PERGAMON INTERNATIONAL LIBRARY
of Science, Technology, Engineering and Social Studies
The 1000-volume original paperback library in aid of education, industrial training and the enjoyment of leisure
Publisher: Robert Maxwell, MC

Nuclear Reactor Kinetics and Control

Other Pergamon Titles of Interest

WILLIAMS	Random Processes in Nuclear Reactors
WILLIAMS	An International Symposium on Reactor Noise
HOWE	Advanced Converters and Near Breeders
MASSIMO	The Physics of High Temperature Reactors
MURRAY	Nuclear Energy
CEC	Pulsed Fusion Reactors
KALLFELZ and KARAM	Advanced Reactors: Physics, Design and Economics
HUNT	Fission, Fusion and the Energy Crisis
SILVENNOINEN	Reactor Fuel Management
KARAM and MORGAN	Environmental Impact of Nuclear Power Plants
RUST and WEAVER	Nuclear Power Safety
KARAM and MORGAN	Energy and the Environment Cost–Benefit Analysis
ASHLEY *el al.*	Energy and the Environment—A Risk–Benefit Approach
ZALESKI	Nuclear Energy Maturity—12 volume set

Nuclear Reactor Kinetics and Control

by

JEFFERY LEWINS
PhD (Cantab), PhD (MIT)

Warden, Hughes Parry Hall
and
Honorary Lecturer, University College London

PERGAMON PRESS
OXFORD · NEW YORK · TORONTO · SYDNEY
PARIS · FRANKFURT

U.K.	Pergamon Press Ltd., Headington Hill Hall, Oxford OX3 0BW, England
U.S.A.	Pergamon Press Inc., Maxwell House, Fairview Park, Elmsford, New York 10523, U.S.A.
CANADA	Pergamon of Canada Ltd., 75 The East Mall, Toronto, Ontario, Canada
AUSTRALIA	Pergamon Press (Aust.) Pty. Ltd., 19a Boundary Street, Rushcutters Bay, N.S.W. 2011, Australia
FRANCE	Pergamon Press SARL, 24 rue des Ecoles, 75240 Paris, Cedex 05, France
FEDERAL REPUBLIC OF GERMANY	Pergamon Press GmbH, 6242 Kronberg-Taunus, Pferdstrasse 1, Federal Republic of Germany

First edition 1978

Library of Congress Cataloging in Publication Data

Lewins, Jeffery.
Nuclear reactor kinetics and control.

Includes bibliographies.
1. Nuclear reactors—Control. 2. Nuclear reactor kinetics.
I. Title.
TK9202.L43 1977 621.48'35 77–8107

ISBN 0–08–021682–X Hard Cover
ISBN 0–08–021681–1 Flexi Cover

Printed Offset Litho in Great Britain by Cox & Wyman Ltd, Fakenham, Norfolk

"Oh, Kitty, how nice it would be if we could only get through into Looking-glass House!"

Contents

Preface

This is a textbook for undergraduates—engineers, physicists and perhaps applied mathematicians—who are studying nuclear engineering; it is also suitable for postgraduates taking an MSc course, say, in this discipline. The central core of the book lies in the application of classical control methods in the frequency space to the dynamic processes of a nuclear reactor. I am persuaded, however, to extend this treatment in two ways: first, to show the reactor as part of an integrated plant including ultimately the load, in the form of an electrical grid, and, second, that non-classical methods should be given some part in a university text. This latter extension covers, therefore, material on fluctuation phenomena (reactor noise), digital computer control and state space optimisation methods. I am not blind to the criticism made by practising nuclear engineers of the usefulness of these theories, but I myself believe that once direct digital control is accepted, it will be natural to incorporate the other developments and therefore that they should be anticipated with simple theory in a text of this sort. Ideas of probability and safety, also discussed in the text, seem to me to be inseparable from a proper appreciation of reactor control and need no apologia.

I would like, of course, to have been able to develop an account of practical applications even further in nearly every section of the book, but space obviously limits such ambitions. That this leads to problems of what to include and what to exclude is a truism of any technical writing; a perusal of the introduction and of the problems and references for further reading in each chapter may partly protect me from the criticism on the part of those who would like more and those who would like less.

An instructor using this book as a text for a one-year course will naturally be selective in the emphasis he gives to various topics; the layout and sequence of the book has in mind the contribution it might make to a team design study which commonly forms part of a final year or MSc course. The first half of the book, Chapters 1 through 5, form a short course, largely based on the frequency space of classical linear design, though Chapter 2, like others marked (†) may be omitted on a first reading. The second half of the text then contains further topics.

The good instructor will also surely see that his students visit a power reactor to see for themselves the scale of the enterprise, and he will no doubt enliven his lectures with practical computations in digital and analogue laboratories. I have found it useful—and also fun—to include some practical illustrations of applied probabilities, and not just poker either, when discussing stochastic questions as in Chapters 6 and 7, placed at a stage in the course when students have developed confidence and can argue the point. Chapter 9, Analogue Computing, can be taken at any time after Chapter 3.

I am conscious that this text on nuclear reactor kinetics and control is entirely about fission reactors and leaves me open to the charge of ignoring fusion reactors. My critics here will be those, I hope, engaged in developing and demonstrating the feasibility of the fusion reactor and who will, therefore, be the appropriate authors for a text to fill this lacuna.

In writing about fission reactors I have tried to give fair coverage to the major reactor systems now in commercial use throughout the world. We can admire the enterprise which has enabled the USA to support not just one but four great construction companies for light water reactors, with offshoots in many other countries. From the United States would come, I believe, an acknowledgement that the world's largest single generating utility, the UK Central Electricity Generating Board, has valuable experience in gas cooled reactors. All will admire the technical purity and devotion of the Canadian heavy water reactor programme. I also discuss control of fast reactors as these are certainly on the commercial horizon with prototypes of more than 1000 MW(th) in four countries. I have put less emphasis on reactors for marine propulsion though there is, of course, unrivalled nuclear experience in the world's naval programmes.

A more serious omission, enforced by space limits, is that of the design and operation of instrumentation, nuclear and non-nuclear, on whose operation control and safety of any plant depends. I must assume other courses or reading are available to make good this deficiency. Similarly, I have thought it unnecessary to go into details of digital computer programming, which must surely be a part of any over-all engineering training, and I have confined myself to a discussion of some of the principles only of employing digital computers in this field.

Classical control methods are well established, known to designers and operators in many disciplines and widely made use of in nuclear engineering in almost all plant built so far. I think one can envisage a digital computer development that is likely to change this pattern, and for that reason I have tried to anticipate where direct digital control (DDC) can make its mark in various discussions throughout the text. Similarly, it seems right to employ SI units (even if these take longer to be adopted than DDC throughout the profession!).

Authors may be allowed, I hope, one or two peculiarities of notation or nomenclature in a field where Elgar's remarks on musical plagiarism are apposite ("of course it sounds the same; we've all of us got only eight notes"). I avoid parenthesis () for multiplication where it might be confused with functional argument and use square brakets [] for multiplication instead, except in Chapter 9 where brakets have a conventional use in analogue computing and in Chapter 6 for expectation functions, where, for example, $<n(n-1)>$ has a well-established meaning not liable to confusion. As for nomenclature, I have stretched the reader's tolerance by introducing three changes to convention for what I at least take to be good reason. Since Dirac's delta is not a function, it seems better to call it for what it is, a distribution. Although "control programme" for the relationship of system pressure and temperatures with load or system power is well established in the USA, it seems to me an uninstructive name, liable to confusion with digital control. I therefore prefer the term pressure/temperature–load function (or PTL). For my third, I have to admit personal inconsistency. Twenty years ago I advocated the use of the "generation time" to simplify neutron kinetics equations. But I overlooked the need which has become more prominent with noise or fluctuation studies, for a different concept introduced by Hurwitz called the generation time. Now I find myself having to suggest that my own pet, still I think highly useful, should be given another name, the "reproduction time". I have also found that there can be much misunderstanding of correlation functions unless they are carefully defined from the practical viewpoint of someone analysing data. At the expense of carrying a normalisation factor, I have used in Chapter 6 the strictly defined correlation function as well as the covariance function. This seems to me, on the one hand, to allow a relation to rigorous mathematical texts on time series while, on the other, it makes the physical interpretation of the Rossi–α experiment somewhat easier. Here (as in the discussion of xenon poisoning in Chapter 4) I have simplified the development in the chapter somewhat, leaving the more detailed, rigorous and less easily understood argument for development in the problems.

With these and other idiosyncrasies, my typists have been patient. I thank my wife for her work on an early draft and comfort through the labour, Jill Shorrocks for typing the final manuscript and Pergamon Press for the finished product. The manuscript also benefited from a critical reading by old friends, friendly enough to be truly critical. Mistakes that remain were put there by me of course. Much of what I have learnt has come in similar friendly engagement with my students in England, America and more recently in Egypt where I would particularly thank Dr. Saad Fadilah for his careful reading. I have many others to thank, only a few of whom receive a pale version of glory in the form of chapter references, good through 1977. My debt is to the many who in thirty-five years have striven to control natural forces for the benefit and security of their fellow men. Above and beyond their technical achievements, I would rate the philosophy they have brought to reactor control and safety as being an intellectual credit to civilised man; we are all in their debt.

HUGHES PARRY HALL, 1977

Acknowledgements

The following diagrams, etc., have been reproduced or adapted by courtesy of the sources shown:

Fig. 4.6 Argonne National Laboratory

Fig. 5.1 Central Electricity Generating Board

Fig. 5.11 General Electric Company

Fig. 5.12 Westinghouse Electric Corporation

Fig. 5.13 British Nuclear Energy Society

Fig. 5.15 International Atomic Energy Agency

Fig. 5.16 Nuclear Power Company

Fig. 5.19 Nuclear Power Company

Fig. 5.23 UK Atomic Energy Authority

Fig. 6.3 University of London Research Reactor

Fig. 6.4 American Nuclear Society

Fig. 6.5 American Nuclear Society

Fig. 7.1 Central Electricity Generating Board

End foldout: Kraftwerke Union

Calculations were undertaken using equipment of Queen Mary College and University College, London, and the digital computing facilities of the University of London.

Acknowledgements

CHAPTER 1

Introductory Review

Nuclear power reactors are an established part of an advanced industrial economy and are to be found in many parts of the world. They offer, on the one hand, a source of power, particularly electrical power, whose raw materials in the form of uranium ores are cheap and have a correspondingly smaller impact on the environment in their winning than fossil fuels. On the other hand, the release of nuclear energy is accompanied by intense and long-lasting radiation whose containment is of fundamental importance. Thus the control of nuclear reactors, the subject of this text, is a matter not only for the day-to-day operation of plant in an efficient manner but also for the study of possible fault conditions, the specification of control systems of adequate reliability and the analysis of experimental information that may have a direct or indirect bearing on these matters.

This book is written as a text therefore in that part of nuclear engineering that can be called reactor kinetics and control with a view to a reasonably broad coverage. It goes beyond an elementary introduction to the subject written in qualitative terms, but it is neither a specialist monograph on the analysis of some specific aspect nor will it substitute for professional experience and practice. Rather it attempts to prepare the student using this book to make good use of subsequent experience.

The remainder of this first chapter is devoted to a review of some but by no means all of the preparatory material utilised in the subsequent chapters. The well-prepared reader will have studied differential calculus through linear differential equations and no review is attempted here. He will also, we hope, have had a course in automatic control and will have taken (or at least be taking simultaneously) a course in reactor physics to understand the conventional diffusion theory description, in multi-group theory, of the distribution of neutrons and the specification of a self-sustaining or critical reactor. It is desirable, too, that he is prepared in heat transfer, enough to appreciate the basic engineering design of a reactor as a plant for turning fission energy into electrical energy, and is taking a practical course in instrumentation.

Readers coming to the book at different levels of preparation will make use of the review material in different ways. For the expert it will serve only to establish notation and nomenclature if this is needed at all. The reader who is seriously underprepared will find that the review material is necessarily condensed and to that extent indigestible; he is advised, however, to "ruminate" upon it before he rushes past the *hors d'ouvres* to the tastier menu in subsequent chapters lest it provokes indigestion if consumed too quickly. Our "average" reader we suppose to be taking a final year in a degree course in nuclear engineering and will be able to proceed steadily through all the chapters.

The elements to be reviewed here cover the treatment of systems of ordinary differential equations using Laplace and Fourier transforms and the subsequent analysis on the frequency space to determine stability and transient response, particularly under automatic control with feedback, using the methods of Nyquist, Bode and Evans. We also review the fundamentals of probability theory with a view to their employment in chapters 6 and 7. While everyone has had some exposure to the ideas of probability, there is still much to be desired in the understanding brought to this deep subject, and the review material is specifically directed to areas where mistakes may be prevalent.

Exercises and problems at the end of this and subsequent chapters will help to consolidate the material and will also introduce new ideas for which there is no space in the main text. Thus the problems vary in difficulty and some are open-ended topics with no concise examination-room answer. For some, as in real life, not all the data necessary to answer the problem are available in the problem statement or even within the cover of this book.

Subsequent chapters are sufficiently well contained, it is hoped, to be read independently, but they follow a natural order perhaps in starting with the neutronics equations and the data appropriate to them for low power reactor behaviour. Chapter 3 provides a range of solutions for these low power cases before extending consideration in Chapter 4 to normal operational behaviour of the system at power, using the linear and frequency domain methods.

Up to this point, the emphasis is largely quantitative and analytical. Chapter 5, however, is largely qualitative in describing the major reactor systems now in commercial use. Chapters 6 and 7 turn to a range of matters whose connecting theme is that of randomness where probability considerations are paramount whether this is a study of the correlation in stochastic processes within the reactor itself or of the safety and reliability of the control system. In Chapter 8 we return to analytical methods for the study of safety and startup problems in a nonlinear description of reactor behaviour in the time dependent state space.

In Chapter 9 analogue computing is introduced since for the nuclear engineering student we envisage using this text it is a natural place to acquaint him with this method of solving a range of equations, a technique that enables us to illustrate the method with a number of important cases in nuclear engineering. That digital computers are not given the same prominence is quite the opposite of decrying their importance to reactor kinetics and control; on the contrary, there is not enough space in such a text to give full coverage to the role of digital computers for both numerical analysis and data handling in reactor control, and we are relieved by the thought that this must of itself form a major part of any current engineering training. Implications for direct digital control (DDC) are touched upon in Chapter 6.

Very largely, this text covers "lumped" kinetics behaviour and not space dependent behaviour. It might even be sufficient reason to say that a full space and energy dependent treatment is too difficult or too long for such a text, but there are two more rationalisations that can be advanced to justify the limitation further. The first is that one has to start somewhere and it is logical to start by a treatment that takes the reactor "as a whole" whether this is a pedagogic viewpoint or whether this is a starting point for more elaborate treatments to extend the spatial treatment in some consistent fashion without jumping immediately into partial integro-differential equations in seven-dimensional Boltzmann space and all the rest of it. Secondly, whatever elaborate models are found necessary in particular cases, the techniques will require testing; one battery of tests requires elaborate codes to reduce to known solutions in special cases. The kinetics and control engineer must at least pass through the stage of developing his "feel" for the discipline from specialised, albeit idealised, cases that he can manipulate analytically. That there should be such an intermediate stage is perhaps well demonstrated by comparing the excellent introductory book by Tyror and Vaughan [8] with the review of dynamic safety codes for light water reactors [9]. Knowles [13] may also be helpful as a short introduction to the place of a power station in a utility grid system, with special reference to nuclear power. Henry [14] and Weisman [15] provide excellent texts on reactor theory and design methods.

Laplace and Fourier Transforms

Much of the analysis of reactor kinetics and control can be done in the form of systems of ordinary differential equations, i.e. equations and their boundary conditions, for dependent variables x_1, x_2, etc., dependent on the independent variable time t. By suitably defining $x_2=\dot{x}_1$, etc., these can always be written as sets of *first* order equations, n such equations implying an nth order system since elimination in the other direction would produce a single nth order equation. Thus

$$\left.\begin{aligned} \frac{dx_1}{dt} &= a_{11}x_1 + a_{12}x_2 + \cdots + s_1; \quad x_1(0) = x_{10} \\ \frac{dx_2}{dt} &= a_{21}x_1 + a_{22}x_2 + \cdots + s_2; \quad x_2(0) = x_{20} \end{aligned}\right\} \tag{1.1}$$

$$\text{etc., i.e. } \frac{dx_i}{dt} = \sum_{i'=1}^{i'=n} a_{ii'}x_{i'} + s_i; \quad x_i(0) = x_{i0}; \quad i = 1, 2 \cdots$$

There is an implication in this notation that the system is linear, but in general the coefficients $a_{ii'}$ might be functions of the $x_{i'}$. In the simplest case the coefficients, including the independent source terms s_i, are constant and the Laplace transform method provides a straightforward way of solving the system.

The Laplace transform is a technique to replace the operation of integration (with its general solution and subsequent fitting of boundary conditions) with an operation of algebraic manipulation, rather like logarithms replace multiplication by a process of addition in the logarithmic transform space. Just as the results can be transformed back by identifying the result as an entry in the log table in reverse, we expect to identify the algebraic result as the transform of a known function of time which is then identified as the solution to the original problem.

The one-sided Laplace transform $L(f(t))$ of a function of time is defined via an integral over all time which has the effect of replacing time dependence with dependence on a new or transform variable p:

$$L(f(t)) \equiv \bar{f}(p) = \int_0^\infty f(t)\, e^{-pt} dt \tag{1.2}$$

It is seen that the exponential weighting is likely to make this integral converge even for $f(t)$ increasing with t. For the time being, indeed, we can suppose p to be a real number, p_0 say, sufficiently large as to dominate any exponential behaviour in $f(t)$.

The advantage of the Laplace transform in the case of constant coefficients is evident since $L(ax(t)) = a\bar{x}(p)$. The applications to differential equations comes by considering the transform of the differential via an integration by parts:

$$\int_0^\infty \frac{dx}{dt}\, e^{-pt} dt = [xe^{-pt}]_0^\infty + \int_0^\infty pxe^{-pt} dt = p\bar{x}(p) - x_0 \tag{1.3}$$

for all reasonable $x(t)$. Thus the system of linear equations with constant coefficients reduces to an algebraic system:

$$p\bar{x}_i - \sum_{i'=1}^{n} a_{ii'}\bar{x}_{i'} = \bar{s}_i + x_{i0} \tag{1.4}$$

with the transformed source separated from the summation. We can expect to find the algebraic matrix inverse and be able to write

$$\bar{x}_i = \sum_{i'=1}^{n} G_{ii'}(p)\, [\bar{s}_{i'} + x_{i'0}] \tag{1.5}$$

or $\bar{x} = G(p)[\bar{s} + x_0]$ in matrix form.

The right hand side divides naturally into two significantly different parts: a driving term due to the source and initial conditions, usually called the excitation function, and the term $G_{ii'}(p)$ that gives the system behaviour modifying the excitation function to yield the system output or response $\bar{x}_i$. In particular, we expect that if a system is to be *stable*, the inherent response with zero driving terms must of itself display a stable nature and not lead to increasing values of $x(t)$. $G_{ii'}(p)$ relates to the *transfer* function for the ith output variable in terms of the i'th input.

SAMPLE TRANSFORMS

Simple functions of time lead to Laplace transforms for which a short list is given in Table 1.1. The transforms of the exponential are readily found. By specialising this result to $\lambda = 0$, imaginary, etc., several other results are obtained. Note that $1/p$ is strictly a step function transform, where the time dependent function rises to 1 at time zero having been zero beforehand. The first entry is for the (Dirac) delta distribution—it is not strictly a function—defined as being zero everywhere except around $t = 0$ and of such a magnitude that the integral over $\delta(t)$ is normalised to unity. Its transform is readily obtained direct from such a definition; note also the implication

$$\int_{-\infty}^{\infty} f(t)\, \delta(t) dt = f(0) \tag{1.6}$$

TABLE 1.1.

Short table of Laplace and Fourier transforms

$f(t)$; $\lambda \equiv 1/\tau$	$\bar{f}(p)$	$\bar{f}(j\omega)$; $j^2 = -1$
$\delta(t)$	1	1
$h(t)$ step function	$\frac{1}{p}$	$\frac{1}{j\omega} = -j/\omega$
$e^{-\lambda t}$ or $e^{-t/\tau}$	$\frac{1}{\lambda + p}$	$\frac{\tau}{1 + j\omega\tau}$
$te^{-\lambda t}$	$\frac{1}{(\lambda + p)^2}$	$\left(\frac{\tau}{1 + j\omega\tau}\right)^2$
$\sin \gamma t$	$\frac{\gamma}{\gamma^2 + p^2}$	$\frac{\gamma}{\gamma^2 - \omega^2}$
$\cos \gamma t$	$\frac{p}{\gamma^2 + p^2}$	$\frac{j\omega}{\gamma^2 - \omega^2}$
$e^{-\lambda t}f(t)$	$\bar{f}(p + \lambda)$	$\bar{f}(j\omega + \lambda)$
$f(t-\tau)$	$e^{p\tau}\bar{f}(p)$	$e^{-j\omega\tau}\bar{f}(j\omega)$
$\lim_{t\to 0} f(t)$	$\lim_{p\to\infty} p\bar{f}(p)$	For well-behaved $p\bar{f}(p)$
$\lim_{t\to\infty} f(t)$	$\lim_{p\to 0} p\bar{f}(p)$	

With the help of this distribution initial conditions can be expressed as equivalent sources. In particular, consider the response of the system for one output x to an impulse source (i.e. one describable via the Dirac distribution and no other initial condition.) We have $\bar{x}(p) = G(p)\bar{s}(p) \to G(p)$, where $G(p)$ is the *system* function (as opposed to $s(p)$ and the initial condition providing a driving term or excitation function). We have the important result that $G(p)$ is the response to an impulse source.

Note also the general "shifting" relations of Table 1.1 together with the limiting value relationships useful for steady state displacements, etc.

If the system of equations leads, as it often does, to rational polynomial fractions $g(p)/f(p)$, where both $g(p)$ and $f(p)$ are polynomials in p, the inverse can generally be found via the method of partial fractions. Let $f(p) = 0$ lead to roots p_i such that we can write $f(p) = [p - p_1]\,[p - p_2] \cdots [p - p_i] \cdots [p - p_n]$. Three results are useful:

(a) If all the roots p_i are distinct, consider the residue R_i where

$$R_i(p) = \frac{g(p)}{f(p)}[p - p_i]$$

Then

$$\bar{x}(p) = \frac{g(p)}{f(p)} = \sum_i^n \frac{R_i}{p - p_i} \tag{1.7}$$

where the $R_i(p)$ are evaluated at $p=p_i$. The result is recognised as the sum of a number of time dependent exponential terms and their coefficients:

$$x(t) = \sum_i^n R_i e^{p_i t} \tag{1.8}$$

It is also worth noting that for the same case of distinct roots, $R_i(p_i)=g(p_i)/f'(p_i)$, where the derivative of f(p) is taken with respect to p before substituting $p=p_i$. These results lead to the well known "cover up" rule for finding partial fractions, for example

$$\frac{1}{[p-2][p+3]} = \frac{1}{5}\frac{1}{p-2} - \frac{1}{5}\frac{1}{p+3} \rightarrow \frac{1}{5}\,[e^{2t} - e^{-3t}] \tag{1.9}$$

(b) If the roots are repeated, say one root repeated once, the expansion in partial fractions is of second order of the form

$$\frac{A_i}{p+p_i} + \frac{B_i}{[p+p_i]^2}, + \sum_{i' \neq i} \frac{C_{i'}}{p+p_{i'}}$$

Whilst formulae for residues are available, it is probably easier to use a direct equation of this form with the original to determine the expansion coefficients A, B, etc.

(c) If the coefficients of the original equations are real, then any complex roots must occur in conjugate pairs of the form

$$p_i = r_i \pm j\omega_i \text{ (with } j = \sqrt{-1}\text{) or in the polar form } A_i e^{\pm j\theta i}$$

These complex pairs then lead to real trigonometrical functions. It is often convenient to note the implication of a *shift* of the Laplace transform variable as being equivalent to the transform of the unshifted function an exponential of the time shift, e.g.

$$\frac{p}{p^2+4p+5} = \frac{p}{[p+2]^2+1} = \frac{p+2-2}{[p+2]^2+1} \rightarrow e^{-2t}[\cos t - 2\sin t] \tag{1.10}$$

FOURIER TRANSFORMS

So far we considered the Laplace transform parameter p as a real number chosen to dominate the real exponentials of the function $f(t)$. More generally, it can be considered a complex number, $p=p_0+j\omega$, where p_0 is chosen to meet this necessity. If the time dependence of $f(t)$ is such that it does not grow exponentially (remains constant or falls), then p_0 may be put to zero. The Laplace transform with $p=j\omega$ is called the Fourier transform of $f(t)$ and its existence is subject to this restriction on the behaviour or $f(t)$. However, the Fourier transform can be obtained also from the idea of the Fourier series representation of a periodic function, and this probably shows its nature more clearly.

Suppose a function is to be represented over an interval $-\frac{1}{2}T$ to $\frac{1}{2}T$ and is assumed periodic outside this range. We may try to represent it in a set of the periodic trigonometrical functions, sin nwt and cos nwt with $w=2\pi/T$. The terms are given by taking $n=0, 1, 2, 3$, etc., to infinity. Thus

$$f(t) \simeq a_0 + \sum_{n=1}^{\infty} [a_n \cos nwt + b_n \sin nwt] \tag{1.11}$$

The expansion coefficients may be found by using the orthogonality properties of the trigonometric functions over the range T, i.e. multiply by one of the expansion functions and integrate over the range; all other functions on the right hand side when integrated are zero. Thus

$$a_0 = \frac{1}{T} \int_{-\frac{1}{2}T}^{\frac{1}{2}T} f(t)\, dt \qquad \text{mean value}$$

$$a_n = \frac{2}{T} \int_{-\frac{1}{2}T}^{\frac{1}{2}T} \cos nwt\, f(t)\, dt \quad \text{symmetric}$$

$$b_n = \frac{2}{T} \int_{-\frac{1}{2}T}^{\frac{1}{2}T} \sin nwt\, f(t)\, dt \quad \text{antisymmetric}$$

With these coefficients, the function is represented in the "best" way. If the original function is continuous, this representation is indeed exact, an equality. If discontinuous, it is best in the sense of minimising the integral of the square error over the range. Of course it may call for all the infinite terms, $n \to \infty$ to achieve these properties.

A more compact representation comes from using the imaginary exponential forms of the trigonometric functions, $e^{jnwt} = \cos nwt + j \sin nwt$. Noting the symmetry involved we write

$$f(t) \simeq \sum_{-\infty}^{\infty} \bar{f}(nw)\, e^{jnwt}$$

where the complex expansion coefficients $\bar{f}(nw)$ are given by

$$\bar{f}(nw) = \frac{1}{T} \int_{-\frac{1}{2}T}^{\frac{1}{2}T} f(t) e^{-jnwt}\, dt \tag{1.12}$$

To pass to a continuous representation, the Fourier transform, over an infinite period, $T = (2\pi/w) \to \infty$, we let nw be the continuous variable ω to obtain

$$\bar{f}(\omega) = \int_{-\infty}^{\infty} e^{-j\omega t} f(t) dt; \quad f(t) = \frac{1}{2\pi} \int_{-\infty}^{\infty} e^{j\omega t} \bar{f}(\omega) d\omega \tag{1.13}$$

This latter shows a formal *inversion* for the Fourier transform. There is an analogous integral inversion (the Bromwich integral) for the Laplace transform though we shall not use it. Some writers will change the normalization of eqn. (1.13) so that the term $\sqrt{2\pi}$ appears symmetrically in both forms. Note that the Fourier transform of a continuous spectrum would have Dirac delta distributions added to represent discrete frequency components.

TRANSFER FUNCTIONS

Keeping in mind that initial conditions can be written with the source terms of our equations (via the Dirac delta distribution), our original problem, eqn. (1.1) in the Laplace transform space can be represented as

$$\bar{x}(p) = G(p)\bar{s}(p) \tag{1.14}$$

The situation is shown graphically in Fig. 1.1.

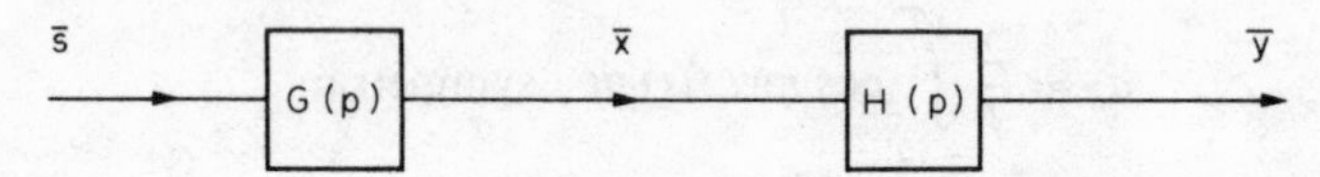

FIG. 1.1. Block transfer functions.

Of course, the output of one part of the system may well provide the input for the next element of the system and so on, as implied in the figure. For the linear systems with constant coefficients we have an important simplification when we use Laplace (or Fourier) transforms in such chains of systems: the transform of the over-all system is simply the product of the transforms of the individual elements. In particular

$$y(p) = G(p)H(p)\bar{s}(p) \tag{1.15}$$

This result, sometimes called the convolution or folding theorem, is established as follows for a single variable system. If we have compound systems, where the response of one element becomes the input of the next, how may these be expressed in terms of $G(p)$ or $G(j\omega)$? Consider first the time dependent solution to an impulse source $g(t)$ called Green's function. Due to the supposed linearity, the response to any source can be obtained by adding (integration) all the elementary solutions in the form

$$x(t) = \int_0^t g(t-\tau)s(\tau)d\tau \quad (t \geqslant 0) \tag{1.16}$$

where the integral is taken up to t, the latest time that a source can affect the response. If we seek to take the Laplace transform, it is convenient to extend this upper limit to ∞ which we may do artificially by introducing a step function $h(t)$, which is zero before $t=0$ and unity for $t>0$. Thus

$$L(x(t)) = \int_0^\infty e^{-pt}[\int_0^\infty h(t-\tau)g(t-\tau)s(\tau)d\tau]dt$$

$$= \int_0^\infty s(\tau)e^{-p\tau}[\int_\tau^\infty e^{-p[t-\tau]}g(t-\tau)dt]d\tau = \bar{g}(p)\int_0^\infty e^{-p\tau}s(\tau)d\tau = G(p)\bar{s}(p) \tag{1.17}$$

where a change in the order of integration, valid for well-behaved functions g and s, enables the Laplace transform of the convolution integral to be expressed as the product of Laplace transforms. In one sense, this is a result we had already anticipated in eqn. (1.5), but more generally this is important as showing that where we have a sequence of elements, the over-all transfer function can be obtained simply as the product of the elements in series.

We have seen that the algebraic manipulation of the transformed equations where there is an impulse (Dirac delta distribution) source leads to expressions of the form $\bar{x}(p)=G(p)$, where $G(p)$ may be called the system transfer function. The more general case will involve arbitrary initial conditions and driving functions (sources). Just as $G(p)$ is the system response to an impulse source, we may see that $G(j\omega)$ is the system response to a sinusoidal excitation of unit amplitude at frequency ω, an observation that makes it possible to determine $G(j\omega)$ experimentally.

We use the terminology of the transform $G(p)$ and the transfer function $G(j\omega)$ somewhat interchangeably since for our purposes we simply let $p \to j\omega$. Particularly when "linearisation" of the model is required and we deal with small departures of the output x as a result of small departures of the input s around the steady state, we can also write the transfer function as $G(p)=\delta\bar{x}(p)/\delta\bar{s}(p)$. When more than one input variable is present a partial differential or thermodynamic notation may be used such as

$$G_1(p) = \left(\frac{\partial\bar{x}}{\partial s_1}\right)_{s_2,\, s_3\, \cdots}$$

showing the remaining input variables s_2, s_3, , , $\cdot$ kept constant.

FEEDBACK AND CONTROL

Many systems have an *inherent feedback* term in their response. Thus the equation $\dot{x}=\lambda x+s$ leads to a Laplace transform of simple lag form

$$\frac{\bar{x}}{\bar{s}} = G(p) = \frac{1}{p-\lambda}$$

(ignoring initial conditions). But this itself could be graphed or split up as in Fig. 1.2.

We note from elementary arguments that λ must be negative for stability (compare the dying away of radioactive decay). We may wish to follow a positive feedback convention of the type shown in the figure to represent inherent feedback.

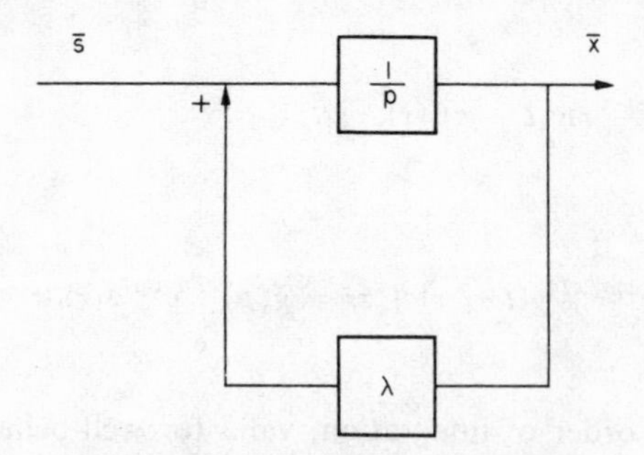

FIG. 1.2. Simple inherent feedback.

Often the output of a system in response to different loads or excitation functions is not satisfactory and the concepts of automatic control are bound up with the idea of modifying the system output by providing, via a feedback term, a correction excitation signal that depends on the output, or rather on the departure of the output from some desired level.

Thus we distinguish between open loop control and closed loop control. A good example of open loop control is a toaster whose time can be set in advance but which does not respond automatically to the degree of browning of the toast. For an example of closed loop control, consider a household heating system which is normally provided with a feedback controller (thermostat) so that the actual temperature of the rooms or their departure from some desired temperature varies the fuel supply to the furnace.

In another example (Fig. 1.3) the response of a turbine generator to different electrical

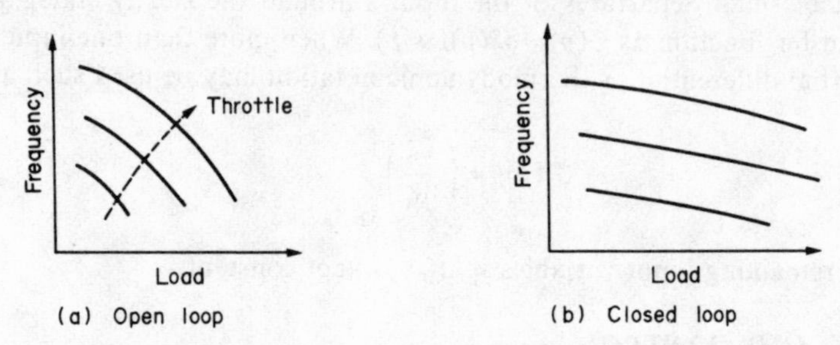

FIG. 1.3. Turbine generator droop characteristics and control.

loads is shown. Without control action there is a noticeable droop in the speed at fixed throttle and higher loads and hence an unacceptable drop in electrical frequency as the load varies. With feedback control to vary the throttle opening with change of speed, through a simple proportional controller perhaps, we may expect to flatten out this undesirable droop. To flatten it still further, using just proportional control, would call for bigger control action or gain for a given speed discrepancy and we would be right to anticipate that there will be a limit to the efficacy of such a simple system, brought about by the onset of dynamic instability in which a small fluctuation in the response of the system is magnified (resonance) by too high a gain, to the point of producing even larger swings in the response.

Just as the sign of the feedback term in the equation represented in Fig. 1.2 must be negative for stability or the analogous decay of the error rather than growth, so feedback control must be negative to promote the reduction of the error between demand and performance. It may be advantageous therefore to use a negative feedback convention as used in the first example of Fig. 1.4.

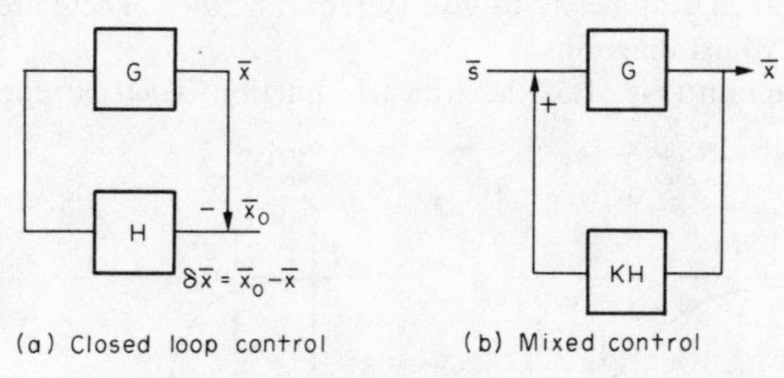

FIG. 1.4. Feedback control systems.

The two examples are not, on the face of it, much different (other than in negative and positive feedback convention) but they bear a different interpretation. In the first, Fig. 1.4(a), we have a closed loop control system with the comparison of a state x against a demand x_0. The resulting difference signal is passed around the loop so that from $\bar{x} = HG[\bar{x}_0 - \bar{x}]$ we have

$$\bar{x} = \frac{HG}{1 + HG}\bar{x}_0 \quad \text{(negative feedback)} \tag{1.18}$$

In Fig. 1.4(b), however, suppose there was no feedback via KH (where K is a scalar or gain). We can regard $\bar{s}$ as a pre-programmed or open loop controller (feed forward control) that leads to an output $\bar{x}$ with forward transfer connection $\bar{x} = G\bar{s}$. With feedback (positive convention) we have

$$\bar{x}(j\omega) = \frac{G}{1 - KHG}\bar{s}(j\omega) \equiv F(j\omega)\bar{s} \quad \text{(positive feedback)} \tag{1.19}$$

defining an equivalent closed loop transfer function F.

In the positive convention, the term $1 - KHG$ is aptly called the return difference while KHG is the open loop gain. If $p \to j\omega$, $\bar{s}(j\omega)$—or its real part—represents an oscillating input at frequency ω and $\bar{x}(j\omega)$—or again its real part—represents an oscillating output at the same frequency but with modified gain and phase. The possibility of dynamic instability is suggested by the form of the return difference, i.e. if $1 - KGH \to 0$ for some ω, unacceptable hunting or oscillations at large amplitudes would occur, again a resonance situation. This interpretation is given closer inspection in a subsequent section and only in a naive sense is it a matter of determining the approach of the open loop gain KGH to the point 1, 0 (positive convention) in the complex Fourier transform plane, or the point -1, 0 (negative convention).

We have now seen how to build up system transfer functions from the elementary transfer functions, either chaining $G(p)H(p)$, etc., for open loop response or, as in eqn. (1.19), for closed loop systems.

GRAPHICAL REPRESENTATION

Transfer functions or Fourier transforms may be represented graphically and also conveniently combined graphically to give system functions. There are two representations, Bode and Nyquist diagrams.

In the Bode diagram (Fig. 1.5) the transfer function $(Gj\omega)$ is represented in polar

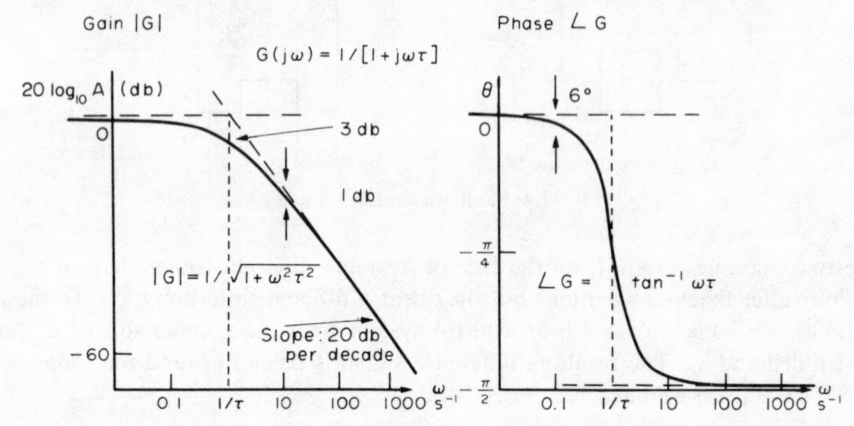

FIG. 1.5. The simple lag as a bode diagram with break frequency construction.

form through its amplitude or gain A and phase θ as a complex number $Ae^{j\theta}$. Gain and phase are then plotted against the frequency (either in radians/s or hertz, i.e. cycles/s) on a logarithmic scale.† Since the combination of successive forward transfer functions requires the *addition* of phases but the *product* of amplitudes, it becomes helpful to plot the amplitudes on a logarithmic scale so that they may be added graphically for this purpose. It is convenient therefore, to plot the amplitude in decibels, i.e. $A(\omega)$ in db= $20 \log_{10}|G(j\omega)|$.‡

Figure 1.5 shows the simple lag plotted as a Bode diagram. It also shows the important approximate construction technique of the "break frequency" method in which asymptotes of G are established for large and small ω. These asymptotes intercept at the frequency $\omega_c \equiv 1/\tau$ (break frequency) and the figures show the departure of the exact result from the asymptote around the break frequency in this example.

Bode diagrams are convenient, then, for the compounding of successive transfer functions. Also note the useful approximation that 6 db corresponds to a factor of two.

An alternative representation is the Nyquist diagram in which $G(j\omega)$ is plotted as an Argand diagram in radial coordinates. (Occasionally a polar plot of log A is useful.) The same simple lag is given as a Nyquist diagram in Fig. 1.6. Note that for negative ω the diagram is completed as a mirror image.

†We shall generally use the angular frequency ω in radians/s which is, of course, 2π greater than the cycles in hertz (Hz).

‡One bel (after Graham Bell) implies a tenfold increase in power taken to go as the square of the amplitude; hence 1 bel $= 2 \log_{10} A$.

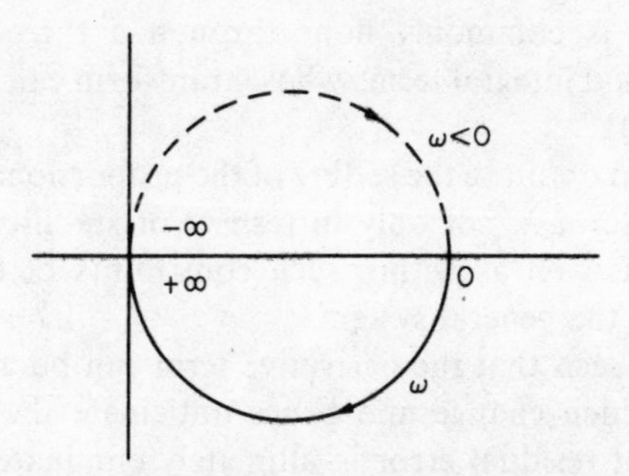

FIG. 1.6. Nyquist diagram for simple lag.

As a second example of the use of a Nyquist diagram we may discuss the representation of transport delays as occur in the flow of a coolant in a pipe. If this takes, say θ s and there is no mixing, then the transform is shifted by a factor $e^{-p\theta}$. When some mixing takes place, as in simple models of heat exchangers, a common approximation is a combination of transport delay and simple lag in the form $G(p)=e^{-p\theta}/[1+p\tau]$. The exponential term is irrational in p and causes some difficulties in analytic representation. The Padé approximants [10] are useful rational polynomial approximations for the exponential in varying orders. One of the simplest, P_{11}, is given by

$$P_{11}=\frac{1-\frac{1}{2}p\theta}{1+\frac{1}{2}p\theta}$$

and the Padé approximants are useful in several other situations.

In the case of a transport lag, certain frequencies will suffer a 180° or π phase shift along the pipe, and this would change, for example, negative feedback to positive feedback with implications for stability. Figure 1.7 illustrates the approximation for the

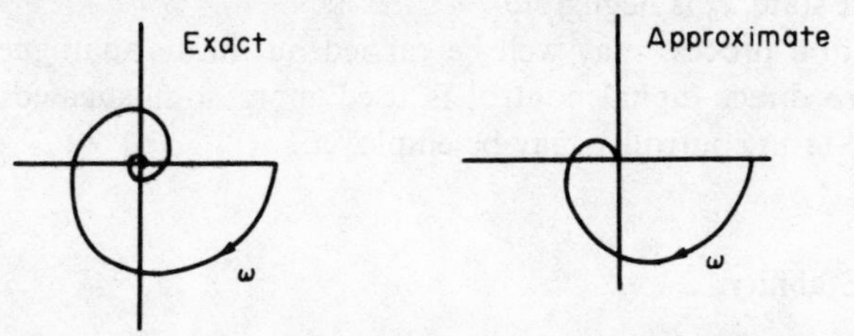

FIG. 1.7. Exact and Padé approximant for simple lag with transport delay.

heat exchanger as Nyquist diagrams both exactly and using the P_{11} approximant. It is seen that the approximant leads to a reasonable representation of the change of phase at the lowest frequency but not for higher frequencies (which would require higher order polynomial approximants).

Conventional (or analogue) feedback controllers take the measured signal, after comparison with a demand or set point signal, shape this difference signal and apply the result to a point of control. The process of measuring the signal and amplification introduce characteristics to be allowed for through their appropriate transfer functions.

The *shaping* of the signal to give desired characteristics is a means of giving effect to a desired control law and is commonly done through a three-term controller having proportional, derivative and integral terms whose transform can be represented therefore as $G_c(p)=K[1+\mu p+(\lambda/p)]$.

The designer will seek to optimise the setting of the proportional gain K, the differential term μ and the integral term λ, not only in respect of stability but also against some objective cost function as well as within such constraints of temperatures, pressures, etc., placed upon him by the general system.

Qualitatively it can be seen that the derivative term can be used to increase the controller response to a sudden change and hence anticipate the necessary control. The integral term ensures that residual error is ultimately eliminated. The size of μ and λ, however, will certainly be restricted by stability considerations.

Objective functions against which the controller terms are to be optimised may involve the variation about a desired state of one or more state variables. To illustrate by means of a single variable, Table 1.2 lists three typical criteria measuring the error

TABLE 1.2.

Objective error functions

IT	Integral time	$\int_0^\tau dt\ (=\tau)$
ISE	Integral square error	$\int_0^\tau [x-x_0]^2 dt$
ITSE	Integral time weighted square error	$\int_0^\tau t[x-x_0]^2 dt$

following a disturbance, where τ is a somewhat arbitrary time at which the departure x from the desired state x_0 is negligible.

The optimisation process may well be carried out in an analogue simulation of the plant, but where direct digital control is used more sophisticated control laws than simply the three-term controller may be employed.

Linear System Stability

Consider a linear system operating in steady state at an operating point (of which there is only the one in a linear system) determined of course by equating the rates of change $\dot{x}_i$ to zero. Inevitably in a real system there will be small disturbances and we study the resulting transients. If these disturbances, whatever their initial form, die out, the system is stable. If they grow or even if only some of them grow, the linearity implies that they will go on growing without bound; the system is unstable. There is the possibility of a behaviour on the *limit* of stability, where an oscillatory behaviour is maintained whose amplitude neither increases nor decreases from the initial disturbance.

The behaviour of systems with simple poles, of any finite order, in the Laplace transform representation illustrates these different possibilities, as in Fig. 1.8.

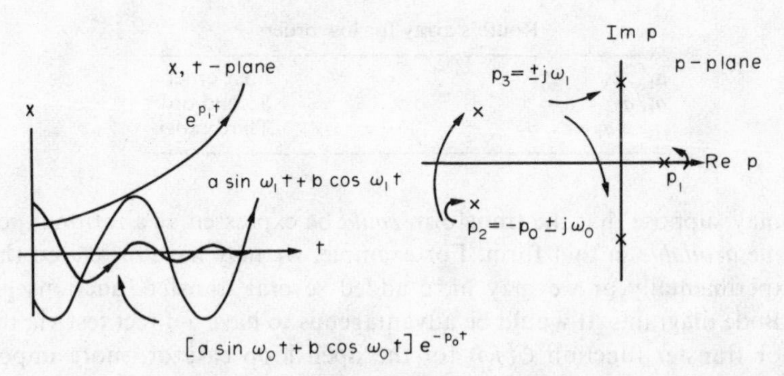

FIG 1.8. Time and Laplace transform representation for stability.

Any system whose Laplace transform $G(p)$ leads to a pole in the right half-plane is unstable because the corresponding time dependent behaviour corresponds to *increasing* exponentials whether or not these exponentials are coupled with oscillatory terms. Single poles on the imaginary axis correspond to the limit of stability, while poles in the left hand plane indicate a stable behaviour with decreasing (though possibly oscillatory) behaviour. A pole at the origin is a little more complicated since a single pole corresponds to the step function $h(t)=1$ for $t>0$, while repeated poles on the imaginary axis indicate further terms in t, t^2, etc., that would denote instability.

Of course the Laplace transform $G(p)$ may have terms in it other than simple poles of finite order, but it may be stated as a general result of complex analysis that if there are poles in the right half p-plane of any form, the time dependent behaviour shows an increase without bound and the system is unstable.

If we attack the question of the stability of a system head on from this point of view, we would try to *find* all the poles in $G(p)$ and see whether any lie in the right half-plane. But this is as difficult as finding the complete time dependent behaviour. We might approach the question obliquely with more efficiency by considering merely the question of whether any poles *exist* in the right half p-plane without actually finding their number or location. If we limit ourselves to transforms $G(p)$ that are rational polynomials of the form $g(p)/f(p)$, then there are several techniques for determining stability from this oblique viewpoint.

If the numerator $g(p)$ is bounded, poles in $G(p)$ arise only by virtue of the zeros of $f(p)$, i.e. the roots of the equation $f(p)=0$. If the polynomial form $f(p)$ is known explicitly, then Routh's criteria test for the existence of poles in the right half p-plane is via the roots of $f(p)$. Suppose $f(p)$ to be in the form

$$f(p) = p^n + a_1 p^{n-1} + \cdots a_n = 0$$

We shall not give the full criteria for arbitrary n but state it explicitly for $n=1$, 2 and 3 in Table 1.3. (See problem 1.12 and ref. 1 for extensions.) Clearly for $f(p)=p+a_1$ we have the trivial result that $p_1=-a_1$, so that if a_1 is negative the system is unstable with the pole lying in the right half plane. The criterion requires the row of numbers each to be greater than zero for stability, terminated as shown for the various orders 1, 2, 3.

However, Routh's approach in any order requires knowledge of the polynomial form.

TABLE 1.3.

Routh's array for low orders

a_1	First order
a_1, a_2	Second order
a_1, $a_1a_2-a_3$, a_3	Third order

Whilst we may suppose that the transform *could* be expressed as a rational polynomial, it may not be *available* in that form. For example, we may have measured the transfer function experimentally or we may have added several transfer functions graphically from their Bode diagrams. It would be advantageous to have a direct test via the Fourier transform or transfer function $G(j\omega)$ for the open loop case or, more important, via $1+G(j\omega)H(j\omega)$ in the case of feedback where stability is more suspect. The latter case is available via the Nyquist criterion for stability.

NYQUIST STABILITY CRITERION (negative feedback convention)

We are examining the behaviour of the feedback expression $\bar{x}=G/[1+GH]\bar{s}$ which may again be written as $\bar{x}=[\bar{g}/\bar{f}]\bar{s}=F(p)\bar{s}$, say. We assume that the over-all transfer function $F(p)$ is again a rational polynomial and, furthermore, we assume that the order of the denominator is greater than that of the numerator, or in more physical terms that the system is such as to attenuate high frequencies. There will be poles in the right half p-plane if either (a) $g(p)$ has poles in the right half plane or (b) $f(p)$, i.e. $1+GH$, has zeros in the right half plane. To start with we assume that the open loop system was stable so that there were no poles from $g(p)=GH$ and we are concerned only with the zeros of $1+GH$. If GH has no poles then $1+GH$ has no poles and we are dealing with a polynomial function that can be entirely represented as the product of its factors $p-p_i$ leading to the poles in $F(p)$ at the roots of $f(p)$. Consider individual roots leading to terms such as $1/[p-p_i]$ in $F(p)$. Suppose (Fig. 1.9) we construct a loop around such a pole in the right half-plane.

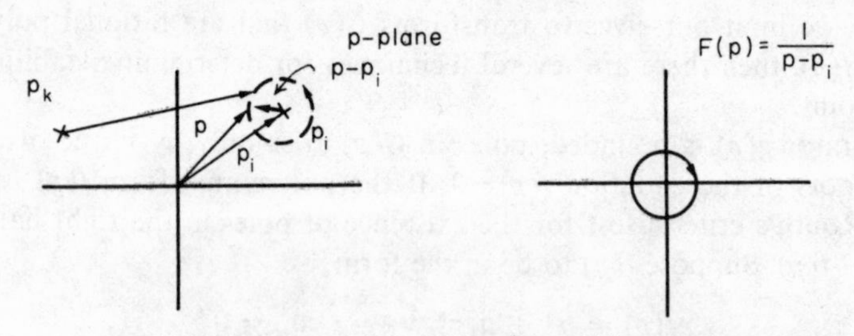

FIG 1.9. Construction of Nyquist stability criterion.

The vector p from the origin traces such a loop enclosing a zero of $\bar{f}\,(p)$ and a pole of $F(p)$. The complex number $p-p_i$ is shown as a vector and as p traces the loop, $p-p_i$ is seen to rotate. That is, the polar form $p-p_i=Re^{j\theta}$ rotates such that the number returns to the same R value but with an angle θ increased by 2π. The corresponding pole $1/[p-p_i]$ is $1/Re^{-j\theta}$ and this also rotates, in the opposite direction, by -2π. If, on the other hand, there is no pole p_i within the loop (compare the pole at p_k), then there is no net rotation. Therefore we plot the complex number $F(p)=1/[p-p_i]$ as p passes round the loop; $F(p)$ will show a rotation if p_i lies inside the loop.

If we extend the loop to cover the whole of the right half p-plane save possibly for any poles on the imaginary axis whose effect must be separately determined, we have tested the whole of the right half p-plane against the presence of a pole that might lie in it. If $F(p)$ rotates, there is a pole. If it does not rotate, but θ returns to its original value and not 2π different, then no pole was contained. This construction of the Nyquist criterion to test $1+GH$ is shown in Fig. 1.10.

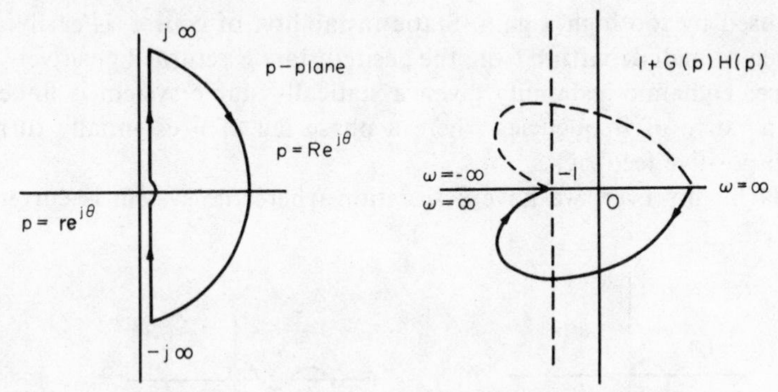

FIG 1.10. Nyquist criterion using the transfer function.

The loop so indicated passes up the imaginary axis of the p-plane, and so in the $1+GH$ plane we are actually studying the behaviour of the transfer function $1+G(j\omega)$ $H(j\omega)$ for ω from $-\infty$ to ∞. The completion of the loop in the p-plane calls for a semi-circle at large radius, but by our assumption of low pass, attenuated high frequencies, the transfer function goes to the origin for large magnitudes of ω. We have the distinct advantage of determining stability in terms of the rotation of the (closed) transfer function.

Rotation is a question of drawing a vector from the origin whose tip traces the completed transfer function loop in the Nyquist diagram. Rather than plot $1+GH$ and determine the rotation about the origin, it is common to plot GH alone and determine the rotation about the point $-1, 0$ on the real axis. *Es egal.*†

If $1+GH$ has more than one pole and is of the form $1/[p-p_1][p-p_2]\cdots$, etc., then the polar form is $Re^{jn\theta}$ and the rotations caused by the loop will be $2n\pi$, where n is the number of poles within the loop, i.e. in the right half p-plane. Thus the number of rotations counts the number of zeros in the return difference $1+GH$ and hence the number of poles of the transfer function $G/1+GH$ in the right half p-plane. If, contrary to assumption, the open loop system had been unstable, there would be right half poles of G and hence in the return difference, of the form $p-p_i=Ae^{j\theta}$, leading to positive rotations in the opposite sense of the zeros investigated. This would have tended to cancel the count and invalidated the result; hence the assumption. In the more general case this effect can be accounted for if we know the number of poles of the open loop

†For a *positive* feedback convention, the diagram is simply rotated 180° or π, since to examine $1-KGH$ we investigate rotations about $+1, 0$.

transfer function and require a corresponding number of rotations of the closed loop transfer function in the Nyquist diagram accordingly.

Figure 1.11 illustrates two possible situations assuming negative feedback in the form of Nyquist diagrams. Figure 1.11(a) itself indicates stability. If the feedback system depends on a gain or parameter K as $1+KGH$ for the return difference, then the same diagram for GH will serve subject to a simple radial scaling according to the factor $1/K$. It is likely, therefore, that at low K the system will be stable as shown in Fig. 1.11(a), while at increasing K or with increasing feedback the system will be unstable. It is convenient to distinguish, therefore, between *static* instability, which is essentially a need for the feedback term to be negative, and dynamic instability, the onset of the instability caused by too high a gain. Static instability, of course, is easily understood as arising when a small departure from the desired state is returned positively to increase the divergence. Dynamic instability given a statically stable system is linked with the existence of a range of frequencies where a phase lag of π essentially turns negative feedback into positive feedback.

In Fig. 1.11(b), however, we have a situation where the system is currently stable

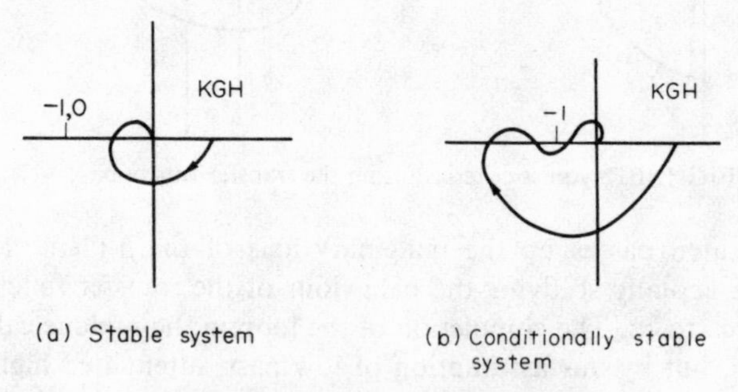

FIG. 1.11. Stability in Nyquist diagrams.

but the *decrease* as well as the increase of a gain K would drive the system unstable. Such a situation is conventionally called conditionally stable. Its disadvantage comes from the possibility that when combined with other stable systems the combination may be (unexpectedly) unstable because of an over-all *reduction* of gain.

ROOT LOCUS METHOD

The Nyquist criterion has the advantage of working with the graphical transfer function data but it does not give detailed information directly on the transient behaviour (though we will quote design rules for phase and stability margin). Such detail requires knowledge of the poles of $G(p)/[1+G(p)H(p)]$ throughout the p-plane. Furthermore, the situation often involves an adjustable parameter, e.g. the gain of a proportional feedback controller, and it is desirable to see the effect of different values of this parameter directly.

Evans[4] (and independently Westcott) suggested a graphical method for studying the roots of the return difference $1+GH$ as a function of a linear parameter. The variation of the roots with the parameter leads to a root locus in the p-plane. Suppose we can write the transfer function in the form of rational polynomials using this time the assumption of *positive* feedback

$$\bar{x} = \frac{G}{1 - GH}\bar{s} = \frac{h(p)}{f(p) - \mu g(p)}\bar{s} \equiv F(p)\bar{s} \tag{1.20}$$

where f, g and h are polynomials of the form $f(p)=p^n+\cdots$, $g(p)=p^m+\cdots$ and μ is a real parameter.† Then the poles arise solely from the zeros of $f-\mu g$. Note again that the expression is not written for the conventional difference of a feedback control system and the actual sign of μ must be allowed for.

We assume that the polynomials can be written in terms of their roots as $f(p)=[p-p_1][p-p_2]\cdots$ and $g(p)=[p-z_1][p-z_2]\cdots$ so that we have at a pole of $F(p)$:

$$\frac{[p - p_1]\,[p - p_2]\cdots[p - p_n]}{[p - z_1]\,[p - z_2]\cdots[p - z_m]} = \mu \tag{1.21}$$

The left hand side is the product/division of complex numbers whose *phases* must add or subtract to the phase of the right hand side, which is $(2k+1)\pi$ for $\mu<0$ and $2k\pi$ for $\mu>0$, where $k=0, 1, 2\cdots$. We also note that when μ is zero, the required poles are the p_i, poles of the *open loop* transfer function, and as μ becomes large in magnitude the roots are the zeros z_i of the *open loop* transfer function.

Without proof we state the construction rules that follow from these observations:

(1) Plot the poles p_i and the zeros z_i of the *open loop* transfer function.
(2) n loci, symmetric about the real axis, emerge from the n poles p_i.
(3) Every zero z_i is a terminator for a locus; the remaining n–m loci terminate at "infinity" either along the real axis or along asymptotes that intersect the real axis at $[\sum p_i-\sum z_i]/[n-m]$, at angles $(2k+1)\pi/[n-m]$ (if $\mu<0$) or $2k\pi/[n-m]$ (if $\mu>0$).
(4) A locus lies on the real axis if the sum of zeros and poles to its right (including the pole it came from) is odd ($\mu<0$) or even ($\mu>0$).
(5) Loci join or leave the real axis vertically at a point p_c, where $\sum 1/[p_c-p_i]=\sum 1/[p_c-z_i]$.

The following example of a simple second order system can, of course, be solved exactly as a quadratic but illustrates the application of the rules. We take $G(p)=1/p$ and $H(p)=\mu/[p+\tau]$. We seek, therefore, the zeros of $p^2+\tau p-\mu=0$. There are no open loop zeros z_i: $m=0$. There are two open loop poles: $p_1=0$ and $p_2=-\tau$; $n=2$. Rules (3) and (5) give the same point, $-\frac{1}{2}\tau$. It is seen (Fig. 1.12) that the nature of the transient response and stability depend substantially on the sign of the parameter.

We shall employ the method in Chapter 4 for a system of more complexity. It is to be remarked, however, that even more complicated systems very commonly are well approximated by a second or third order system in the range of interest, which is usually close to the imaginary axis. Thus simple graphical methods are indeed helpful for a preliminary analysis, and with the availability of computer aid design it can be expected

†This is not the gain of the controller, due to the form chosen for the polynomials, but is the gain times a suitable product of the time constants of the transfer functions.

that even more complicated systems will yield to a root locus study of the effect of varying one or even more parameters.

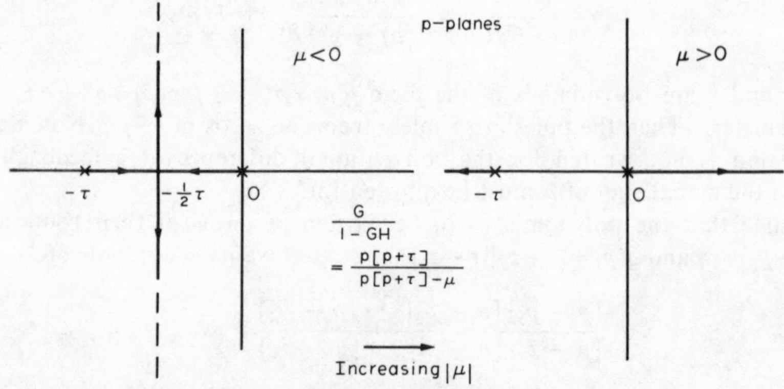

FIG. 1.12. Elementary root locus example.

STABILITY MARGINS

The Nyquist criterion itself determines a limit of stability, sustained oscillations. A satisfactory system must be offset from this limit with sufficient damping of transients to ensure they die out in acceptable times. Fortunately, the Nyquist diagram serving the Nyquist criterion can itself be utilised to determine approximate transient conditions in the form of stability margins, at least if we omit conditionally stable systems. Figure 1.13 illustrates the conventional designer's guide to ensure adequate transient behaviour of a stable system. Phase and gain margin definitions are shown.

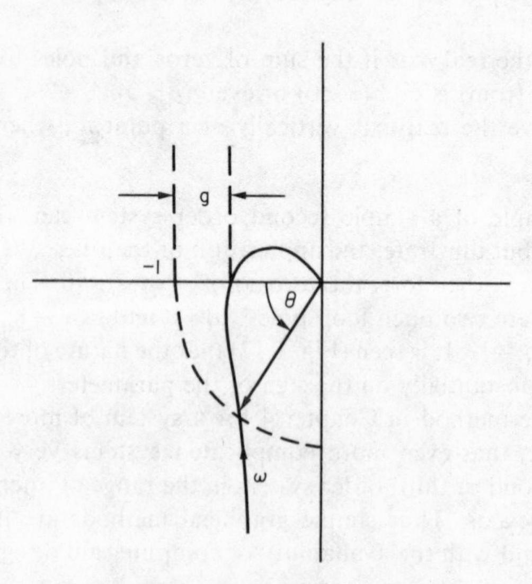

FIG. 1.13. Stability margins.

The setting of margins is open to some choice. Widely used margins in control practice are to take a gain margin g of 8 db and a phase margin θ of 35°. However, in the design of a nuclear power station for a life of, say, 30 years and to allow for the deterioration of components, it might be wiser to adopt values of 12 db and 50° respectively in order to see that temperature transients are well damped and do not lead to overshoots above critical temperatures, etc.

Review of Probability Theory

The reader will, of course, have studied probability theory to be used in chapters 6 and 7 in studies of reactor fluctuations and system reliability. This section briefly reviews the notation and concepts needed for those chapters with emphasis on some of the pitfalls that occur when we leave deterministic studies and consider fluctuations and variations. Wonnacott [5] and Feller[7] are good introductory texts for the reader needing further preliminary study. Certainly probability has been the graveyard for the reputation of many mathematicians, including d'Alembert and Leibniz; holes are still being dug in nuclear engineering. The first difficulty, perhaps, is in defining probability itself.

FOUR DEFINITIONS OF PROBABILITY

An early definition of probability centred about the outcome of events being "equally likely" and the corresponding idea of a probability space of such equally likely results. Thus the probability p of a coin turning up heads is commonly taken as $\frac{1}{2}$ since the complementary probability $p' = 1 - p$ of turning up tails is equally likely and the total probability, corresponding to a certainty of one event or the other, is one. We try to argue that there are two states and as we have no reason to distinguish them they are equally likely if the coin is "fair". One conceptual difficulty is that we do distinguish them and if we refine this distinction until it disappears, we are talking vacuously since we cannot tell which is heads and which is tails. Problems in quantum mechanics, the Gibbs paradox in thermodynamics, etc., stem from this difficulty.

We must also be careful in discussing compound states and their likelihood, such as tossing two coins (either two coins simultaneously or the same coin twice in succession; is there any difference?). The apparent states are both heads, both tails, or mixed. This, however, obscures the fact that the third state has two forms (HT or TH) and, after a little thought, we might be convinced that there are four "equally likely" states. Overall, this concept as a definition has the unfortunate defect of circular argument: what is "equally likely" if not "equally probable"?

A second approach, favoured here, uses observed frequencies. If over n trials there are m favourable outcomes, we might say the probability of the favourable outcome per trial is the frequency m/n. Here further difficulties become apparent. After another trial it is possible (even "likely") that the new m/n has changed. Does this mean that the "real" probability has changed? We meet this objection by saying that in the first place it is certainly possible—we don't know enough about the circumstances of the experiment. But even so, it may be that there is an underlying probability as yet unknown for which the observed frequency is the best estimate we have. The powerful central limit theorem or law of large numbers represents an argument to justify this idea with its support for obtaining improved estimates of the underlying probability by conducting more trials. We meet the difficulty in fact by contemplating the hypothesis that the underlying probability is constant and asking whether the observed fluctuations in frequencies can reasonably be expected on the basis of this hypothesis or not. We can often disprove the hypothesis, never prove it. This is where science differs from mathematics.

An associated difficulty is really a difficulty of another sort and arises from the question: How does the outcome of the next trial depend on the outcome of the last trial (or the last two, three, etc.)? In its classical form, this is the string of successive heads: is the next outcome more likely to be a tail to make the observed frequency tend to the theoretical $\frac{1}{2}$ or a head because the observed frequencies suggest a biased coin?

These difficulties are enough to spur the mathematician into a third definition, the axiomatic treatment of probability. Here the approach is to let p be something subject to laws and operations with rigorously deducible consequences. The mathematician can be satisfied; it is somebody else's problem to say whether this has anything to do with "probabilities".

Finally, a fourth definition centres on subjective gambling: the probability of an event is what you are willing to bet on the outcome. This definition will not be pursued, but it has interesting applications in nuclear engineering when discussing the possible outcome of severe accidents and the "probability" that members of the public, say, are willing to accept increasingly severe consequences.

Here we treat probability largely in accordance with the second definition, that the best estimate of the probability of an event is the frequency at which it occurs.

CLASSIFICATION OF PROBABILITY SEQUENCES

The probability of an isolated event is not of great interest and the interest turns more on compound events. The first of these is perhaps essentially linked with the frequency definition. Suppose we have a sequence of events $\{x_i: i=1, 2, 3, \cdots, n\}$, n finite or infinite. The summation over the set leads to our observed frequency of a particular result x. Is there any connection, however, between the order of events in the sequence? It is helpful to contemplate a classification based on successively complex connections.

(a) RANDOM PROCESSES

In this classification there is *no* dependence of the outcome of a trial on the *sequence* of previous trials, only on the average over the set of trials. It is in practice not easy to produce events purely randomly capable of satisfying the many stringent tests that this classification should pass. Thus the numbers offered as "random" in a computer are certainly nothing of the sort. Great effort is put into the gambling machine ERNIE to try to select prizes in such an independent way (so that people buying serially numbered tickets, for example, are "fairly" treated). The best source of random events is probably the radioactive decay leading to α-particles if the time scale of observation is short (hours) compared to the mean life (thousands of years). Radioactive decay seems to satisfy most stringent tests that the decay of one atom is uninfluenced by the decay of another of the same species. We can return in a moment to discuss some of the probability distributions falling into this random category.†

(b) MARKOV PROCESSES

These refer to situations where the outcome of the next trial is dependent on the last trial but not on earlier trials. In other words, to specify the probability of the future state we need to know the present state but not any earlier history. Markov processes are also known, obviously, as chain processes and have an immediate example in the fission chain reaction: the probability of seeing a neutron from fission is enhanced if we recently saw a neutron from fission that could have led to a continuing chain reaction. If the earlier event showed no neutron, there may be no neutrons to continue the chain. Markov chains may be absorbing (with the possibility of terminating) or non-absorbing, e.g. the daily state of the weather, dependent to some extent on yesterday's weather, going on for ever. In the special circumstance that the non-absorbing chain eventually repeats itself, we may have an ergodic Markov chain.

Without going further into the hierarchy of connectedness, consider an example where the state of a system is one of six results to be determined by throwing a conventional die. As such, the probability of transferring to the next state is independent of the current and past states; we have a random system. If, however, the transfer probability is modified by ignoring the result of the die if it leaves the state unchanged and throwing a replacement until a result leads to a change of state, then the transfer probabilities are dependent on the current state—a Markov process. If the restriction is extended further so that the die is discounted of the current and preceding results, the transfer probability now depends on the last as well as the present—a non-Markovian stochastic process.

This illustrates how compounding one level of process may lead to a process further removed from random. Note also that the Markov property is analogous to the deterministic principle of, say, mechanics, that the future can be determined from the present state without regard to the past except for the extension to a probabilistically determined future.

PROBABILITIES AND DISTRIBUTION FUNCTIONS

When the various outcomes are discrete we may readily envisage the probability of a particular outcome x from the set of possible outcomes X as p_x with values for each outcome whether these are finite or denumerably infinite. There is a natural requirement that *some* outcome is inevitable and we will have the normalization $\sum p_x = 1$. We are also interested in *continuous* distributions. This might be a distribution in time, the interval, say, before a specified event occurred. It might be a distribution in the form of a continuous signal recorded, say, along magnetic tape (in analogue—not digital form), the outcome being some varying magnitude. Here the probability of seeing an event *exactly* at value x is negligible. We can talk, however, of the probability distribution

†Emphasis is sometimes added to the definition of this description by talking of *purely* random, independently random, etc. We shall use the term random itself for this special category and reserve the term stochastic for more general non-deterministic phenomenon that may nevertheless be dependent on sequence and hence not random.

function (or PDF) via the probability $f(x)\delta x$ of seeing an outcome in the band δx about x, thus establishing $f(x)$ as the PDF. Of course we have the same normalization that $\int_{-\infty}^{\infty} f(x)dx = 1$.

Note that we write the PDF as $f(x)$ not $p(x)$ since $f(x)$ is not a probability as such. It has the dimensions of inverse x and only $f(x)\delta x$ is a (non-dimensional) probability.

The PDF indicates the probabiluty of observing a value "around" x. Related to this is the cumulative distribution function (CDF) to give the probability of seeing a value anywhere up to x:

$$F(x) = \int_{-\infty}^{x} f(x')dx' \tag{1.22}$$

From the normalization equation, $F(\infty) = 1$. Of course, in many situations $x<0$ is not meaningful and $f(x)$: $x<0$ is zero.

MOMENTS

The distribution of observations, with discrete probabilities or continuous PDF, is usefully summarised by the moments of the distribution and thereupon generally written in the two forms

$$i\text{th moment: } <x^i> \equiv \sum_x x^i p_x \quad \text{or} \quad \equiv \int_{-\infty}^{\infty} x^i f(x)dx \tag{1.23}$$

Special cases are:

(a) Zeroth moment: normalization to certainty

$$<1> = 1 = \sum_x p_x, \quad = \int_{-\infty}^{\infty} f(x)dx$$

(b) First moment: the mean

$$<x> = \sum_x xp_x, \quad = \int_{-\infty}^{\infty} xf(x)dx$$

(c) Second moment:

$$<x^2> = \sum_x x^2 p_x, \quad = \int_{-\infty}^{\infty} x^2 f(x)dx$$

There are various ways of writing the expectation of an event affected by fluctuation and we have, for example,

$$E(g(x)) = <g(x)> = \bar{g} = \sum_x g(x)p_x, \quad = \int_{-\infty}^{\infty} g(x)f(x)dx \tag{1.24}$$

Here we shall normally reserve the use of the overbar for the *observed* mean, etc.

The reader will also be familiar with the observational measures of spread and the corresponding expectations of mean and variance and standard deviation (Table 1.4).

TABLE 1.4.

Measures of spread

	Observed	Expected
Mean:	$\bar{x} = \sum_{i=1}^{i=n} x_i/n, \quad \int_{-\frac{1}{2}T}^{\frac{1}{2}T} x(t)dt/T$	$\langle x\rangle = \Sigma\, x p_x, \quad \int_{-\infty}^{\infty} x f(x) dx$
Variance:	$\sigma_0^2 = \sum_{1}^{n} [x_i - \bar{x}]^2/n, \quad \int_{-\frac{1}{2}T}^{\frac{1}{2}T} [x-\bar{x}]^2 dt/T$	$\sigma^2 = \langle (x-\bar{x})^2\rangle = \langle x^2\rangle - \langle x\rangle^2$
Standard deviation:	$\lvert\sigma_0\rvert$	$\lvert\sigma\rvert$

PRIMARY LAWS OF PROBABILITY

The two primary laws for finding probabilities of compound events may be written using the notation of set theory. Let Ø be the null set, U the universal set (no event and all events), let A' represent the complement of event A (i.e. NOT A) and $\cup$ the union of events, $\cap$ the intersection of events.

(1) The probability of observing at least one of two events, A and B, is

$$p(A \cup B) = p(A) + p(B) - p(A \cap B) \tag{1.25}$$

There are several corollaries arising from $p(\emptyset) = 0$ and $p(U) = 1$, a statement that the probability of *no* event in the set is zero and the probability of *some* event in the set is unity. If the events are mutually exclusive then $A \cap B$ is null and the corresponding probability is zero. Thus for *mutually exclusive* events

$$p(A \cup B) = p(A) + p(B)$$

Similarly, the probability of seeing an event A or not seeing it is

$$p(A \cup A') = p(A) + p(A') = p(U) = 1$$

The second relationship is really a definition of conditional probability, $p(B|A)$ or the probability of B *given* the occurrence of A:

$$p(A \cap B) = p(B)p(A|B) \tag{1.26}$$

As a corollary

$$p(A)p(B|A) = p(B)p(A|B) \tag{1.27}$$

Only if B is drawn from a random distribution independent of the distribution of A will the conditional probability $p(B|A)$ be the same as the probability $p(B)$.

It is difficult to overemphasise the significance of this restriction on using $p(B)$ for $p(B|A)$. If safety is being evaluated, for example, a misunderstanding on this point may give a totally spurious confidence for the safety and reliability of a compound system if it is not recognised that the failure of one safety device may be dependent on the failure of the other. This point will be taken up in detail, of course, in Chapter 7.

EXAMPLES OF RANDOM PROBABILTIY DISTRIBUTIONS

The binomial distribution for the likelihood p_x of seeing exactly x occurrences of repeated identical events in n trials, where the probability per event is p, can be expressed as

$$p_x = {}^nC_x p^x[1 - p]^{n-x}; \quad <x> = pn; \quad \sigma^2 = npp' \tag{1.28}$$

where, of course, x is integer, ${}^nC_x$ the combinatorial coefficient which can be written $n!/x![n-x]!$ and $1-p$ is the complementary probability of not seeing one event in a trial. Two limiting cases of the number of trials are of interest:

(a) *Gaussian distribution* (law of large numbers). Suppose the number of trials grows very large and the results closer together, going over to a continuous distribution. We obtain

$$f(x) = \frac{1}{\sigma\sqrt{2\pi}} \exp\left\{-\tfrac{1}{2}\left[\frac{x - <x>}{\sigma}\right]^2\right\} \tag{1.29}$$

Some writers reserve the term normal distribution for the case $<x> = 0$, $\sigma = 1$, so that $f(x) \equiv (2\pi)^{-\frac{1}{2}} e^{-\frac{1}{2}x^2}$.

(b) *Poisson distribution* (law of small numbers). Suppose the number of trials grows very large but the probability of success p diminishes such that the expected rate of successes $<x> = np$, remains constant. We obtain

$$p_x = \frac{<x>^x}{x!} e^{-\langle x \rangle} \tag{1.30}$$

and for this special case $\sigma^2(= <x^2> - <x>^2) = <x>$ so that the variance-to-mean ratio (VTMR) sometimes called the relative variance (rel. var.), is *unity* for a Poisson distribution (and thus makes a useful test for a random Poisson distribution).

Problems for Chapter 1

1.1 Regard the Laplace transform variable p as a complex number and use the Bromwich integral

$$x(t) = \frac{1}{2\pi j} \int_{p_0 - j\infty}^{p_0 + j\infty} \bar{x}(p) e^{pt} dp$$

to find the inversions of Table 1. 1 direct.

1.2. Use Cauchy's thorem

$$\oint f(z)dz = 2\pi j \Sigma \text{Res}(z_i)$$

to find the inversions direct.

1.3. Given $\bar{x}(p) = 1/p[p + \lambda]$, find the inversion by (a) partial fractions and Table 1.1 and (b) by direct integration.

1.4. Show that for a simple pole, $R_i(p_i) = g(p_i)/f'(p_i)$.

1.5. Represent the Dirac delta distribution $\delta(t)$ as a Fourier series taking a fundamental and two harmonics only in a graphical representation. What is the Fourier transform of $\delta(t - \tau)$?

1.6. The spectral power density (using the analogue of power having no phase, only magnitude, whilst current has phase and amplitude) is obtained from a Fourier transform as the square of the amplitude. Show that the power spectral density of the time dependent function cos $\omega_0 t$ is $\omega^2/[\omega^2_0 - \omega^2]^2$. Put $\omega - \omega = \delta\omega$ and find simple approximations for the spectral power density for the cases (a) $\omega \ll \omega_0$, (b)$\omega \simeq \omega_0$, and (c)$\omega \gg \omega_0$.

1.7. Find the (approximate) values in db for magnitudes of: (a) 1/10, (b) 1/2, and (c) 8. For the simple lag (Fig. 1.5) find the corrections to the break frequency asymptotes at (i) half/double the break frequency, and (ii) tenth/ten times the break frequency for amplitude and phase. Derive similar information to help plot a simple lead: $1 + j\omega\tau$.

1.8. Sketch Bode and Nyquist diagrams for simple lead $(1 + j\omega\tau)$ and integrating element $(1/j\omega\tau)$. Show analytically that the simple lag has semicircular form in the Nyquist diagram. Construct the heat exchanger approximation $e^{-p\theta}/[1+p\tau]$ in Bode diagrams, using the P_{11} approximation, for $\theta = \frac{1}{2}\tau$.

1.9. We may well need to introduce a phase lead to compensate for lags that tend to bring about instability. These are difficult to realise in practice from a pure differential term $p \rightarrow j\omega$ and more commonly we employ a control element with lead and lag of the form

$$G(j\omega) = \frac{1+j\omega\tau_1}{1+j\omega\tau_2} \quad (\tau_1 > \tau_2)$$

Sketch Nyquist and Bode diagrams.

1.10. A second order damped system in normal form has the equation

$$\frac{d^2x}{dt^2} + 2\zeta\omega\frac{dx}{dt} + \omega_n{}^2x = S$$

where $\zeta > 0$ is the damping factor and ω_n a natural frequency. Find the transfer function and hence plot Bode and Nyquist diagrams for a range of damping factors $\zeta < 1$, $\zeta = 1$, and $\zeta > 1$. Sketch the transient response for large and small damping factors.

1.11. The 2 × 2 system is shown in diagram A. Find a matrix and vector representation of the input

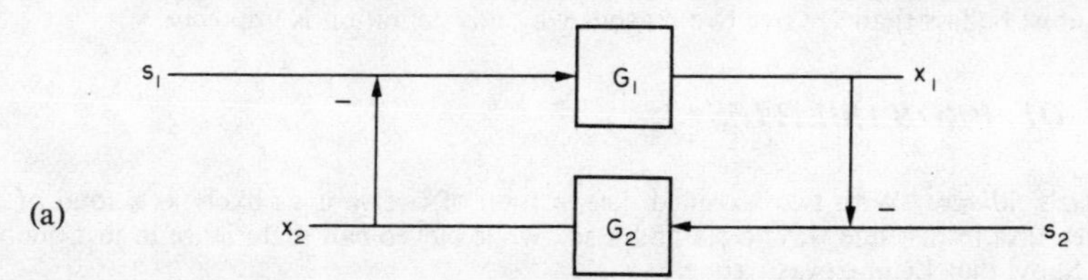

and output and hence the generalised matrix transfer function $G_{ii'}$. In diagram B two feedback loops in parallel are shown. Show that the forward and *one* feedback loops can be represented as the equivalent transfer $G_1 = G/[1 - GH_1]$ and hence write the over-all transform for the two loop system. Show how this can be generalised to a three-loop system, and on *n*-loop system.

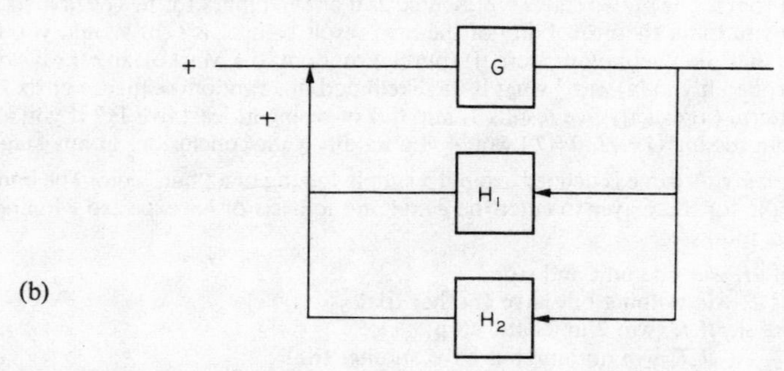

1.12. For the following transfer functions find *K*-values for which the system is stable: Assume negative feedback.

(a) $G = \dfrac{1}{[1+\Lambda p][1+\tau p]}$; $H = \dfrac{K}{1+\theta p}$ for $\theta = 2\tau = 4\Lambda$

(b) $G = \dfrac{1+0.2p}{p[1+1.43p][1+0.0066p]}$, $H = \dfrac{K[1+0.4p]}{[1+2.5p][1+0.02p]}$

1.13. In applying Routh's criterion, the column of numbers formed from the coefficients Table 1.3 must all be positive for a stable system. Apply the criterion to the following cases:

(a) p^3-p.
(b) p^3+p^2-4p-4.
(c) $p^3+6p^2+11p+6$.

More generally the criterion may be applied to the equation $a_0p^n+a_1p^{n-1}+\ldots a_n=0$ by considering determinants obtained from the array of coefficients, an n × n matrix:

$$\begin{pmatrix} a_1 & a_0 & 0 & 0 & \ldots & & \\ a_3 & a_2 & a_1 & a_0 & \ldots & & \\ a_5 & a_4 & a_3 & a_2 & a_1 & a_0 & \\ \ldots & & & & & & a_n \end{pmatrix}$$

An element in this scheme whose index lies outside the range 0, n is taken as zero. The determinant of this array and the successive minors formed of order $n-1$, $n-2$, etc., must all be greater than zero for all the roots to be within the left half-plane. Show that this rule gives rise to Table 1.2 and extend the table to fourth order. What direct conclusions can you draw if (i) the last coefficient of the polynomial a_n is positive (negative), or (ii) all the coefficients have the same sign (some have opposite signs)?

1.14. *Closure of the Nyquist diagram.* Suppose the transfer function has a term in $1/p$. Sketch the transfer function and see that it does not have $G(o^-)=G(o^+)$. It is therefore necessary to "close" the transfer function loop for the purpose of a Nyquist test of stability. To do this consider the loop of small radius around the origin as shown in Fig. 1.10 in polar form as $re^{j\theta}$, where r tends to zero and θ passes from $-\frac{1}{2}\pi$ to $+\frac{1}{2}\pi$. What happens to the term in $1/p$ for these values? Hence "close" the loop. Using this analysis, determine whether the case $1+GH=1+K[p+\lambda]/p[p+\gamma]$ is stable or unstable for (a) $K>0$, (b) $K<0$, and with $0<\gamma<\lambda$.

1.15. Stability might be defined in terms of the open loop gain assuming negative feedback, $KG(\omega)H(\omega)$, that for any ω such that the phase of the open loop gain is π radians or 180°, the magnitude of the open loop gain must be less than 1. Give two reasons why this definition is imprecise.

PROBLEMS OF PROBABILITIES

1.16 (a) *Leibniz's fallacy.* "With two six-sided dice, a total of twelve is as likely as a total of eleven. For 12 can materialise in just one way: a six and a six, while eleven can materialise in just one way: a six and a five." Show that Leibniz was wrong.

(b) *d'Alembert's fallacy.* "To get one head with two tosses of a coin, heads on the first toss makes the second unnecessary. So there are three cases: H then TH and TT of which only the last is unfavourable. So the probability of getting at least one head is $\frac{2}{3}$." Do you agree?
Hint: Use a *sample space* of mutually exclusive events with equal probabilities.

1.17. The outcome of a trial is one of two states represented as 0 or 1 in binary form. The first six results are 101111. (a) What do you think the probability of the next result being 1 is? (b) Would you modify this if you knew for sure that the fluctuations were (i) (purely) random, (ii) Markovian? (c) If you took the hypothesis that the probability in (a) was $\frac{1}{2}$ what is the likelihood in a random sequence of six trials of seeing (i) exactly this pattern, (ii) exactly five results 1, and (iii) of seeing at least five 1s? If you are now told that the trials are coin tossing ($1=H$, $0=T$) would you modify your conclusions at any stage?

1.18. *St. Petersburg Paradox.* A game is defined around a simple tossing of a "fair" coin. The immediate object is to find a fair stake for the player to enter the game on the basis of his expected winnings. The winnings are arrived at as follows:

On the first result, if H, win one unit and stop;
if T, win nothing but have another trial.
If taking a second result, if H, win 2 units and stop;
if T, win nothing but have another trial.

The game is repeated indefinitely with the return for H *doubling* each time until an H stops the game and payment is made.

(a) Draw an event tree to show the possible paths for the result.
(b) Find the expected return from the game in units.
(c) Propose a fair stake equal to this expected return.

Would you play? What relevance does this have to safety studies?

1.19. *Crown and anchor* (or AceyDeucey). A board of six entries corresponding to three six-sided dice. Bets are placed on the board and payment made (with return of stake) according to the number of dice showing the bet times the stake. For example, bet one unit on "3" and two dice show 3 so receive stake and two units. Is this a fair game (or why is it illegal in the services?) Comment on the analysis that since there are three chances of winning for six squares, you must win half the time.

1.20. Suppose buses arrive randomly with uniform probability distribution function at a mean rate $b\ s^{-1}$. What proportion of the time will you have to wait longer than the mean time between buses, $1/b$? Is this affected by the time you arrive at the bus stop as evidenced by the number of people waiting? Is the model of random bus arrivals realistic?

1.21. In a group of 200 apparently similar reactor operators there are three to each of whom three accidents occurred last year out of a total of 100 reported accidents to the operators as a group. The operations manager is considering suspension of these three men from reactor operations on the grounds that they are "accident prone". On the basis of Poisson statistics advise the manager whether he can expect to decrease the human error rate by such a move.

1.22. This question serves to introduce the material of Chapter 6. A casual labourer has the choice of working in either of two locations. He goes to work by train, catching either the north- or south-going train according to which train arrives first at the station. The worker keeps no regular time and so may be said to arrive at the station at any point in the hour, uniformly at random. Nevertheless, employment records show that he works five times more often in the north than in the south. There are as many trains in a day going north as south. On the basis of a deterministic schedule-keeping time-table, deduce something about the train times to account for the random output (work place) no longer being uniformly distributed.

Suppose there was no deterministic train time-table but the service consisted of one train which travelled backwards and forwards (shuttle) from end to end without pause at either terminus. Deduce something about the position of the *worker*'s station to account for the observed distribution of the random output.

1.23. (a) Consider a random event with a probability distribution function $f(x)$. What is the joint probability of one result in δx_1 about x_1 *and* another in δx_2 about x_2? What is the triple event probability?

(b) In another, Markov system, the probability of a value in δx_1 about x_1 *and* a value in δx_2 is given by $g(x_1,x_2)\delta x_1 \delta x_2$. What is the probability of a result in δx about x? What is the triple event probability?

(c) In a higher order system the triple event probability must be specified for complete knowledge in the form $h(x_1,x_2,x_3)\delta x_1 \delta x_2 \delta x_3$. What are the double and single event probabilities?

References for Chapter 1

1. M. F. Gardener and J. L. Barnes, *Transients in Linear Systems* (1) Wiley, New York, (1st edn.), 1942.
2. A. Pollard, *Process Control*, Heinemann, London, 1971.
3. R. V. Churchill, *Complex Variables and Applications*, McGraw-Hill, New York, 1960.
4. W. R. Evans, *Control System Dynamics*, McGraw-Hill, New York, 1954.
5. T. H. Wonnacott and R. J. Wonnacott, *Introductory Statistics* (2nd edn.), Wiley, New York, 1972.
6. R. E. Barlow and F. Proscan, *The Mathematical Theory of Reliability*, Wiley, New York, 1965.
7. W. Feller, *An Introduction to Probability Theory* (2nd edn.), Wiley, New York, 1957.
8. J. C. Tyror and R. I. Vaughan, *An Introduction to the Neutron Kinetics of Nuclear Power Reactors*, Pergamon, Oxford, 1970.
9. S. Fabic, Review of the existing codes for loss-of-coolant accidents, *Adv. in Nucl. Sci. and Tech.* 10. Plenum, New York, 1977.
10. P. M. Hammond and C. L. Seebeck, *An. Maths. Monthly* **56**, 243 (1949).
11. W. R. Evans, Graphical analysis of control systems, *AIEE* **67**, 547 (1948).
12. E. J. Wright, Digital computer generation of root locs, *I. Jl Mech. EE* **2** (3) 41–49 (1974).
13. J. B. Knowles, Principles of nuclear power station control, *Jl BNES* **15**, 225 (1976).
14. A. Henry, *Nuclear-Reactor Analysis*, MIT, Cambridge, 1975
15. J. Weisman (ed.), *Elements of Nuclear Reactor Design*, Elsevier, Amsterdam, 1976.

CHAPTER 2

Neutron and Precursor Equations†

In the fission process, nuclear fuels such as ^{235}U are fissioned by neutrons. This produces immediate γ-radiation, high speed fission fragments and usually between two and three high speed or fast neutrons. The majority of the heating effect comes from the kinetic energy of the fragments or fission products. The time taken for the fission itself is negligible compared with the time taken for one neutron to continue the chain process by fissioning the next fuel nucleus, this latter time being of the order of 10^{-3} to 10^{-6} s, depending on reactor type.

These times are so short that control of a nuclear power plant would appear to be difficult. Fortunately, the balance of the chain reaction, in which for a steady production of power it is necessary that there should be one and only one neutron released from fission causing a further fission, is assisted by the phenomenon of *delayed neutrons*. The fission fragments themselves are radioactive. Their decay schemes lead to a release of energy and thus a so called after-heat that continues to arise for some time after the fission process has stopped. Some of these schemes include a decay process that releases a neutron, and since there is a time lag before this step occurs, the result is a delayed neutron.

The fission fragment decay is governed by probability so that the decay times vary. Nevertheless, average or mean decay times can be identified and these turn out to have lengths of the order of seconds and thus very much longer than the characteristic times of the direct or prompt fissioning process. Even though the proportion of neutrons released via the decay of fission fragments is small, less than 1% of the total neutron release, this much-extended time scale makes the control of a nuclear power plant relatively easy. It is the purpose of this chapter to see how the physics of the fissioning process leads to equations that describe, in part, the reactor behaviour as a function of time.

The description of the time dependent behaviour of a nuclear reactor is generally so complicated as to motivate us to separate these matters and the solution of the corresponding time dependent equations from the description of the behaviour of the reactor (and its associated plant) in space, neutron energy, temperature, etc. Our ambition in this chapter is to limit the problem to dealing with a set of ordinary differential equations to describe the behaviour of neutrons and the delayed neutron precursors of the *form*

†This chapter is directed to evaluating form and data for the "lumped" model equations used subsequently; it may be omitted on a first reading. Chapter 3 recapitulates the essential data and equations.

$$\frac{dn}{dt} = \frac{\rho - \beta}{\Lambda} n + \sum_{i=1}^{i=I} \lambda_i c_i + s \quad i = 1, 2, 3, \cdots, I \tag{2.1}$$

$$\frac{dc_i}{dt} = \frac{\beta_i}{\Lambda} n - \lambda_i c_i \tag{2.2}$$

with their associated initial conditions $n(0)$ and $c_i(0)$.

In the simplest possible case, these symbols can be understood as follows: the dependent variables are $n(t)$ (the number of neutrons) and $c_i(t)$ (the number of precursors of the ith type), all varying generally with the time t.

Coefficients of these equations can be understood to signify:

- ρ reactivity (non-dimensional), a measure of net production probability.
- β_i fraction of precursors of the ith type produced in fission compared to the total production of (prompt) neutrons and precursors of all types.
- λ_i mean decay rate of precursors of ith type (s^{-1}).
- Λ neutron reproduction time (s).
- s independent source rate (neutron/s).
- β total delayed neutron fraction $= \sum_i \beta_i$.

In principle all coefficients might be time dependent but in practice one would normally suppose that only the reactivity $\rho(t)$ is likely to vary with time and it is common to solve the equations parametrically with constant values of all the other coefficients. Indeed, the equations have been specially manipulated into just this form to allow just such a simplification, where $\rho(t)$ expresses the essential time dependency of the difference between production and removal rates per neutron present in the reactor. It is also the coefficient that links the kinetics equations (2.1) and (2.2) with the other significant processes in the reactor such as energy balances and consequent temperature changes.

To quote simple equations is not to justify them. This chapter is therefore devoted to reviewing suitable values of the physical constants to be used in the equations and to the justification whereby a full description in space, energy and time may be reduced to the *form* of eqns. (2.1) and (2.2) and to some extent, when this is not possible. We shall see that such a reduction introduces *effective* values of both the dependent variables and some of the coefficients that arise from averaging over the space, energy, etc., involved. The chapter concludes with some practical methods for finding these effective values.

Physical Data

A useful compilation of physical data for reactor kinetics is Keepin's monograph[1] to which we could add the valuable review by Tuttle[2] as well as the older *Handbook of Reactor Physics*[7]. We draw heavily on these compilations, and reference can also be made to the recent work of Besant *et al.*[10] and Walker and Weaver[13]. Evaluated data in computer compatible form is available (ENDF/B, etc.)[16].

Delayed neutron precursors are formed as fission fragments. These fragments either emit a neutron directly after some characteristic radioactive decay period or through a sequence of decays. The yields of delayed neutrons β as a fraction of the mean number of neutrons emitted in fission $\bar{\nu}$ (prompt and delayed) is less than 1% for all fissionable isotopes, but the decay or hold-up times in the form of precursors are large compared to the time taken for one prompt neutron to produce another in fission, i.e. precursor delay times $\tau = 1/\lambda$ between $\frac{1}{3}$ s and 80 s. This leads to the significant influence of delayed neutrons on the rate of response of a reactor, at least in normal circumstances when reactivity changes are small so that the delayed neutrons are a principal part of the difference between production and removal.

Thirty to forty precursor chains have been identified chemically, some only tentatively, and the most important come from the halogens bromine and iodine. There is no reason to suppose that each link in the chain ending with a delayed neutron is governed by any other than the normal laws of radioactive decay and the corresponding distribution of precursors and subsequent neutrons about the mean implied by mean times $1/\lambda$. But the experimental difficulties of direct identification of precursors with short mean lives are substantial. The generally accepted approach is to fit observed dynamic behaviour in reactors (ideally in "clean", easily analysed, geometrically simple assemblies) to an assumed model of single precursor elements that decay directly to delayed neutrons, i.e. without going through a sequence in the decay chain. The number of such equivalent precursors (or groups) is itself determined from the "best" fit of the data to the model.

It is almost universally found† that for a single fissionable isotope a best fit is achieved using six of the pseudo-precursors in the sense that five gives a poorer predictive fit to $n(t)$ and seven lead to larger uncertainties in the fitted standard deviations of the parameters. Thus six groups are used, even though some of these groups must clearly contain more than one physical species.

The source of data in "clean" assemblies implies that the data in large part come from fast fissionable assemblies even though the results may be used subsequently in application to thermal power reactors. This is justified not only by the greater precision of data from fast assemblies (built specifically for this purpose in many cases) but also by the observation that the *relative* data appear to be quite insensitive to the energy of fissioning neutrons and are sensitive only to the absolute yields or values which may be separately determined for fast and thermal fission.

ABSOLUTE AND RELATIVE YIELDS

The convenient parameters in the neutron dynamic equations are the (mean) decay rates λ_i and the delayed neutron fractions β_i of the precursor groups, the fraction of the total yield of neutrons in fission $\bar{\nu}$ that leads to a neutron in the ith group. However, the immediately available experimental information is reported in terms of the *absolute* neutron yield per fission rather than the delayed neutron fraction. The group *relative* yield a_i is also employed and the connection between these parameters can be displayed as:

$$\text{absolute yield per fission} = \beta\bar{\nu}$$
$$\text{group relative yield } a_i = \beta_i/\Sigma\beta_i = \beta_i/\beta$$

†Besant *et al.*[10] have recently suggested an approach taking a more direct physical representation, but the simple six-group interpretation is now well established.

Table 2.1 gives recommended absolute yields $\beta\bar{\nu}$ for a total over all groups in the main fissionable isotopes (after Tuttle and Keepin). The present author believes that in particular the standard deviations quoted are optimistic, but there is a difficulty in making piecemeal changes without degrading the consequent data, so perhaps these are the best to be used in practice. Differences in the yield of ^{235}U and ^{239}Pu are especially notable.

TABLE 2.1.

Recommended absolute delayed neutron yields from neutron induced fission[1, 2]

Isotope	$\beta\bar{\nu}$	SD	%SD
^{232}Th	0.0545	± 0.0011	±2.1
^{233}U	0.00698	0.00013	1.9
^{234}U	0.0106	0.0012	11.3
^{235}U	0.01698	0.00020	1.2
^{236}U	0.0231	0.0026	11.3
^{238}U	0.04508	0.00060	1.3
^{239}Pu	0.00655	0.00012	1.8
^{240}Pu	0.0096	0.0011	11.5
^{241}Pu	0.0160	0.0016	10

Values in the table are an average, where appropriate of fast and thermal fissionable assembly data. Besant *et al.*[10] indicates lower values for the more common isotopes, while Walker[13] proposes the values $0.0165 \pm 10\%$ for thermal and $0.0157 \pm 10\%$ for fast fission of ^{235}U which might imply a 5% reduction in the Table 2.1 value.

The (total) fractional yield β is obtained by dividing the information in Table 2.1 by $\bar{\nu}$-values, and at this point it is essential to employ $\bar{\nu}$-data appropriate to the energy spectrum of the reactor under consideration.

Table 2.2 gives a series of recomended $\bar{\nu}$-values for the mean neutron yield from fission

TABLE 2.2

Recommended values of mean neutron yield from fission (to include delayed neutrons)

(a) Thermal fission, after Hanna *et al.*[3]

Isotope:	^{233}U	^{235}U	^{239}Pu	^{241}Pu
$\bar{\nu}$	2.2866	2.4229	2.8799	2.934
SD	±0.0069	0.0066	0.0090	0.012

(b) Fast fission in ^{235}U spectra, after Keepin[1] and ref. 7

Isotope:	^{232}Th	^{233}U	^{235}U	^{238}U	^{239}Pu	^{240}Pu	^{241}Pu
$\bar{\nu}$	2.44	2.62	2.57	2.79	3.09	3.32	2.99
SD	±0.15	0.05	0.04	0.10	0.06	0.14	0.06

(to include delayed neutrons) for typical fast and thermal spectra. Keepin[1] gives further details of the variations of $\bar{\nu}$-values with the energy of neutron causing fission. The high $\bar{\nu}$-value in ^{239}Pu (important for breeding) only accentuates its low $\beta\bar{\nu}$-value.

The detail of individual groups is quoted in Table 2.3 in the form of decay rates and *relative* yields a_i with Tuttle's standard deviations as they arise from the fitting process. These, again, are based on fast assembly data and may be converted to fractional yields using the appropriate combination of data in Tables 2.1 and 2.2. The relative yields are substantially independent of the energy of fissioning neutrons and also of the absolute value due to the fitting of dynamic data rather than use of absolute measurements.

TABLE 2.3.
Group relative yields and decay constants

Group	i	Relative yield a_i	Decay constant $\lambda_i(s^{-1})$
^{232}Th	1	0.034 ± 0.003	0.0124 ± 0.0003
	2	0.150 ± 0.007	0.0334 ± 0.0016
	3	0.155 ± 0.031	0.121 ± 0.007
	4	0.446 ± 0·022	0.321 ± 0.016
	5	0.172 ± 0.019	1.21 ± 0.13
	6	0.043 ± 0.009	3.29 ± 0.44
^{233}U	1	0.086 ± 0.004	0.0126 ± 0.0006
	2	0.274 ± 0.007	0.0334 ± 0.0021
	3	0.227 ± 0.052	0.131 ± 0.007
	4	0.317 ± 0.016	0.302 ± 0.036
	5	0.073 ± 0.021	1.27 ± 0.39
	6	0.023 ± 0.010	3.13 ± 1.00
^{235}U	1	0.038 ± 0.004	0.0127 ± 0.0003
	2	0.213 ± 0.007	0.0317 ± 0.0012
	3	0.188 ± 0.024	0.115 ± 0.004
	4	0.407 ± 0.010	0.311 ± 0.012
	5	0.128 ± 0.012	1.40 ± 0.12
	6	0.026 ± 0.004	3.87 ± 0.55
^{238}U	1	0·013 ± 0.001	0.0132 ± 0.0004
	2	0.137 ± 0.003	0.0321 ± 0.0009
	3	0.162 ± 0.030	0.139 ± 0.007
	4	0.388 ± 0.018	0.358 ± 0.021
	5	0.225 ± 0.019	1.41 ± 0.10
	6	0.075 ± 0.007	4.02 ± 0.32
^{239}Pu	1	0.038 ± 0.004	0.0129 ± 0.0003
	2	0.280 ± 0.006	0.0311 ± 0.0007
	3	0.216 ± 0.027	0.134 ± 0.004
	4	0.328 ± 0.015	0.331 ± 0.018
	5	0.103 ± 0.013	1.26 ± 0.17
	6	0.035 ± 0.007	3.21 ± 0.38
^{240}Pu	1	0.028 ± 0.004	0.0129 ± 0.0006
	2	0.273 ± 0.006	0.0313 ± 0.0007
	3	0.192 ± 0.078	0.135 ± 0.016
	4	0.350 ± 0.030	0.333 ± 0.046
	5	0.128 ± 0.027	1.36 ± 0.30
	6	0.029 ± 0.009	4.04 ± 1.16
^{241}Pu	1	0.010 ± 0.003	0.0128 ± 0.0002
	2	0.229 ± 0.006	0.0299 ± 0.0006
	3	0.173 ± 0.025	0.124 ± 0.013
	4	0.390 ± 0.050	0.352 ± 0.018
	5	0.182 ± 0.019	1.61 ± 0.15
	6	0.016 ± 0.005	3.47 ± 1.7

The variations of yields between isotopes are expected in terms of the systematics of nuclear fission and have been well correlated with models of the fission process. Further inspection of Table 2.3, however, shows a variation in λ_i between isotopes. This serves to underline the fact that the six groups are not explicit species with a unique physical λ_i. If we have a design in which, for example, ^{235}U, ^{238}U and several plutonium isotopes are present, the slightly different λ_i between isotopes presents a complication. We would not wish to carry, say, thirty explicit, separate groups in the equations. One method of locally averaging over I' isotopes, say, might be to write (with $\tau_i = 1/\lambda_i$)

$$\beta_i = \sum_{i'}^{I'} \beta_{ii'}; \quad \tau_i = \sum_{i'}^{I'} \beta_{ii'}\tau_{ii'}/\beta_i \tag{2.3}$$

in each group i before taking spatial effects into account (see also ref. 17).

FRACTIONAL YIELDS AND TIME CONSTANTS

A desired form of the delayed neutron data for calculations of the dynamic behaviour from first principles is as β_i and λ_i. Table 2.4 therefore combines the preceding data and may be regarded as a working source for precursor data. The data is given for thermal and for fast fission in a ^{235}U spectrum. Mean times $\tau_i = 1/\lambda_i$ are given and are

TABLE 2.4.

Working delayed neutron data

(a) Thermal fission

	Group	β_i	λ_i (s^{-1})	τ_i (s)
Isotope ^{233}U	1	0.000241	0.00126	79.37
$\beta = 0.00281$	2	0.000769	0.0334	29.94
$\pm$ 0.00005	3	0.000637	0.131	7.63
	4	0.000890	0.302	3.31
$\beta/\lambda = 0.050$ s	5	0.000205	1.27	0.787
	6	0.000065	3.13	0.319
Isotope ^{235}U				
	1	0.000266	0.0127	78.74
$\beta = 0.00700$	2	0.001492	0.0317	31.55
$\pm$ 0.00008	3	0.001317	0.115	8.70
	4	0.002851	0.311	3.22
$\beta/\lambda = 0.089$ s	5	0.000897	1.40	0.714
	6	0.000182	3.87	0.258
Isotope ^{239}Pu				
	1	0.000086	0.0129	77.52
$\beta = 0.00227$	2	0.000637	0.0311	32.15
$\pm$ 0.00004	3	0.000491	0.134	7.46
	4	0.000746	0.331	3.02
$\beta/\lambda = 0.033$ s	5	0.000234	1.26	0.794
	6	0.000080	3.21	0.312
Isotope ^{241}Pu				
	1	0.000054	0.0128	78.13
$\beta = 0.00545$	2	0.001249	0.0299	33.44
$\pm$ 0.00054	3	0.000943	0.124	8.06
	4	0.002127	0.352	2.84
$\beta/\lambda = 0.060$ s	5	0.000993	1.61	0.621
	6	0.000087	3.47	0.288

(b) Fast fission (over ^{235}U spectrum)

	Group	β_i	λ_i (s^{-1})	τ_i (s)
Isotope ^{232}Th	1	0.000759	0.0124	80.65
	2	0.003350	0.0334	29.94
$\beta = 0.0223$	3	0.003462	0.121	8.26
$\pm$ 0.0014	4	0.009962	0.321	3.12
	5	0.003842	1.21	0.826
$\beta/\lambda = 0.225$ s	6	0.000960	3.29	0.304
Isotope ^{233}U	1	0.000229	0.0126	79.37
	2	0.000732	0.0334	29.94
$\beta = 0.00266$	3	0.000605	0.131	7.63
$\pm$ 0.00007	4	0.000845	0.302	3.31
	5	0.000194	1.27	0.787
$\beta/\lambda = 0.048$ s	6	0.000061	3.13	0.319
Isotope ^{235}U	1	0.000251	0.0127	78.74
	2	0.001406	0.0317	31.55
$\beta = 0.00660$	3	0.001241	0.115	8.70
$\pm$ 0.00013	4	0.002687	0.311	3.22
	5	0.000845	1.40	0.714
$\beta/\lambda = 0.084$ s	6	0.000172	3.87	0.258
Isotope ^{238}U	1	0.000210	0.0132	75.76
	2	0.002214	0.0321	31.15
$\beta = 0.0161$	3	0.002618	0.139	7.19
$\pm$ 0.00062	4	0.006270	0.358	2.79
	5	0.003635	1.41	0.709
$\beta/\lambda = 0.124$ s	6	0.001212	4.02	0.249
Isotope ^{239}Pu	1	0.000081	0.0129	77.52
	2	0.000594	0.0311	32.15
$\beta = 0.00212$	3	0.000458	0.134	7.46
$\pm$ 0.00010	4	0.000695	0.331	3.02
	5	0.000218	1.26	0.794
$\beta/\lambda = 0.031$ s	6	0.000074	3.21	0.312
Isotope ^{240}Pu	1	0.000081	0.0129	77.52
	2	0.000789	0.0313	31.95
$\beta = 0.00289$	3	0.000555	0.135	7.41
$\pm$ 0.00035	4	0.001012	0.333	3.00
	5	0.000370	1.36	0.735
$\beta/\lambda = 0.0389$ s	6	0.000084	4.04	0.247
Isotope ^{241}Pu	1	0.000054	0.0128	78.13
	2	0.001246	0.0299	33.44
$\beta = 0.00544$	3	0.000941	0.124	8.06
$\pm$ 0.00055	4	0.002122	0.352	2.84
	5	0.000990	1.61	0.621
$\beta/\lambda = 0.062$ s	6	0.000087	3.47	0.288

seen to span a range from a few to some 80 s. The total fractional yield β and the sum β_i/λ_i are also given for each isotope. Quoted standard deviations for the former are obtained from the usual combination of assumed independent deviations of the $\beta\bar{\nu}$ and $\bar{\nu}$ data of Tables 2.1 and 2.2. The internal six-group data is so highly correlated by the fitting procedure, however, that for standard deviations within isotopes it is better to refer directly to Table 2.3.

ENERGY SPECTRA

Prompt fission spectra are well established and could be taken† to follow Maxwellian distribution with a probability distribution function

$$f(E) = \frac{2}{\sqrt{\pi}} \frac{1}{kT} \left[\frac{E}{kT} \right]^{\frac{1}{2}} e^{-E/kT} \tag{2.4}$$

where k is Boltzmann's constant (0.0000138 aJ/K), T is the spectrum temperature (K) and E is the detailed energy of the distribution. The equivalent mean energy is given by

$$\bar{E} = \frac{3}{2} kT.$$

The spectrum of delayed neutrons is not so well measured but it is clear that their mean energy is appreciably below that of prompt neutrons leading to a different "effectiveness" compared to prompt neutrons. The delayed spectra certainly show peaks at the lower energies; nevertheless, for practical purposes one will often assume the same Maxwellian distribution for averaging.

Table 2.5 gives mean energies and corresponding temperatures for prompt and delayed neutrons (after Keepin[1]). The figures for the delayed neutrons are suspect, but newly

TABLE 2.5.

Neutron energy spectra, after Keepin[1]

(a) Prompt neutron fission spectrum parameters (MeV)

Isotope:	^{233}U	^{235}U	^{239}U
kT	1.31	1.29	1.41
$\bar{E}$	1.965	1.935	2.11

(b) Delayed neutron fission spectrum parameters (MeV)

Isotope:	^{232}Th	^{233}U	^{235}U	^{238}U	^{239}Pu	^{240}Pu
kT	0.37	0.26	0.29	0.37	0.27	0.28
$\bar{E}$	0.49	0.39	0.43	0.49	0.40	0.42

N.B.—1 MeV (kT) is 0.160219 pJ, equivalent to about 10^{11} K.

published data (see the review by Walker and Weaver[13]) are in disagreement; thus Eccleston and Woodruff would decrease values in Table 2.5(b) by about 15% while Evans and Krick, also 1976 values, would increase them by some 20%. One has to say that this important matter is not yet resolved.

† An older form is Watt's curve, $0.484 \sinh \sqrt{2E} e^{-E}$. ($E$ in MeV)

SPREAD OF NEUTRON YIELDS

The variation of ν around its mean $\bar{\nu}$ is needed in considering stochastics or reactor noise (Chapter 7). Diven's original measurements[11] of the distribution of yields shows that either a binomial fit

$$p_\nu = {}^mC_\nu \left[\frac{\bar{\nu}}{m}\right]^\nu \left[1 - \frac{\bar{\nu}}{m}\right]^{m-\nu}$$

is suitable, with m the *maximum* number of neutrons from a fission ($m=5$ for ^{235}U) or more generally for neutron induced fissionable isotopes, a Gaussian fit of the form[15]:

$$p_\nu = \tfrac{1}{2}[\text{erf}(x^+) - \text{erf}(x^-)]$$

with $x^+ = [\nu - \bar{\nu} \pm \tfrac{1}{2}]/1.08$. Figure 2.1 shows the ^{235}U distribution of p_ν.

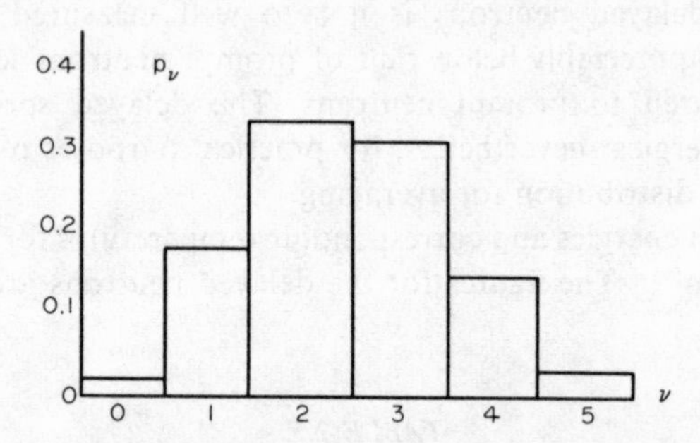

FIG. 2.1. Distribution of neutrons in fission from ^{235}U.

The binomial fit is not exact for all isotopes and a better experimental measure of the distribution is Diven's D parameter, $D = [\overline{\nu^2} - \bar{\nu}^2]/\bar{\nu}$ as given in Table 2.6. Note, however, from Fig. 2.1 that some fissions yield *no* neutrons but that the peak probability arises

TABLE 2.6.

Diven parameter for the distribution of neutron yields in fission[11]

Isotope:	^{233}U	^{235}U	^{239}Pu	^{240}Pu
D	0.786	0.795	0.815	0.807
SD	0.013	0.007	0.017	0.008

for ν close to $\bar{\nu}$. Since the Diven parameter is fundamental to much of the new developments in reactor noise, the author feels that this would be a useful area for experimentalists to confirm or update the values given above. They should, anyway, be understood as measuring the spread of the *prompt* yield, i.e. strictly of $\nu[1-\beta]$ in our notation. It may easily be shown that D is related to a binomial distribution as $D = 1 - 1/m$, so that the departure of D-values from unity is itself a measure of the departure of the fission distribution from a Poisson distribution, where D would be unity ($m \to \infty$).

SPECIAL CASES

Before turning to the consequences of spatial and energy variations on the physical data, we should point out a few special cases that must occasionally be accounted for, leading to β_i and λ_i values in addition to those just tabulated.

Thus as well as conventional fission, we may have $(n,2n)$ or $(n,3n)$ reactions as in the splitting (fissioning) of Be^9. When this is significant we have to decide whether to include the extra source of neutrons in the production term of the neutron balance or as a negative removal term. If the former, the fission spectrum, fractional yields, etc., will in principle be changed. The choice is arbitrary but must be made consistently. (Indeed, in some versions of transport theory we talk about a yield of neutrons including (n,n) and (n,n') from scattering.)

In a heavy water reactor there is a noticeable (γ,n) reaction in deuterium. In addition to providing a few additional prompt neutrons (which might be negligible) this mechanism will lead to additional delayed neutrons from the γ-radiation of the fission products (after heat). The inclusion of what are essentially additional delayed neutron groups is complicated by the dependence of the yield on the structural design of the reactor and consequently the access of the γ-rays to the deuterium. Specific studies, however, indicate that a further six to nine groups may be needed for a representation, with small yields compared to the standard six groups and generally longer mean lives. They cannot be quoted in any generality† due to the shielding effect of structures and cladding, but should be allowed for in specific calculations.

Other complications arise with circulating fuel reactors. Part of these complications are dealt with under the effective value approach of the next section, but if some of the fuel is taken outside the reactor in times short enough for the delayed neutrons to be carried out with it, then, of course, the yield of delayed neutrons within the reactor is affected.

Effective Values

We now look more closely at the origin of eqns. (2.1) and (2.2) and to what extent they are meaningful when reactor properties vary in space and energy and not only time. Clearly if the detailed description is to be reduced to a set of equations dependent on time only, then somehow we are introducing average properties, averaged over space and energy. We can beg the question of *how* to average only when properties are uniform in space and energy and are therefore independent of the method of averaging. But if the absorption cross-section is high in the core, low in the reflector, for example, what is the "best" average absorption cross-section?

The detailed description of the neutrons and precursors in the reactor will involve (expected) density distributions, functions of position r, energy E and time t. Indeed, some descriptions of neutron behaviour would have us consider velocity (direction as well as speed), but for our purpose a variation in position r and energy E will illustrate the theory sufficiently. We therefore consider density functions $N(r,E,t)$ and $C_i(r.t)$ as follows: $N(r,E,t)dr\,dE$ is the (expected) number of neutrons at time t found in a small

† See problem 2.1 for some representative figures.

element of volume dr around r and in the energy band dE around E. Here $drdE$ is not so small as to make fluctuations evident but is small enough to say that N is uniform across it. Similarly, $C_i(r,t)dr$ is the number of precursors of the ith type† found in dr about r at time t.

Most models‡ of neutron and precursor behaviour lend themselves to balance or accounting equations that have the following *form*:

$$\frac{\partial N}{\partial t} = [1 - \beta]\chi_p FN - RN + \sum_i \lambda_i \chi_i c_i + S \tag{2.5}$$

$$\frac{\partial C_i}{\partial t} = \beta_i FN - \lambda_i C_i \tag{2.6}$$

Here the left hand sides are the resultant rates of change of densities. The right hand side contains a series of production and removal terms. The first is a prompt production from fission term, where $[1-\beta]$ gives the fraction of fission production that is prompt neutron, $\chi_p(E)$ gives the distribution of prompt neutrons in energy and FN gives the fission production rate. In a simple model, $FN = \nu\Sigma^f vN$, with v the neutron speed and Σ a macroscopic cross-section. This production term is supplemented by the term giving the production of delayed neutrons from precursor decay and this includes a similar spectrum distribution, $\chi_i(E)$ this time, for the distribution of delayed neutrons. Of course every neutron has to be emitted with some energy so that the $\chi(E)$ are normalised such that $\int_0^\infty \chi(E)dE = 1$.

It will be convenient to write a total production from fission term as

$$PN = [1 - \beta]\chi_p FN + \sum_i \beta_i \chi_i FN \tag{2.7}$$

which can be interpreted as the production term if there were no hold up of precursors before decay (or else interpreted as the neutron equation if the precursor balance is in the steady state allowing for the substitution $\lambda_i \chi_i C_i = \beta_i \chi_i FN$). Obviously in many static or steady state approximations the role of the delayed neutrons with their slightly different spectrum χ_i is ignored in view of their small β_i, and the approximation $PN \simeq \chi_p FN$ is made.

To continue with eqn. (2.5), RN is a removal term, covering both absorption, leakage and the transfer of neutrons between energies in scattering. The detail of this term depends heavily on the model employed.

$S(r,E,t)$ is an independent source density. This is sometimes referred to as an "external" source as if it were necessarily outside the reactor. The essence of the term, however, is that it is an expression for the neutrons arising independent of the neutrons already present and causing fission.

λ_i, β_i and $\beta = \sum_i \beta_i$ are the physical parameters of the previous section.

Note that the operators P $(=\chi F)$, R, etc., may generally be differential, algebraic, integral or a mixture of all three without changing the general concept of these balance equations.

† To save complicating the notation we do not show a separate summation over i' isotopes. We use i here to count all the possible precursor types.

‡ Familiarity is assumed with the equations of reactor physics as given, for example, by Henry.[6]

The reduction of these equations to the desired set of ordinary differential equations has to resolve certain conflicting factors:

(i) we would like the result to be conceptually identifiable with the simple forms in n, c_i and the parameters of eqns. (2.1) and (2.2);

(ii) we want to avoid the necessity of having to *solve* eqns. (2.5) and (2.6), particularly in the time dependent case. (Indeed, if we could solve them exactly we would have all the information already that could possibly be extracted from eqns. (2.1) and (2.2);

(iii) we would like "easy" ways to calculate the coefficients in the resulting form of eqns. (2.1) and (2.2) while including in them detailed description inherent in eqns. (2.5) and (2.6).

Perhaps not surprisingly these requirements conflict and it will be (i) above that will suffer most in the compromise to achieve all the aims in some measure.

THE LINEAR WEIGHTING

How then shall we average the detailed description of the process involved in these equations? One appealing way is to resort to total numbers of events per second by means of an integration over energy and volume. Of course we do not know the shape of the functions $N(r,E,t)$ and $C_i(r,E,t)$, but we might go on to suppose that we could approximate for them in terms of the shape in r and E of the solution to a similarly designed reactor in steady state (so that time does not appear in this reference reactor equation and its equations are correspondingly easier to solve). We can refer to this steady state, reference reactor as an unperturbed reactor.

Our initial concept then makes no approximation as yet but simply forms such terms as†

$$\bar{n}' = \iint N'(r,E,t)dr\,dE \qquad \text{total number of neutrons} \tag{2.8}$$

and

$$\bar{R}\bar{n}' = \iint R'(r,E,t)N'(r,E,t)dr\,dE \quad \text{total removal rate} \tag{2.9}$$

where N' is the actual true solution. We then have an appealing definition of an average removal rate per neutron present as

$$\bar{R}'(t) = \iint RN'dr\,dE / \iint N'dr\,dE \tag{2.10}$$

and the approximation enters when we attempt to use an unperturbed or reference value $N(r,E)$ for N'. Although N' is time dependent in general, of course the magnitude tends to cancel top and bottom in such expressions.

The primary question is whether this approximation is accurate enough to admit the substitution N for N', since it certainly leads to ordinary differential equations in time dependent variables and coefficients. We can show, however, that the approximation *fails*.

Write a general balance equation of the form

$$\frac{\partial N'}{\partial t} = [P' - R']N' + S' \tag{2.11}$$

† Integration over a total reactor volume and energy range assumed.

with which we associate an unperturbed reference reactor in steady state with equation

$$0 = [P - R]N \tag{2.12}$$

The change in properties, etc., can be written $\delta P = P' - P$ and $\delta R = R' - R$. The change in density or solution can be written $\delta N = N' - N$. We try to develop eqn. (2.11) in a way that will allow us to make use of eqn. (2.12) after integrating over the reactor:

$$\begin{aligned} \frac{d\bar{n}'}{dt} = \iint \frac{\partial N'}{\partial t}\, dr\, dE = \iint [P-R]N + [P-R]\delta N \\ + [\delta P - \delta R]N + [\delta P - \delta R]\delta N + S'\,dr\,dE \end{aligned} \tag{2.13}$$

On the right hand side the first term cancels in view of eqn. (2.12). The third term will represent the effect of the perturbation in properties but based, as we wish, on the unperturbed density. Note that the integrals will only be required to be carried out for this term in the region where the perturbation actually exists. This is attractive and allows the experimentalist and the analyst to focus on the effect of, say, a sample in the reactor. The fourth term may be negligible if both the perturbation in properties and the resultant perturbation in densities are (relatively) small, and we may feel justified in neglecting this term. If not, we must indeed approximate for the perturbed density N'. The last term in S', however, is known.

But we are left with the uncomfortable term in $[P-R]\delta N$. This does not vanish and represents the consequence of the density change through the rest of the reactor. Even though δN may be small in the rest of the reactor, outside the perturbation, its total effect in this integral form may be large and is certainly likely to be of the same order of magnitude as the third term. There is no justification for ignoring it.

To illustrate this point, consider Fig. 2.2 which shows in an elementary diffusion model the flux distribution in a one-dimensional "slab" reactor suffering a gross perturbation consisting of the introduction of a strong control "sheet" in the mid-plane.

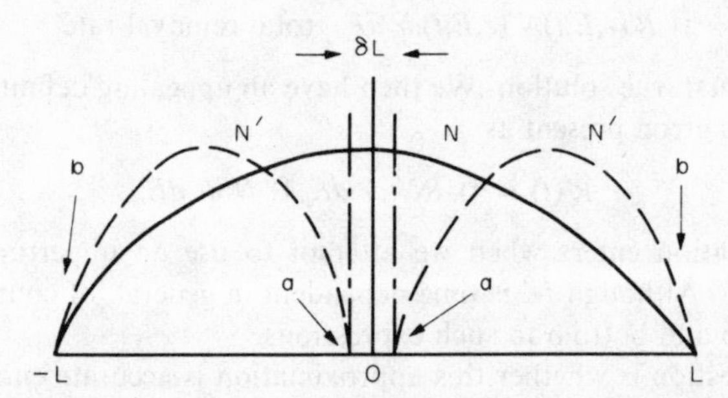

FIG. 2.2. The reactor perturbed by a black control sheet.

No one would sensibly suppose that the effect of this "black" absorber could be evaluated by using the unperturbed density and the perturbation of properties; one might more reasonably, however, try to estimate the effect of the control sheet by evaluating the perturbed density shape around the control sheet and hence the rate at which neutrons are being absorbed in the control sheet via the current of neutrons at $a-a$. Let us suppose one does this in some accurate approximation. Would it adequately describe the new balance of neutron reaction rates? No, because it ignores the additional leakage of neutrons from the outer faces $b-b$ of the reactor, i.e. the consequent effect of the perturbation in density shape throughout the original reactor. It is readily established in diffusion theory that these two effects are of the same order of magnitude, and to compute the effect of the control sheet solely in terms of the perturbation in properties by virtue of determining the neutrons absorbed in the control sheet is to seriously underestimate the total effect of the control material—by 33% in this example.

THE FLUX WEIGHTING

We may generally say that the effect of a perturbation in properties (large or small) is as much indirect, via the consequent density change throughout the reactor, as it is direct, in the perturbation itself. What is wanted, therefore, is a way of eliminating the troublesome term in $[P-R]\delta N$ when we form our averages.

This aim may be achieved by introducing a weighting function into our equations before integrating and specially selecting that weighting function to serve the desired purpose. Certainly we can introduce any weighting we like before integration over space and energy, and the result will be coefficients or averages dependent on time only, but, of course, these weighted averages will no longer have the simple meaning of the total reaction rate, etc., that we first had when the weighting function was uniform or unity.

We now show that for a simple one-group diffusion theory model, the desirable weighting function is simply the density (or flux) solution itself in the unperturbed reactor, solution of the neutron balance equation and with the same boundary and interface conditions. The example is based on one-dimensional theory and can readily be extended to show the result is true in two and three dimensions.

The unperturbed equation for the flux ($\phi = vN$), where we assume source-free steady state properties, and hence a steady state equation, is

$$D\frac{d^2\phi}{dx^2} + \nu\Sigma^f\phi - \Sigma^a\phi = 0 \tag{2.14}$$

with D the diffusion coefficient.

Suppose the perturbation consists of adding a small amount of absorber of thickness δL with additional cross-section $\delta\Sigma^a$ at the centre line with the geometry of Fig. 2.2 (but this time with a less severe perturbation). The perturbed equation is then

$$D\frac{d^2\phi'}{dx^2} + [\nu\Sigma^f - \Sigma^a - \delta\Sigma^a]\phi' = \frac{1}{v}\frac{\partial\phi'}{\partial t} \tag{2.15}$$

Form from eqn. (2.15) the integral weighted with ϕ:

$$\int_{-L}^{L} \phi D\frac{d^2\phi'}{dx^2} + \phi[\nu\Sigma^f - \Sigma^a - \delta\Sigma^a]\phi' dx = \frac{1}{v}\int_{-L}^{L}\phi\frac{\partial\phi'}{\partial t}dx \tag{2.16}$$

To deal with the first term, integrate by parts twice using the boundary conditions that $\phi'(\pm L)=\phi(\pm L)=0$. Since the unperturbed flux is time independent,

$$\int_{-L}^{L} \frac{d}{dx}\left[\phi D \frac{d\phi'}{dx}\right] dx - \int_{-L}^{L} \frac{d\phi}{dx} D \frac{d\phi'}{dx} + \phi[\nu\Sigma^f - \Sigma^a - \delta\Sigma^a]\phi' dx = \frac{1}{v}\frac{d}{dt}\int_{-L}^{L} \phi\phi' dx \qquad (2.17)$$

and the first term vanishes in view of the boundary conditions. After another integration by parts, we arrive at a stage where the term in ϕ' can be reordered as

$$\int_{-L}^{L} \phi' \left[D \frac{d^2\phi}{dx^2} + \nu\Sigma_f\phi - \Sigma^a\phi\right] dx - \int_{-\Delta L}^{L} \phi\delta\Sigma^a\phi' dx = \frac{1}{v}\frac{d}{dt}\int_{-L}^{L} \phi\phi' dx \qquad (2.18)$$

We see that the bracketed term vanishes identically for *any* ϕ' in view of the equation for ϕ', the selected weighting function. We therefore have

$$-\int_{\Delta L} \phi\delta\Sigma^a\phi' dx = \frac{1}{v}\frac{d}{dt}\int_{-L}^{L} \phi\phi' dx \qquad (2.19)$$

This is the result we have been seeking. The rate of change of our averaged neutron population $\int\phi\phi' dx$ has been related solely to the perturbation in properties and not (explicitly) to the remaining unperturbed properties. Of course these properties appear implicitly through the choice of ϕ as weighting function since this satisfies an equation based on the unperturbed properties.

If $\delta\Sigma^a$ and therefore $\delta\phi'$ are relatively small we can now use the first order approximation with some confidence and substitute ϕ for ϕ'. However, without approximation at this stage we have *weighted* averages:

$$\bar{n}'(t) = \int\frac{\phi\phi'}{v} dx; \int \phi[P' - R'] \frac{\phi'}{v} dx = [\bar{P}' - \bar{R}']\bar{n}'(t) \qquad (2.20)$$

Putting $\qquad \bar{\Lambda}' = 1/\bar{P} \quad \text{and} \quad \bar{\rho}' = [\bar{P}' - \bar{R}']/\bar{P}'$

we could write this *exactly* as

$$\frac{d\bar{n}'}{dt} = \frac{\bar{\rho}'}{\bar{\Lambda}'}\bar{n}' \qquad (2.21)$$

The first order approximation in eqn. (2.21) could be indicated by dropping the primes.

Clearly $\bar{n} = 1/v\int\phi^2 dx$ is no longer the simple neutron population. Neutrons in the centre, where ϕ is high, are counted more than neutrons at the edge, where ϕ is low. This is quite reasonable, describing the relative importance of neutrons to the continued fission process; those in the centre will have a better chance of fissioning instead of leaking out. We return to this point later in the problems.

THE ADJOINT WEIGHTING

At this point we generalise the idea of the "best" weighting function. Unfortunately, for general energy dependent models the density (or flux) itself is not the best weighting function able to eliminate the unwanted terms in the original properties. (This arises through a lack of symmetry between the processes increasing neutron energy, i.e. fission, and the processes decreasing neutron energy, i.e. scattering.) Rather, the best function is called the adjoint density, related to the density through a related equation called the adjoint equation having similar boundary conditions and interface conditions. Its solution, the adjoint function, is written $N^+(r,E)$. It is time independent if we base the perturbation on a reference reactor that is critical in steady state. Only in the one-energy group case do the two equations coalesce, and N^+ has the same shape as N or ϕ.

The adjoint equation uses adjoint operators, related to the operators such as P, R or F and denoted here by P^*, R^* or F^* respectively. If M is an operator, we are seeking a way of changing the order of terms such as N^+MN to NM^*N^+. In general, this cannot be done but the adjoint operator allows such a commutation, *on the average*, when integrated over all space and energy. In the one-group case the diffusion equation is self-adjoint under this integration, making use of the boundary conditions, etc. In the multi-group diffusion case it can readily be shown that the appropriate adjoint operator for M is formed as the transpose M^T (thus allowing for the asymmetry of the multi-group case). In the multi-group diffusion theory, of course, integration over a continuous energy is replaced by summation over groups which is readily expressed in a matrix and vector notation.

To proceed in general, however, consider eqns. (2.5) and (2.6) in the unperturbed steady state with no source. We may eliminate $C_i(r)$ to leave

$$0 = [1-\beta]\chi_p FN + \sum_i \beta_i \chi_i FN - RN = PN - RN \tag{2.22}$$

The equation adjoint to this is taken to define N^+:

$$0 = [1-\beta][\chi_p F]^* N^+ + \sum_i \beta_i [\chi_i F]^* N^+ - R^* N^+ = P^* N^+ - R^* N^+ \tag{2.23}$$

Now suppose the system is perturbed to new properties and the form

$$\frac{\partial N'}{\partial t} = [1-\beta]\chi_p' F' N' - R' N' + \sum_i \lambda_i \chi_i' C_i' + S' \tag{2.24}$$

$$\frac{\partial C_i'}{\partial t} = \beta_i F' N' - \lambda_i C_i' \tag{2.25}$$

Multiply (2.25) by the delayed neutron spectrum $\chi_i'(E)$. Next multiply both equations by $N^+(r,E)$ and integrate over the whole reactor space and energy. There results

$$\begin{aligned}\frac{d}{dt}\iint N^+ N' dr\, dE &= \iint N^+[1-\beta]\chi_p' F' N' dr\, dE - \iint N^+ R' N' dr\, dE \\ &\quad + \sum_i \iint N^+ \lambda_i \chi_i' F' N' dr\, dE + \iint N^+ S' dr\, dE\end{aligned} \tag{2.26}$$

and

$$\frac{d}{dt}\iint N^+ \chi_i' C_i' dr\, dE = \iint N^+ \beta_i \chi_i' F' N' dr\, dE - \iint N^+ \lambda_i \chi_i' C_i' dr\, dE \tag{2.27}$$

Recollect that in

$$\bar{n}' = \iint N^+(r,E)N'(r,E,t)dr\,dE$$

the term N^+ is time independent. Then $\bar{n}'$ can be considered as an average of the neutron population at any time. Similarly for

$$\bar{c}_i' = \iint N^+\chi_i' C_i' dr\,dE$$

which may be considered an average precursor population, including a weighting according to the contribution to be made by delayed neutrons based upon their energy spectrum. Define the following average properties or coefficients as in Table 2.7. It

TABLE 2.7.
General effective reactor kinetics coefficients

$\bar{P}' = \iint N^+P'N'dr\,dE / \iint N^+N'dr\,dE$	Mean production rate
$\bar{\beta}_i' = \beta_i \iint N^+\chi_i' F'N'dr\,dE / \iint N^+PN'dr\,dE$	Effective delayed neutron fraction
$\bar{R}' = \iint N^+R'N'dr\,dE / \iint N^+N'dr\,dE$	Mean removal rate
$\bar{n}' = \iint N^+N'dr\,dE;\quad \bar{c}_i' = \iint N^+\chi_i' C_i' dr\,dE$	Weighted neutron and precursor population
$\bar{s}' = \iint N^+S'dr\,dE$	Weighted source rate
$\bar{\rho}' = [\bar{P}' - \bar{R}']/\bar{P}'$	Reactivity
$\bar{\Lambda}' = 1/P'$	Reproduction time

follows from these definitions that the reduced eqns. (2.5) and (2.6) take the desired *form*:

$$\frac{d\bar{n}'}{dt} = \frac{\bar{\rho}' - \bar{\beta}'}{\bar{\Lambda}'}\bar{n}' + \sum_i \lambda_i \bar{c}_i' + \bar{s}' \tag{2.28}$$

$$\frac{d\bar{c}_i'}{dt} = \frac{\bar{\beta}_i'}{\bar{\Lambda}'}\bar{n}' - \lambda_i \bar{c}_i' \tag{2.29}$$

Definitions of effective values $\bar{n}'$, $\bar{c}_i'$, $\bar{\beta}_i'$ and $\bar{s}'$ are conceptually clear and need little comment. The reactivity is based on the difference between weighted production and removal, normalised to weighted production rates. The reproduction time $\bar{\Lambda}'$ is the reciprocal of the weighted production rate associated with fission. There is an arbitrary element in such definitions; fission includes events producing no neutrons, one neutron as well as two or more neutrons and thus includes events that might not be thought of as production.

So far, the reduction is exact, depending on the actual or perturbed density N'. From the property of the adjoint operators and subsequent appeal to eqn. (2.23), we have

$$\bar{\rho}' = \frac{\bar{P}' - \bar{R}'}{\bar{P}'} = \frac{\iint N^+[\delta P - \delta R]N'dr\,dE}{\iint N^+P'N'dr\,dE} \tag{2.30}$$

with the desirable property of relating $\bar{\rho}'$ directly to the perturbation in properties.

Can we now drop the prime and use N for N'? The answer, unfortunately, has to be equivocal. If the perturbations are small so that the "*shape*" in r and E of N' changes only a little, then this approximation will be successful. Of course the overall magnitude or time dependency cancels in the average coefficients just defined. If not, then not.

What can be said with precision is that only an adjoint weighting will admit the first order approximation with any hope of success, though there is admittedly a choice in reference reactor and adjoint function used. We have now seen the origin of the "effective" values for the various coefficients because such terms as $\bar{\Lambda}$, $\bar{\rho}$, $\bar{\beta}_i$ and $\beta = \Sigma\beta_i$ are weighted values. In particular, $\bar{\beta}_i$ differs from β_i and we may usefully write an effectiveness factor γ_i as

$$\gamma_i = \frac{\bar{\beta}_i}{\beta_i} = \frac{\iint N^+ \chi_i F N \, dr \, dE}{\iint N^+ P N \, dr \, dE} \tag{2.31}$$

In a moment we shall give some explicit models for calculating adjoints and effectiveness factors, but we turn first to the relation of these effective kinetics parameters to the so-called static reactivity calculations. At this point let us just note the additional arbitrary element in defining effective coefficients, the reference to an arbitrary reactor in steady state with no source, chosen as the reference reactor.

Static Reactivity Calculations

It is easier to compute the neutron distribution in a steady state, static model, than it is in a general time dependent model. Thus the precursor densities can be eliminated in the steady state and, indeed, the difference between delayed and prompt neutron spectra is ignored in many steady state codes. Since many such computation schemes for static solution of the density distribution are available, it is not surprising that we should seek to obtain the distributions N and N^+ for kinetics calculations from the results of steady state calculations.

In a static calculation, time independent cross-sections, etc., are assumed. The properties of the system investigated may not really define a critical system, capable of being in the steady state with no source. In which case some adjustment or criticality reset mechanism must be proposed. This is an artificial adjustment of some parameter in the system and the amount of adjustment needed is some sort of measure of the departure of the system from critical.

There are several ways this departure could be measured, depending on the critically reset mechanism. Perhaps we chose to introduce a detailed representation of the control rod mechanism. Perhaps instead we chose to add an amount of absorber uniformly throughout the whole reactor; in this case, if we add a so-called $1/v$ cross-section to the flux equation, we are adding a constant term, the same for every position and energy to the neutron balance. The strength of this term (sometimes called the α-reset mechanism) is the measure of the departure of the system from criticality.

But the nature of the fission equations is such that a more common criticality reset choice is the so-called λ-reset where a uniform adjustment of the fission production term is assumed, the same at every position and energy. The static equations can be represented therefore as

$$0 = \lambda P N_s - R N_s \tag{2.32}$$

where λ is a uniform multiplier or eigenvalue and we are careful to show that the resulting static density N_s is not identical with either N or N' of the kinetic models.

If the reactor is really critical, λ equals unity. If the reactor is subcritical, production must be increased to obtain this artificial balance and $\lambda > 1$. If supercritical, $\lambda < 1$. Thus λ is a measure of the departure from criticality. Unless the system is critical and $\lambda = 1$, the measure is not unique and the density N_s is not N. Experience with the solution of steady state equations shows that this form of criticality reset is a most convenient one.

In that case, we seek to use the computed λ-value to find the reactivity $\bar{\rho}$ for a kinetics calculation. Correspondingly, we must use consistent values for the remaining kinetics parameters if the resulting kinetic equations are to be meaningful. This consistency is necessary however small the perturbation and however good at first order approximation, using N_s for N', may be, since the weighted reactivity depends on its weighting function as well as the first order approximation.

If we use $N^+{}_s$ as the weighting function taken from the adjoint equation corresponding to eqn. (2.32),

$$\bar{\rho} = \frac{\iint N_s^+ \delta PN \, dr \, dE}{\iint N_s^+ PN \, dr \, dE} = [1 - \lambda] \frac{\iint N_s^+ PN \, dr \, dE}{\iint N_s^+ PN \, dr \, dE} = 1 - \lambda \tag{2.33}$$

justifying the view that $1 - \lambda$ can be used as a reactivity. Strictly, all other kinetics parameters must be calculated with the same weighting function $N_s^+(r,E)$.

Since the point is not universally understood we can express these conclusions in another way. In the simplest possible case the kinetics equations separate exactly into functions of time only (with corresponding ordinary differential equations) and functions of the remaining variables such as position and energy. This simple case is limited to properties independent of time and (as seen in the problem) even here there is a complication that *one* separated solution is insufficient, and in principle an infinite set is wanted. The alternative is to reduce the equations to ordinary differential equations using a space and energy dependent weighting function W, say. This may always be done formally, employing also the (unknown) density distribution N' which may be changing in shape over r and E as time changes.

If the weighting W is selected as unity, then we have a simple count of the number of neutrons or the total reaction rate in the system giving us a simple interpretation of the average or reduced parameters in the reduced equations. If, for example, in a Monte Carlo calculation the development of the neutron reactions is followed and such a count is made, the reactivity is just such a total reaction measure. There is no reason to suppose that the reactivity calculated from a Monte Carlo solution on this basis agrees with the reactivity from a static calculation with λ-reset or a perturbation calculation with adjoint weighting.

The attraction of the adjoint weighting is the scope we have for making the first order approximation for the shape of N'. The attraction of the λ-reset or static calculation is the availability of codes for computing λ and hence $\rho_s = 1 - \lambda$. In all cases we must be consistent in using the remaining parameters in a kinetic calculation based on a consistent weighting function. The only time the different representations will give the same value of reactivity is when the system is actually critical and we find we are computing $\rho = 0$. Away from critical, different definitions of reactivity will have different scaling factors.

Practical Calculations of Kinetics Parameters

If the criticality and flux distribution calculations for a reactor design warrant an elaborate digital computer analysis, it is likely that similar effort is needed for calculation of effective kinetics parameters. We start with some simple models to give the physical "feel" of the various expressions.

REACTIVITY PERTURBATION TERM

We have given the one-group diffusion expression for a reactivity change due to a perturbation in absorption cross-section. For more general perturbations it is convenient to carry out an integration by parts of the diffusion or leakage term to avoid having to find the perturbation $\nabla\,\delta D\nabla$.

Making use of boundary and interface conditions we have

$$\bar{\rho} = \frac{-\int [\nabla\phi^2]\delta D\,dr + \int \phi^2[\delta[\nu\Sigma^f] - \delta\Sigma^a]\,dr}{\int \phi^2\nu\Sigma^f\,dr} \tag{2.34}$$

One-group diffusion theory is of limited value. Multi-group theory is better. If $M\phi=0$ defines the multi-group equations in matrix and vector form, the required adjoint operator is the transpose of M, i.e. $M^*=M^T$. Thus in a two-group model

$$\begin{pmatrix} \nabla\cdot D_1\nabla - \Sigma_1, & \nu\Sigma^f \\ p\Sigma_1, & \nabla\cdot D_2\nabla - \Sigma_2 \end{pmatrix}\begin{pmatrix}\phi_1\\ \phi_2\end{pmatrix} \equiv M\phi = \begin{pmatrix}0\\0\end{pmatrix} = 0 \tag{2.35}$$

leading to adjoint equations†

$$\begin{pmatrix} \nabla\cdot D_1\nabla - \Sigma_1, & p\Sigma_1 \\ \nu\Sigma^f, & \nabla\cdot D_2\nabla - \Sigma_2 \end{pmatrix}\begin{pmatrix}N_1^+\\ N_2^+\end{pmatrix} = M^T N^+ = 0 \tag{2.36}$$

It is not difficult to show from the theory of simultaneous equations that the criticality of eqns. (2.35) is related to the determinant of the equations and the adjoint equations have the same determinant. The adjoint system is therefore critical when the forward or flux system is critical. These equations are supplemented by boundary and interface conditions analogous to those applied to the flux. Thus if $\phi(L)=0$, take $N^+(L)=0$. If $\nabla\phi|_L=0$, take $\nabla N^+|_L=0$, and, finally, if $k\nabla\phi+\phi|_L=0$, take $k\nabla N^+ + N^+|_L=0$.

REPRODUCTION AND LIFETIME

The expression for the neutron reproduction time using the flux in a two-group model is

$$\bar{\Lambda} = \frac{\int N_1^+\phi_1/v_1 + N_2^+\phi_2/v_2\,dr}{\int N_1^+\nu\Sigma^f\phi_2\,dr} \tag{2.37}$$

where we assume no fast fission. The integrals are taken over all the volume in principle except that the denominator will be evaluated over the core only and not the reflector, where $\Sigma^f=0$.

† If we use densities instead of fluxes in eqns. (2.35) then the group speeds will cancel out in the transposed adjoint equations, and for that reason we write an adjoint density N^+ rather than an adjoint flux.

Multi-group diffusion theory leads to analogous expressions and continuous energy models call for integrations over energy as well as space.

EFFECTIVE PRECURSOR FRACTIONS

To evaluate $\bar{\beta}_i$ or γ_i we might take a three-group diffusion model where the first two groups describe the fast flux of prompt and delayed neutrons, respectively, caused by thermal fission, and the third group a common thermal group into which the two fast groups separately transfer or scatter. The equations of this model are

$$\begin{pmatrix} \nabla\cdot D_1\nabla - \Sigma_1, & 0, & [1-\beta]\nu\Sigma^f \\ 0, & \nabla\cdot D_2\nabla - \Sigma_2, & \beta\nu\Sigma^f \\ p_1\Sigma_1, & p_2\Sigma_2, & \nabla\cdot D_3\nabla - \Sigma_3 \end{pmatrix} \begin{pmatrix} \phi_1 \\ \phi_2 \\ \phi_3 \end{pmatrix} = 0 \tag{2.38}$$

Resonance escape probabilities p_1 and p_2 for prompt and delayed neutrons respectively will be essentially the same in view of the energies of both sets of neutrons from fission lying above the major resonance region, but they have been distinguished here for clarity. The significant physical effect in a thermal reactor is the additional likelihood of fast leakage by prompt neutrons before they are slowed down to the $\frac{1}{2}$ MeV with which delayed neutrons appear. In this sense we expect a prompt neutron to be less effective than a delayed neutron in achieving the status of a thermal neutron and continuing the chain process.

The appropriate adjoint equation takes the transpose of the matrix of coefficients of eqn. (2.38) and the resulting vector solution $N^+ = (N_1^+, N_2^+, N_3^+)$ used in the expression

$$\gamma_i = \frac{\int N_2^+\nu\Sigma^f\phi_3\,dr}{[1-\beta]\int N_1^+\nu\Sigma^f\phi_3\,dr + \beta\int N_2^+\nu\Sigma^f\phi_3\,dr} \simeq \frac{\int_c N_2^+\Sigma^f\phi_3\,dr}{\int_c N_1^+\Sigma^f\phi_3\,dr} \tag{2.39}$$

and its approximation in integrals over the core, again assuming thermal fission only. If the reactor is uniform and bare, however (or can be described for the present purpose via reflector savings), so that we may substitute $-B^2\phi$ for $\nabla^2\phi$ in each group, we may develop a result to illustrate the significance of eqn. (2.39). We find that ϕ_1, ϕ_2 and ϕ_3 can now be eliminated from the equations to leave a criticality relation

$$\left\{\frac{[1-\beta]p_1}{1+L_1^2B_2^2} + \frac{\beta p_2}{1+L_2^2B_2}\right\}\nu\Sigma^f - [1+L_3^2B^2]\Sigma_3 = 0 \tag{2.40}$$

where $L_i^2 = D_i/\Sigma_i$ and $1/[1+L_i^2B^2]$ is the ith group non-leakage probability. This equation may be put in the conventional form of a factor formula (with $\eta = \nu\Sigma^f/\Sigma_3$) as

$$\frac{\eta}{1+L_3^2B^2}\left[\frac{[1-\beta]p_1}{1+L_1^2B^2} + \frac{\beta p_2}{1+L_2^2B^2}\right] = 1 \tag{2.41}$$

If delayed and prompt neutrons had the same fission energies then the final bracket would simplify to $p/[1+L_1^2B^2]$ and the expression would become the conventional two-group criticality relation. The full form shows the separate role played by the delayed neutrons and prompt neutrons. We have

$$\gamma = \frac{p_2/[1 + L_2^2B^2]}{\dfrac{[1-\beta]p_1}{1 + L_1^2B^2} + \dfrac{\beta p_2}{1 + L_2^2B^2}} \simeq \frac{p_2}{p_1}\frac{1 + L_1^2B^2}{1 + L_2^2B^2} \tag{2.42}$$

A similar approach in continuous slowing-down theory (Fermi age) leads to a criticality relation

$$\frac{\eta\epsilon f}{1 + L^2B^2}[[1-\beta]p_1 e^{-\tau_1 B^2} + p_2 e^{-\tau_2 B^2}] = 1 \tag{2.43}$$

and correspondingly

$$\gamma \simeq \frac{p_2}{p_1}\exp([\tau_1 - \tau_2]B^2) \tag{2.44}$$

In the above we have developed an expression for γ using one group of precursors only. The knowledge of their mean energies hardly warrants more precision. The resulting $\gamma = \bar{\beta}/\beta$ is largest for small reactors with large, fast leakage probabilities. The University of London research reactor has a value $\gamma = 1.20$ while a large Magnox reactor has $\gamma \sim 1.03$.

When γ_i is assumed the same for each precursor group, the $I+1$ kinetics equations simplify somewhat to become

$$\left.\begin{aligned} \frac{d\bar{n}}{dt} &= \frac{\bar{\rho}/\gamma - \beta}{\bar{\Lambda}/\gamma}\bar{n} + \sum_i \lambda_i \bar{c}_i + \bar{s} \\ \frac{d\bar{c}_i}{dt} &= \frac{\beta_i}{\bar{\Lambda}/\gamma}\bar{n} - \lambda_i \bar{c}_i \end{aligned}\right\} \tag{2.45}$$

We see in eqns. (2.45) that we may solve in terms of the measured β_i rather than $\bar{\beta}_i$ (or in terms of $a_i = \beta_i/\beta$) if at the same time we use a reactivity $\bar{\rho} = \rho/\gamma$ and a reproduction time $\Lambda = \bar{\Lambda}/\gamma$.

STATIC CALCULATIONS

The λ-reset or static calculation method conveniently produces a reactivity $\rho_s = 1 - \lambda$, and if we wish to use this calculation of ρ we need calculate corresponding kinetic parameters. Rather than solve the adjoint static equation for the necessary adjoint weighting function, it is possible to obtain consistent values of the remaining kinetic parameters direct from a static density calculation as follows.

To calculate Λ_s, suppose we had a prompt neutron only system. If this is non-critical it would lead in the separated equations to an eigenvalue or ω-value, the same at every point in space and energy, of amount ρ_s/Λ_s. In a flux representation this eigenvalue is equivalent to a uniform (in space) $1/v$ absorber. We therefore calculate the static reactivity arising from such a uniform $1/v$ absorber of normalisation b, say. The reproduction time is then given by $\Lambda_s = \rho_s/b$.

To compute the effective fractions it is necessary that the static calculation can accommodate the addition of a uniform amount to the fission cross-section with the *delayed neutron spectrum* χ_i. If the addition is of amount $\beta_i \chi_i F$ leading to a computed reactivity ρ_s, then $\gamma_i = \rho_s / \beta_i$. This is perhaps most neatly done in an adaption of conventional iteration schemes (see ref. 14).

Conclusion

In this chapter we have provided not only the fundamental physical data for reactor kinetics but reasons and methods for calculating effective values of the coefficients of the equations. In the simplest cases these equations describe the time dependency of the neutron population, but in general they must be understood to describe weighted populations of neutrons and precursors. In doing this we introduced the adjoint functions as the best weighting function in arriving at these space and energy averaged values.

No explicit answer has been given to the question "Can we make the first order approximation for the space and energy dependence of the neutron and precursor densities appearing in these averages?" since there is no simple answer. Some of the problems in this chapter explore situations when the approximation breaks down. For many practical purposes it is an approximation that is made, at least over a limited period of time.

Let us briefly indicate a scheme for solving time dependent equations when we cannot make a simple approximation of the sort desired. To solve the equations in their full generality, in time space and energy, when the cross-sections, i.e. the coefficients of the full equations, themselves vary with time being dependent on the neutron and consequently power and temperature levels, is a formidable task, principally because of the short characteristic time for neutron behaviour (Λ) which calls for correspondingly short time steps. It is expensive to look for such a full solution in the general three-dimensional, time and energy dependent case.[9]

A more realistic approach in general would be to suppose that over a more reasonable time interval the cross-sections remain constant and to join the successive solutions over such time intervals together. At the beginning of an interval a static solution is obtained for the time independent equations and this shape used throughout the interval to produce time dependent equations of the reduced type (eqns. (2.28 and 2.29)) by means of which the neutron levels at the end of the interval are computed. This change is then available in simultaneous calculations of the thermal and hydraulic balances, themselves having a slower natural time scale, from which changed cross-sections are evaluated for the next time interval. Such an approach might be called the quasi-static method (Fig. 2.3). Some allowance is made in the shape equations for the influence of the rate of change as a pseudo-absorption cross-section,[9] using an estimate of the persistent period obtained from a previous step.

It is a matter of some judgement to know, of course, when the reduced equations method, when the quasi-static method or when a full treatment would be appropriate (although infeasible perhaps except in special circumstances such as test calculations). This expertise takes us beyond the scope of the text and into the province of professional design laboratories.

What we have achieved in this chapter is a rational basis for the reactor kinetics equations used for many day-to-day calculations in nuclear engineering.

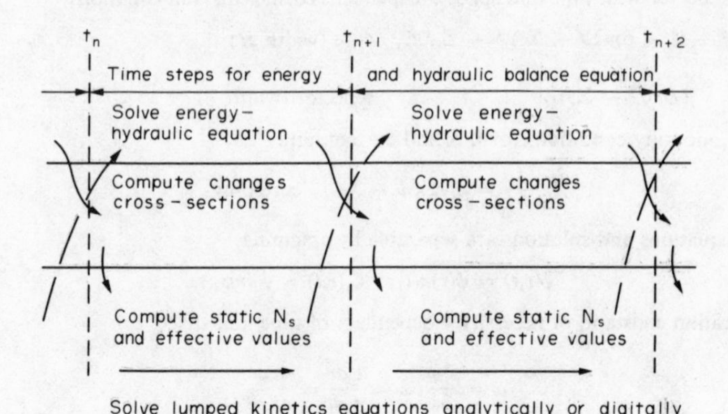

FIG. 2.3. Diagrammatic representation of quasi-static calculations.

Problems for Chapter 2

2.1. Evaluate the so-called effective reactor lifetime that takes the fraction of delayed neutrons into account: $\Lambda + \sum_i \beta_i/\lambda_i$ for the following reactors:

(a) Fast ^{239}Pu. (b) Nat. U-graphite. (c) Slightly enriched D_2O. (d) AGR. Sources of additional data include ref. 7. For (c) include delayed photo-neutrons according to

Group	7	8	9	10	11	12	
β_i	0.00076	0.00023	0.000080	0.000038	0.000023	0.000027	
λ_i	0.28	0.017	0.005	0.002	0.0004	0.0001	(s^{-1})

2.2. Show that the mean life of a precursor is $1/\lambda$ directly from the probability distribution function based on an age-specific failure rate λ (constant).

2.3. Evaluate the Fermi age τ including first flight effects for prompt neutrons and delayed neutrons in (a) H_2O, and (b) D_2O. Find a reactor size and corresponding buckling in a continuous slowing down model that leads to a 10% difference in the non-leakage probability between prompt and delayed neutrons, i.e. in $\exp(-B^2\tau)$.

2.4. A reactor is constructed with nat. U. After operating for a year and a half the ^{239}Pu content has built up to 0.5% while the ^{235}U content has fallen from 0.7 to 0.5% of the original uranium loading. Assuming a bare uniform reactor, find suitable values of β_i and λ_i averaged over the two fissionable isotopes at the beginning and end of this period, using $\gamma = 1.03$.

2.5. Select your own values of coefficients in a two-region, two-group reactor model and evaluate Λ, comparing

(a) $\bar{\Lambda} = \dfrac{\int N_1^+\phi_1/v_1 + N_2^+\phi_2/v_2 dr}{\int N_1^+ \nu\Sigma f \phi_2 dr}$ adjoint weighted

(b) $\Lambda_l = \dfrac{\int \phi_1/v_1 + \phi_2/v_2 dr}{\int \nu\Sigma f \phi_2 dr}$ linear weighted (total reaction rate)

(c) $\Lambda_v = \dfrac{\text{volume reactor}}{\text{volume core}} \dfrac{1/v_1 + 1/v_2}{\nu\Sigma f}$ volumetric average

2.6. *Separable equations in diffusion theory*. Consider the one-group diffusion theory for a uniform core and a uniform reflector with time and space independent coefficients and equations

$$\frac{1}{v}\frac{\partial\phi}{\partial t} = \begin{cases} (D_c\nabla^2 + (1-\beta)\nu\Sigma f - \Sigma_c^a)\phi + \Sigma_i\lambda_i C_i & \text{core (width } H) \\ (Dr\nabla^2 - \Sigma_r^a)\phi & \text{reflector (width } T) \end{cases}$$

with the usual continuity conditions for flux and current, and

$$\frac{\partial C_i}{\partial t} = \beta_i\nu\Sigma f\phi - \lambda_i C_i \quad \text{core only}$$

Show that the equations and solutions are separable by assuming

$$\phi(r,t) = \psi(r)n(t); \quad C_i(r,t) = \psi_i(r)c_i(t)$$

Write a separation constant, of necessity independent of time and space, as

$$\omega = \frac{\partial\phi}{\partial t}/\phi = \frac{1}{n}\frac{dn}{dt} = \frac{1}{c_i}\frac{dc_i}{dt}$$

so that $n(t) = n_0e^{\omega t}$ and $c_i(t) = c_{i0}e^{\omega t}$ are potential solutions.

Note that ω-modes are not the same as α-modes since we are including the precursor equations. Thus distinguish between ω-, α- and λ- modes.

Treat the precursor equation first and hence separate the flux equation.

Make a final separation in the flux equation into the form based on

$$\frac{D_c\nabla^2\psi}{\psi} = \text{constant}, \; B_c^2: \quad \text{core, with } B_c^2 = \frac{\nu\Sigma f - \Sigma_c^a - \omega/v}{D_c}$$

$$\frac{Dr\nabla^2\psi}{\psi} = \text{constant}, \; -\kappa_r^2: \quad \text{reflector, with } \kappa_r^2 = \frac{\Sigma_r^a + \omega/v}{D_r}$$

There will be many such solutions with differing values of ω determined by the criticality relation obtained by the continuity conditions between core and reflector of the form

$$D_cB_c \tan B_c\tfrac{1}{2}H = D_r\kappa_r \coth \kappa_rT$$

We expect to satisfy the original equations with any linear combination of these solutions and thus have in principle a way to match arbitrary initial conditions.

Note, however, that each shape $\psi(r) = \psi_i(r)$ is slightly different since the ω differ even in the set of $I+1$ solutions having a similar flux shape in the sense of having no oscillations in the core, 1 oscillation in the core, 2, etc. Because of this splitting, the flux shape at the beginning of the transient will change and only adopt the lowest mode shape finally when the persisting term dominates the remaining I solutions of the ω-modes in the first group. There is no reason to suppose that the flux shape before the disturbance leading to these ω-values is in the shape $\psi(r,\omega)$.

How many ω-mode solutions would you expect to find grouped with a similar shape in G-group diffusion theory?

2.7. Write the equations adjoint to eqn. (2.38) using the transposed matrix. Specialise to the case of a uniform bare reactor so that
$\nabla{\cdot}D\nabla \rightarrow -DB^2$. Show that the formal expression

$$\frac{\bar{\beta}}{\beta} = \frac{\int N_2^+\nu\Sigma f\phi_3\, dr}{\int [N_1^+[1-\beta] + N_2^+\beta]\nu\Sigma f\phi_3\, dr}$$

reduces to the expression given in eqn. (2.40).

2.8. The adjoint weighting can be given an interpretation as an importance function, a measure of the contribution of a neutron, etc., at a particular location in space energy and time, to some measurable characteristics. These characteristics are legion, leading to many varieties of importance functions,[8] but we are concerned here principally with the ultimate persisting distribution that will remain after transients have died away in a time-constant property reactor.

For a one-dimensional slab reactor with uniform properties (i.e. bare) in the one-group diffusion theory and no precursors, find the set of possible solutions of the form sin Bx_0, sin B_1x, cos B_0x, cos B_1x, etc. What are the expressions for the buckling values B_n in this model? Represent a single neutron at position x_0 by means of a Dirac delta distribution $\delta(x-x_0)$, and if the same neutron is represented by a sum of all the possible solutions, each correctly normalised, of the form

$$\delta(x-x_0) = \sum_i a_i \sin B_i x + c_i \cos B_i x$$

show how to find the expansion coefficients a_i and c_i using the orthogonal properties of the expansion series. Hence find the expansion coefficient for the lowest, persisting term cos B_0x. Thus show that the strength of the lowest term induced by a neutron at x_0 is proportional to cos B_0x_0. Within this normalization, therefore, cos B_0x_0 is the importance of a neutron at x_0 to the ultimate neutron distribution.

References for Chapter 2

1. G. R. Keepin, *Physics of Nuclear Kinetics*, Addison-Wesley, Massachusetts, 1965.
2. R. J. Tuttle, Delayed neutron data for reactor physics analysis, *Nucl. Sci. and Engr.* **56**, 37–71 (1975).
3. G. C. Hanna, C. H. Westcott, H. D. Lammel, B. R. Leonard, J. S. Story and T. M. Attree, *At. Energy Rev.* **7** (4) 3 (1969).
4. G. Ficg, Measurement of delayed neutron spectra, *J. Nucl. Energy* **26**, 585–92 (1972).
5. J. Lewins, *Calculation of the MITR In-Hour Equation*, Internal MIT Report, Mass. Inst. Tech., 1958.
6. A. Henry, *Nuclear Reactor Analysis*, Mass. Inst. Tech., 1975.
7. H. Sodak (ed.) *Reactor Handbook Physics*, III A (2nd edn.), Interscience, New York, 1962.
8. J. Lewins, *Importance: The Adjoint Function*, Pergamon, Oxford, 1965.
9. W. Werner, Solution methods for the space-time dependent neutron diffusion equation, *Adv. Nucl. Sci. and Tech.* **10**, Plenum, New York, 1977.
10. C. B. Besant, P. J. Challa, M. H. McTaggart, P. Tavoularidis and J. G. Williams. Absolute yields and group constants of delayed neutrons, *JBNES 16*, 161–176 (1977).
11. B. C. Diven, H. C. Martin, R. F. Taschek and J. Terrel, Multiplicities of fission neutrons, *Phys. Rev.* **101** (3) 1012–15 (1956).
12. J. Lewins, Reduction of the time-dependent equations for nuclear reactors, *Nucl. Sci and Tech.* **8** 399 (1961).
13. J. Walker and J. Weaver, Reactor kinetics *data, Adv. in Nucl. Sci. and Tech, 11*, Plenum, 1978.
14. D. R. Harris, A method of calculating neutron importance using few group codes, *Trans. Am. Nucl. Soc.* **3**, 396 (1960).
15. J. Terrell, Distribution of fission neutron numbers, *Phys Rev* **108**, 783 (1957).
16. Evaluated Nuclear Data Files available from international centres including Brookhaven Lab., USA and AERE Harwell, UK. For a general introduction, see S. Pearstein, Evaluated nuclear data files, *Adv.* m. *Nucl. Sci.* and *Tech.* **8**, Academic, 1975.
17. J. E. Cahalan and K. O. Ott, Delayed neutron data for fast reactor analysis, *Nucl. Sci. and Engr.* **50**, 208 (1973).

CHAPTER 3

Elementary Solutions of the Kinetics Equations at Low Power

Kinetics Parameters

In the last chapter we discussed the physical data appearing in the kinetics equations and the need to compute effective values of the coefficients if these equations are to take ordinary differential form. The reader coming direct from Chapter 2 may well feel enmeshed in a maze of "effectiveness" and feel that he has lost sight of the meaning of the reactor kinetics equations.

It is appropriate, therefore, to return to fundamental concepts. After such a discussion we derive some exact solutions for limited behaviour of the reactivity and, finally, we discuss some useful approximation methods.

In this chapter we are essentially ignoring everything else in the system except the neutronics equations, e.g. no consideration is given to temperature effects where a rising neutron level would cause a higher energy output, a corresponding temperature change and hence a change in the reactivity. In this chapter, if the reactivity is to vary, it is assumed to vary as a given function of time. So we may be said to be solving the low power equations where temperature effects are not noticeable (albeit for the mean or expected neutron behaviour and not yet allowing for the stochastic or probabilistic description).

For convenience, we restate the neutron kinetics equations as justified in the last chapter. We omit explicit reference to effective or averaged values on the strict understanding that the effective values are now actually being used from here on.

$$\frac{dn}{dt} = \frac{\rho - \beta}{\Lambda} n + \sum_i \lambda_i c_i + s; \quad n(0) = n_0 \tag{3.1}$$

$$\frac{dc}{dt} c_i = \frac{\beta}{\Lambda} \beta_i n - \lambda_i c_i; \qquad c_i(0) = c_{i0} \tag{3.2}$$

Suppose we had no delayed neutrons or their precursors ($\beta_i = \beta = 0$). Then the kinetics equations reduce to

$$\frac{dn}{dt} = \frac{\rho}{\Lambda} n + s; \quad n(0) = n_0 \tag{3.3}$$

For constant coefficients the solution is readily obtained: the Laplace transform would represent "overkill":

$$n(t) = n_0 e^{\rho t/\Lambda} + \frac{s\Lambda}{\rho}[e^{\rho t/\Lambda} - 1] \tag{3.4}$$

Thus Λ/ρ is the effective time constant of the system without delayed neutrons.

REPRODUCTION OR GENERATION TIME

Λ is called the reproduction time (sometimes the generation time†) and is conceptually the mean time for one neutron to produce another prompt neutron or precursor in fission. In the simplest model, therefore, $\Lambda = 1/\nu\Sigma^f v$. Λ is closely related to the neutron lifetime l which is defined as the mean time for a neutron to be *removed* from the reactor. In the same simple model, $l = 1/[\Sigma^a v + DB^2 v]$, where $\Sigma^a v$ gives the rate of loss by absorption and $DB^2 v$ gives the rate of loss by leakage. These are the simplest expressions for the general definitions $\Lambda = 1/\bar{P}$ and $l = 1/\bar{R}$ of Chapter 2.

Clearly in a critical reactor, where production probability equals removal probability, $\Lambda = l$ and typically both range from, say, 1 ms in a heavy water reactor or Magnox through, say, 10 μs in a light water reactor to 0.1 μs in a fast reactor.

If the reactor is *not* critical, $\Lambda \neq l$, but the difference is likely to be small and negligible. We choose to use the reproduction time Λ in preference to the lifetime l because the former is normalised to fission production and this is a natural measure when delayed neutrons are to be represented also as a fraction of fission production. The equations are simpler and the solutions solving with Λ as a constant parameter are more readily obtained. One can, of course, express the equations in terms of l (see the problems for this chapter).

Strictly, neither Λ nor l are constant if we change the configuration of the reactor. But from the simple expressions just given we might say that if the change arises from the insertion of control rods, then Λ is going to change even less than l. Only if the fuel loading is changed will Λ change appreciably. In any case we are straining at gnats in taking note of a change of less than 1% in either Λ or l themselves.

REACTIVITY

It is a very different matter in considering changes in the reactivity ρ since matters have been deliberately arranged to represent the essential and sensitive difference between production and removal rates that drive the kinetics equations. The 1% change that could be ignored in Λ may lead to dangerous rising neutron levels via the reactivity ρ.

Conceptually,

$$\rho = \frac{\text{production rate—removal rate}}{\text{production rate}} \tag{3.5}$$

and in the elementary model, $\rho = [\nu\Sigma^f - \Sigma^a - DB^2]v/\nu\Sigma^f v$.

†Hurwitz had already used generation time, however, for $1/\Sigma^f v$.

Here, again, reactivity has been normalised to fission production rates and we have these three consistent kinetics parameters ρ, Λ and β. When production probabilities equal removal probabilities, the reactor is *critical*, able to support a steady state neutron level without the presence of an independent neutron source; $\rho=0$. For $\rho>0$ we have a supercritical system and for $\rho<0$ a subcritical system. Equation (3.3) shows that the neutron population will rise or fall correspondingly in a source-free system.

The reproduction time Λ is essentially positive, of course. Note that there is an upper limit, $\rho \leqslant 1$ for any sensible definition of production and removal, but this represents a vast overestimate of the maximum reactivity desired or even realisable.

To appreciate the nature of the solutions neglecting delayed neutrons, suppose we have no source, that $\Lambda=0.1$ ms and we have a reactivity of 0.001, i.e. for every 1001 neutrons produced 1000 die, leaving one only to increase the population. Then $n \sim \exp(\rho t/\Lambda) \sim \exp(10t)$. In one second only the neutron population would grow by a factor of more than 20,000. It is likely that the reactor would be damaged by the strains of such an increase and could not sustain this rising period. Even so it will be seen that the task of controlling a reactor that potentially made such large, rapid changes, by either human operator or automatic control, is a formidable one. Fortunately the effect of the *delayed neutrons* in this situation is to replace the exponential behaviour $\exp(10t)$ with a term of the order of $\exp(10^{-4}t)$, which is evidently much easier to control.

There are other measures of the departure from critical in addition to the reactivity. Well known alternatives and their conceptual nature are:

$$\underset{\text{constant}}{\overset{k_{\text{eff}}}{\text{effective multiplication}}} = \frac{\text{production}}{\text{removal}} = \frac{\nu\Sigma^f v}{[\Sigma^a + DB^2]v}$$

$$\underset{\text{constant}}{\overset{k_{\text{ex}}}{\text{excess multiplication}}} = \frac{\text{production—removal}}{\text{removal}}$$

$$= k_{\text{eff}} - 1 \qquad = \frac{\nu\Sigma^f - \Sigma^a - DB^2}{\Sigma^a + DB^2}$$

It follows that $\rho=k_{\text{ex}}/k_{\text{eff}}$ and that $l=\Lambda/(1-\rho)$. Again the kinetics equations can be written in terms of k_{ex} as well as l; the result is more complicated than the forms in ρ and Λ.

REACTIVITY UNITS

We collect several measures of the reactivity, recollecting that fundamentally ρ is a non-dimensional number defined as "(production–removal)/production" and including the weighting function used in averaging these reaction rates. Nevertheless, it is customary to consider so-called static and kinetic measures.

The new units are introduced partly because a "true" reactivity of, say, 0.1 is a very large amount in dynamic terms, and smaller units are convenient.

STATIC MEASURES

These are obtained as the result of calculations of an assumed steady state configuration making some necessary adjustment if the properties do not actually define a critical system.

(a) Per cent reactivity: $\rho_{pc} = 100\rho$
(b) Nile: $\rho_N = \rho_\Delta = 100\rho$
(joke: the Nile has a large delta)
(c) Milli-Nile: $\rho_{mN} = \rho_{m\Delta} = 10^5\rho$
(d) Per cent mille (pcm): $\rho_{pcm} = 10^5\rho$

KINETIC MEASURES

These are based on observed periods in a steadily exponentiating reactor, for example, and are usually related to β (or more properly, β—effective).

(e) Dollar† and cent: $\rho^* = \rho/\beta$
so $\rho^* = \$1$ has $\rho = \beta$
$\rho^* = 1¢$ has $\rho = \beta/100$

(f) Inhour or inverse hour: an amount of reactivity leading to a steady increasing exponential period of one hour. It will be shown that if λ_i are measured in s^{-1},

$$\rho_{in} = \frac{\Lambda + \sum_i \beta_i/\lambda_i}{3600}\rho = \frac{\Lambda + \beta\tau}{3600}\rho$$

For convenience we reproduce in Table 3.1 the delayed neutron working data discussed in Chapter 2, including the appropriate averages β and β/λ defined later in this chapter, for thermal fission in the principle isotopes.

Some Exact Solutions

We now consider the nature of the kinetics equations with delayed neutrons included. It is seen that there are a set of $I+1$ equations with $I+1$ initial conditions. We expect, therefore, $I+1$ independent solutions to be matched to these initial conditions. The equations are *linear* in the dependent variables, n and the c_i, so that solutions can be added together and remain solutions in seeking this "match".

From our discussion we may suppose all the coefficients except possibly ρ and s are independent of time. If ρ and s in a particular case are time independent, we have a constant coefficient system where the Laplace transform method leads to simple algebraic manipulation as a way of getting the answers. If $s(t)$ varies we may still be able to proceed if we can find its transform in simple terms. But if $\rho(t)$ varies we will meet difficulties with the term $\rho(t)n(t)$ and will only be able to find exact solutions in strictly limited

†"Dollar" and "cent" were introduced by L. Slotin who was to die in the first criticality accident. "Inhour" was introduced by E. Fermi in the first reactor at Stagg Field.

TABLE 3.1.

Working delayed neutron data

Thermal fission

	Group	β_i	λ_i (s^{-1})	τ_i (s)
Isotope ^{233}U				
	1	0.000241	0.0126	79.37
$\beta = 0.00281$	2	0.000769	0.0334	29.94
$\pm$ 0.00005	3	0.000637	0.131	7.63
	4	0.000890	0.302	3.31
$\beta/\lambda = 0.050$ s	5	0.000205	1.27	0.787
	6	0.000065	3.13	0.319
Isotope ^{235}U				
	1	0.000266	0.0127	78.74
$\beta = 0.00700$	2	0.001492	0.0317	31.55
$\pm$ 0.00008	3	0.001317	0.115	8.70
	4	0.002851	0.311	3.22
$\beta/\lambda = 0.089$ s	5	0.000897	1.40	0.714
	6	0.000182	3.87	0.258
Isotope ^{239}Pu				
	1	0.000086	0.0129	77.52
$\beta = 0.00227$	2	0.000637	0.0311	32.15
$\pm$ 0.00004	3	0.000491	0.134	7.46
	4	0.000746	0.331	3.02
$\beta/\lambda = 0.033$ s	5	0.000234	1.26	0.794
	6	0.000080	3.21	0.312
Isotope ^{241}Pu				
	1	0.000054	0.0128	78.13
$\beta = 0.00545$	2	0.001249	0.0299	33.44
$\pm$ 0.00054	3	0.000943	0.124	8.06
	4	0.002127	0.352	2.84
$\beta/\lambda = 0.060$ s	5	0.000993	1.61	0.621
	6	0.000087	3.47	0.288

cases (and even then the solutions are exact only in the sense of yielding to numerical integration by quadratures of known functions).

As just stated, the equations are linear in n and the c_i with time dependent coefficients in general, particularly $\rho(t)$. At a later stage we shall seek to couple the reactor kinetics equations for the neutron balance with equations in temperature, etc., from an energy balance, so that instead of $\rho(t)$ being a specified function of time it will, through the additional state equations involving temperature, etc., be coupled to the power and hence to the neutron level $n(t)$. At that point the extended system, described by state variables T, n, etc., can be said to be nonlinear because of the term $\rho(T)n(t)$. However, at low power we suppose $\rho(t)$ is specified independently of n and the terminology "nonlinear" would strictly be incorrect.

We now seek to develop some exact solutions to the low power kinetic equations for the mean neutron level.

SOURCE VARIATIONS

The most elementary problem is likely to be to solve the equations with their initial conditions given that the only variations are caused by the source s. If this is done for the special case that one neutron is added at time zero, then *any* general variation of the source $s(t)$ can be built up from the one source neutron result by suitably summing over the more complicated source function. Indeed, any initial conditions can be represented in this way. For simplicity, we assume that the reactor is in steady state before the source neutron is added so that $n_0 = c_{i0}\lambda_i\Lambda/\beta_i$, and we may take $n_0 = 0$ (one source neutron at time zero is equivalent to one initial neutron). Then

$$\left.\begin{aligned} \frac{dn}{dt} &= \frac{\rho - \beta}{\Lambda} n + \sum_i \lambda_i c_i + \delta(t); \quad n(0) = 0 \\ \frac{dc_i}{dt} &= \frac{\beta_i}{\Lambda} n - \lambda_i c_i; \quad\quad c_i(0) = 0 \end{aligned}\right\} \tag{3.6}$$

where $\delta(t)$, the Dirac delta distribution, represents the source neutron added at $t=0$.

The Laplace transform is represented by an overbar. In the space of the transform variable p:

$$\bar{n}(p) = \int_0^\infty n(t) e^{-pt} dt.$$

$$\left.\begin{aligned} p\bar{n} - n_0 = p\bar{n} &= \frac{\rho - \beta}{\Lambda} \bar{n} + \sum_i \lambda_i \bar{c}_i + 1 \\ p\bar{c}_i - c_{i0} = p\bar{c}_i &= \frac{\beta_i}{\Lambda} \bar{n} - \lambda_i \bar{c}_i \end{aligned}\right\} \tag{3.7}$$

noting that the assumed constant coefficients are simply factored out of the integral of the transformed variables. Then

$$\bar{c}_i = \frac{\beta_i}{\Lambda[\lambda_i + p]} \bar{n} \tag{3.8}$$

$$p\bar{n} = \frac{\rho - \beta}{\Lambda} \bar{n} + \sum_i \frac{\lambda_i \beta_i}{\Lambda[\lambda_i + \rho]} \bar{n} + 1 \tag{3.9}$$

which may be rearranged as

$$\bar{n}(p) = \frac{1}{p + \sum_i \frac{\lambda_i \rho}{\Lambda[\lambda_i + p]} - \frac{\rho}{\Lambda}} \equiv 1/f(p) \tag{3.10}$$

The term in the denominator arises in many applications so it is convenient to denote it by the special term $f(p)$. We see that $f(p)$ could be rationalised by multiplying by the product of all the terms $(\lambda_i + p)$. The result would be a rational polynominal fraction in p with a power of $I+1$ in the numerator and a power of I in the denominator. To proceed further, it would be convenient to express $\bar{n}(p)$ or rather $1/f(p)$ as a sum of partial fractions since we could then recognise the inversion or corresponding time dependency. To do this we expect to write

$$f(p) = \frac{(p - p_0)(p - p_1)\dots(p - p_{i'})\dots(p - p_I)}{(p + \lambda_0 t)\dots(p + \lambda_i)\dots(p + \lambda_I)} = \frac{\prod_{i'}(p - p_{i'})}{\prod_l (p + \lambda_i)} \tag{3.11}$$

in the π or product notation.

Thus we seek $I+1$ roots, $p_{i'}$ of the equation $f(p_{i'})=0$ and with their help find the residues such that we can write

$$\bar{n}(p) = \frac{1}{f(p)} = \sum_{i'=0}^{I} R_{i'}(p_{i'}) \frac{1}{p - p_{i'}} \tag{3.12}$$

Thus much turns on the solution of the equation $f(p)=0$, for the $I+1$ roots p_i'. The general nature of the solution may be obtained by inspection of the equivalent forms

$$p\left[\Lambda + \sum_i \frac{\beta_i}{p + \lambda_i}\right] = \rho = \Lambda p + \beta - \sum_i \frac{\beta_i \lambda_i}{p + \lambda_i} \tag{3.13}$$

regarding ρ as a function of p. Figure 3.1 is built up from the following observations:

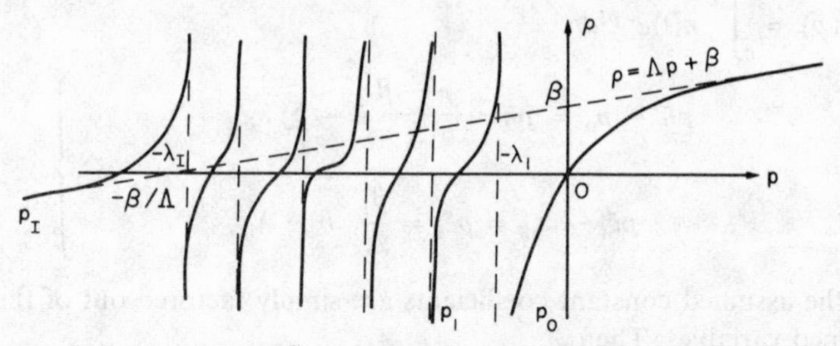

FIG. 3.1. Roots of $f(p)=0$, the inhour equation (diagrammatic).

(i) The equation has a singularity around each $p=-\lambda_i$, where the term $1/[p+\lambda_i]$ changes from "$+\infty$" for p just less than $-\lambda_i$ to "$-\infty$" for p just more than $-\lambda_i$, as seen from the right hand form.

(ii) Again in the right hand form, for large magnitudes of p the summation term is negligible. Therefore these values of p must approach an asymptote given by

$$\rho = \Lambda p + \beta$$

This intersects the ρ-axis at β, the p-axis at $-\beta/\Lambda$ and has a slope of Λ. Since β is small, $\ll 1$, the solution for $\beta < \rho < 1$ (the physical limit of ρ) approaches this asymptote for $\rho \gg \beta$.

(iii) The p-ρ plane has been divided into $I+1$ strips by the well-spaced values of the $-\lambda_i$. Within each strip ρ is a continuous function of p that passes from $-\infty$ to $+\infty$ as p increases through the strip. There is therefore a real root in each strip and this accounts for all $I+1$ roots of the polynomial equation. In particular, there are no complex roots. If in a multi-isotope model there are groups with common values of (some) λ_i there would be repeated roots. However, the ordinary differential equations for such precursors could be aggregated into a single equation so the possibility can be ignored.

(iv) The left hand form shows that there is a root at the origin, $p=0$ for $\rho=0$. The root to the right, for $\rho>0$, will be positive, rising to the asymptote. It is the only positive root and corresponds to a rising exponential in the time dependent solution. For $\rho<0$ all roots are negative and correspond to falling exponentials.

(v) The root p_0 in the right hand strip is largest and whether positive or negative will lead to an exponential ultimately dominating all other exponential terms. It may therefore be called the dominant or persistent root. All the remaining roots correspond to dying transients.

(vi) Since $\rho = p = 0$ is a known solution, we may develop the expression for the slope of the locus of eqn. (3.13) through the origin:

$$\left.\frac{d\rho}{dp}\right|_{p=0} = \Lambda + \left.\sum_i \frac{\beta_i \lambda_i}{[p + \lambda_i]^2}\right|_{p=0} = \Lambda + \sum_i \frac{\beta_i}{\lambda_i} \equiv \Lambda + \sum_i \beta_i \tau_i \tag{3.14}$$

This gives an "average" time constant for small reactivities of $\rho/p_0 = \Lambda + \beta\tau = \Lambda + \Sigma\beta_i\tau_i$. The second summation term dominates Λ. It is rarely less than 100 Λ (heavy water) and often more. There is therefore the dramatic reduction of the effective reactor characteristic time from Λ (when $\beta \ll |\rho|$) to $\Lambda + \beta\tau$ (when $\rho \to 0$). When $\rho > \beta$, however, the reactor is behaving as a prompt multiplying system with a natural characteristic time Λ and a prompt reactivity $\rho_p = \rho - \beta$. This is evidently of potential danger, and much thought in design is exercised to see that it is not possible to reach this stage (except for special research facilities where peculiar care is taken to limit the duration of such a prompt transient to a few Λ). These two extremes lead to dominant or persisting solutions of the form:

(a) prompt behaviour, $\rho > \beta$: $n(t) \to \exp\left(\frac{\rho - \beta}{\Lambda} t\right)$;

(b) delayed behaviour, $\rho \sim 0$: $n(t) \to \exp(\rho t/\beta\tau)$.

(vii) We have now demonstrated the origin of the term "inhour" in the dynamic unit of reactivity. For related reasons Fig. 3.1 is sometimes called the (full) inhour figure based on an "inhour" equation (3.13). However, the usual units are 1/s and not inverse hours. Due to the significance of the dominant root, especially on a rising period with $\rho > 0$, it is also helpful to plot the corresponding period $T_0 = 1/p_0$ as shown in Fig. 3.2. The data is for ^{235}U and shows the 10 s period which is commonly used as a safety criterion against excessive positive reactivity.

For $\rho > 0$, T_0 varies between zero and infinity. Values of T_0 less than 10 s indicate a rapidly rising exponential growth to be avoided (though this is to be distinguished from the transient periods before the persistent period dominates the remainder). If $T_0 = \infty$ the reactor without a source is critical. If the reactor is critical but has a source, however, there is a steady *linear* rise (and no specific exponential period of course).

(viii) It is possible to define an "instantaneous" period $T = n/(dn/dt)$ but it is the *steady* rising period that has a clear interpretation in experimental determinations of ρ against p_0. Whilst negative values of T are meaningful in the sense of implying $\rho < 0$, it is more difficult to wait for the transients to die away in relation to the term in p_0 in order to leave a pure exponential behaviour. Indeed, the fastest *long term* drop in $n(t)$ for $\rho \ll 0$ can still only be at $p_0 = -\lambda_1$. That is, the long term shutdown is governed by the rate at which the slowest decaying precursors give up their delayed neutrons.

We have now developed the *form* of eqn. (3.13). Actual solutions are obtained numeric-

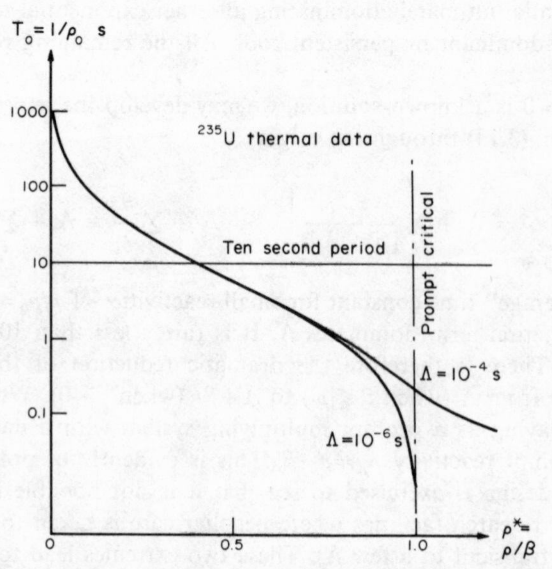

FIG. 3.2. Steady exponentiating positive period against reactivity.

ally (see problems). We return to the solution of the impulse source problem, eqn. (3.12). Since the roots p are distinct, the residues are conveniently given by

$$R_{i'}(p_{i'}) = 1/\left.\frac{df}{dp}\right|_{p=p_{i'}} = 1/\left[1 + \sum_i \frac{\beta_i \lambda_i}{\Lambda[p_{i'} + \lambda_i]^2}\right] \equiv A_{i'} \tag{3.15}$$

As a check, $\sum A_i' = 1$, corresponding to the initial source neutron. The full solution is then

$$n(t) = \sum_{i'} A_{i'} \exp(p_{i'} t) \tag{3.16}$$

where the p_i' are obtained from the equation $f(p) = 0$ and the A_i' from eqn. (3.15). For a stronger impulse source, $s_0\delta(t)$ say, linearity of the equation allows us to multiply the solution of eqn. (3.16) by s_0 to answer the new problem. Indeed, a general source $s(t)$ leads to a convolution solution summing all the inherent impulse sources:

$$n(t) = \sum_{i'} A_{i'} \int_{-\infty}^{t} s(t') \exp(p_{i'}[t - t'])dt' \tag{3.17}$$

Returning again to the impulse source problem, we can consider the dominant root to be smaller in magnitude than any λ_i and develop A_0 approximately so that for sufficiently long times

$$n(t) \to A_0 \exp(p_0 t) \simeq 1 + \left[\sum_i \frac{\beta_i}{\Lambda\lambda_i}\right]^{-1} \exp(p_0 t) \simeq \frac{\Lambda}{\beta\tau} \exp(\rho t/\beta\tau) \tag{3.18}$$

The interest here is in the coefficient $A_0 \simeq \Lambda/\beta\tau$ which again takes the small value $\simeq 10^{-3}$. It is as if the persisting neutron behaviour were given by a persisting exponential starting not with the actual one neutron but only 10^{-3} neutrons. The original neutron has instead gone over into precursors, held back by the slower time characteristics of precursors. Indeed, the concentration of precursors is far higher than that of neutrons in normal circumstances, and in the steady state we have $\sum_i c_{i0} = n_0\beta\tau/\Lambda \simeq 10^3 n_0$.

STEP CHANGE IN REACTIVITY

A similar analysis allows us to develop the solution to a problem where the reactor properties change suddenly at $t=0$ and thereafter remain constant. Thus the coefficients are essentially constant and an exact solution can be found. Indeed, the solution is closely related to the impulse solution with the more detailed explicit initial conditions in n_0 and c_{i0} and with constant source s. Then the Laplace transformed equations are

$$p\bar{n} - n_0 = \frac{\rho - \beta}{\Lambda}\bar{n} + \sum_i \lambda_i \bar{c}_i + \bar{s} \tag{3.19}$$

$$p\bar{c}_i - c_{i0} = \frac{\beta_i}{\Lambda}\bar{n} - \lambda_i \bar{c}_i \tag{3.20}$$

For simplicity we develop the result starting from steady state so that $c_{i0} = [\beta_i/\Lambda\lambda_i]n_0$, leaving the more general result to be followed through as a problem. Similarly, take the source to be absent.

Substitution and rearrangement gives

$$\bar{n}(p) = n_0 \frac{1 + \sum_i \dfrac{\beta_i}{\Lambda[p+\lambda_i]}}{p + p\sum_i \dfrac{\beta_i}{\Lambda[p+\lambda_i]} - \dfrac{\rho}{\Lambda}} = n_0\left[1 + \sum_i \frac{\beta_i/\Lambda}{p+\lambda_i}\right]\Big/ f(p) \tag{3.21}$$

where $f(p)$ is the same expression as in the impulse source case. Thus to solve the problem we require the same roots p_j for the current values of ρ, Λ and the β_i, λ_i and refer again to the inhour relation (Fig. 3.1).

The residues in this problem are not the same, of course, but are given by

$$R_{i'}(p_{i'}) \equiv B_{i'} = n_0 \frac{1 + \sum_i \dfrac{\beta_i/\Lambda}{p_{i'} + \lambda_i}}{\left.\dfrac{df}{dp}\right|_{p=p_{i'}}} = \frac{1 + \sum_i \dfrac{\beta_i/\Lambda}{p_{i'} + \lambda_i}}{1 + \sum_i \dfrac{\beta_i\lambda_i/\Lambda}{[p_{i'} + \lambda_i]^2}} \tag{3.22}$$

As a check, $\sum B_i' = n_0$ in correspondence with the initial condition. The formal solution is then

$$n(t) = \sum_{i'=0}^{i'=I} B_{i'} \exp(p_{i'}t) \tag{3.23}$$

The exponential terms in this result are the same as for the source variation but the expansion coefficients, the residues or $B_{i'}$ differ, reflecting the different initial conditions. It is here interesting to develop the coefficient of the *fastest* transient B_I with a view to determining the change in the neutron level that occurs as this fast transient decays away, leaving the remaining I terms (dominated by the delayed neutron behaviour) to carry the time dependency of n forward. Since $p_I \simeq [\beta+\rho]/\Lambda$ we find

$$B_I \simeq \frac{\rho}{\beta-\rho}\, n_0 \tag{3.24}$$

Thus after a short "prompt" transient, a few Λ-values, this term only has decayed to zero; other terms have not changed noticeably. Since the sum of the $B_{i'}$ is n_0, we find that the neutron level at the end of this prompt transient, usually called the prompt "jump" is $\beta/(\beta-\rho)\, n_0$.

We speak of the prompt jump from n_0 to its immediate value, followed by a slow transient (involving I exponentials) towards the dominant or persistent exponential in p_0.

The step *reduction* of reactivity is particularly interesting as a method, called in this context the rod-drop experiment, of measuring the reactivity worth of a control rod by suddenly inserting it in a critical system and monitoring the subsequent reduction in neutron level. The observed variation is to be fitted to the above theory with a view to finding the reactivity worth of the rod. The experiment is therefore applicable to rods with substantial reactivity worth where it would be unacceptable to *withdraw* the rod to add that amount of reactivity to a critical system (and where therefore the rising exponential period of Fig. 3.2 is inapplicable).

It is worth noting that several groups of delayed neutrons are needed to represent the *decay* of the flux adequately. This follows from the discussion (viii) of Fig. 3.1, that the long term behaviour is governed by the slowest decay group, so that a single group with an average behaviour would not adequately represent the rod-drop results.

Some Approximate Methods

The general discussion so far, and particularly the development of the two exact solutions, will have illustrated the consequences of a system combining widely different time constants, e.g. the typically 10^{-6} s corresponding to prompt neutrons and, say, 10^{-1} s corresponding to the longer-lived precursors. Simultaneous differential equations with this characteristic are said to be "stiff" and stiffness leads to fundamental difficulties in obtaining solutions, by any means.

Exact (analytic) solutions to the reactor kinetics equations are limited to cases where the reactivity varies with time in a simple manner (and the remaining coefficients are constant, of course). In addition to the (piecewise) constant case already dealt with, it is possible to develop exact solutions to three classes of reactivity variation: ramp

($\rho=\rho_0+\gamma t$), inverse ($\rho=\gamma/t$) and exponential ($\rho=\exp[\gamma t]$). More complicated and interesting cases have to be approximated. Even these "exact" solutions call for integration by quadratures, i.e. over known functions of time, and are not explicit in the sense of being reduced to commonly tabulated functions. Even in the constant reactivity case, of course, with, say, three isotopes and six precursor groups, the evaluation of so-called exact solution involves considerable computation.

Methods using analogue computation are inherently approximate but lend themselves to the representation of more complicated reactivity functions. They also suffer from stiffness (see Chapter 9). Methods involving digital computation start by approximating the differential form with a *difference* equation based on $dn/dt \simeq [n(t+h)-n(t)]/h$, where h is some time interval. To find the neutron behaviour over, say, 50 s, we would ideally like to take long intervals h. Unfortunately, the *error* in the difference approximation for a differential equation turns on the *shortest* time constants, and in systems of stiff equations we are driven to take h-intervals less than this shortest time constant. As a result we may have to take unacceptably many short steps to develop the required solution on a digital computer, at least without special devices.

We see then that some forms of approximation are commonly necessary. More expensive exact solutions have a role as a test for cheaper approximate solutions, and once the range of validity is established these approximations are valuable. In this section we develop a few of the analytical approximations, starting with the reduction of the I-precursor equations to a single average precursor equation with the corresponding simplicity of solution and interpretation.

ONE-GROUP PRECURSORS

Some characteristics of a full model with, say, six precursor groups have been established. It would be a useful simplification to consider one representative precursor group instead. The equations would then be

$$\left.\begin{aligned}\frac{dn}{dt} &= \frac{\rho-\beta}{\Lambda}n + \lambda c + s; \quad & n(0) &= n_0\\ \frac{dc}{dt} &= \frac{\beta}{\Lambda}n - \lambda c; & c(0) &= c_0\end{aligned}\right\} \tag{3.25}$$

and the inhour equation reduces to $p^2+[\lambda+[\beta-\rho]/\Lambda]p-\rho\lambda/\Lambda=0$. It remains to specify values of the one-precursor group coefficients, β and λ, and these can be chosen perhaps to reproduce some leading characteristics of the full I-group model.

We cannot expect to match every characteristic feature of a full solution but the following points seem important:

(i) So that the potentially dangerous behaviour around prompt critical is accurately represented, we want the asymptotic line $\rho=\Lambda p+\beta$ of Fig. 3.1 to be preserved. Thus we select a β-value$=\sum_i\beta_i$, and, of course, use the exact values of ρ and Λ since these are unaffected by the proposed simplification.

(ii) Both models provide for a root $p=0$ at the origin when $\rho=0$. We may therefore determine the remaining one-group parameter λ by matching the *slope* through the origin and hence securing a match of the persistent solution for small departures of the reactivity from zero (though not, of course, matching the remaining transients). To do this we take $\beta/\lambda = \sum \beta_i/\lambda_i$ or take τ of the one-group model to be the same τ as developed in eqn. (3.14) for the I-group model.

There is, of course, something arbitrary in the choice of these particular characteristics to be matched. In another problem there may be more interesting features to reproduce in the simple model. For example, the transfer function to be developed in Chapter 4 can be matched over a prescribed frequency band.

This approximation has given us a two-equation system, i.e. of second order. A further approximation makes use of the stiffness of these two equations to reduce the complexity to a single order level (for which solutions for arbitrary reactivity variations are readily found) in one of two ways, depending on whether the reactivity variation is rapid or slow.

CONSTANT PRECURSOR APPROXIMATION

If the reactivity is above prompt critical the neutron level changes rapidly compared to the rate at which precursors decay to yield delayed neutrons. There is a range of rising transients therefore where it is acceptable to assume *constant precursors* as a source in the neutron equation. This is equivalent, of course, to neglecting the rate of change of precursors in the precursor equation which then degenerates from a differential to an algebraic equation.

Thus

$$\frac{dn}{dt} = \frac{\rho - \beta}{\Lambda} n + \frac{\beta}{\Lambda} n_0 + s; \quad n(0) = n_0 \tag{3.26}$$

where n_0 is the initial value of n in the assumed initial steady state. If s is also constant the formal solution as a quadrature is

$$n(t) = \left[n_0 + \left[s + \frac{\beta}{\Lambda} n_0 \right] \int_0^t \exp\left(-\int_0^{t'} \frac{\rho - \beta}{\Lambda} dt'' \right) dt' \right] \exp\left(\int_0^t \frac{\rho - \beta}{\Lambda} dt' \right) \tag{3.27}$$

for general reactivity variations $\rho(t)$. The constant precursor approximation can be used in a full I-group model also, of course, to good effect. Since the values of λ (or λ_i) do not appear, the model can only be valid for fast, predominantly supercritical transients where the hold up of the precursors is immaterial to the development of the neutron density.

In particular, consider a *fast* ramp variation where $\rho(t) = \beta + \mu t$, i.e. jumping from critical to prompt critical and then increasing, μ positive. If the independent source is zero,

$$n(t) = n_0 \exp(\mu t^2/2\Lambda) \left[1 + \frac{\beta}{\Lambda} \int_0^t \exp(-\mu t'^2/2\Lambda) dt' \right] \to n_0 \left[1 + \beta \sqrt{\frac{\pi}{\mu\Lambda}} \right] \exp(\mu t^2/2\Lambda) \tag{3.28}$$

where the final form is related to the error function asymptotic expansion. The nature of the rapidly increasing term in $\mu t^2/2\Lambda$ is evident. Figure 3.3 plots the result and illustrates the severe consequences of such an accident, for the more important case of *delayed* critical ($\rho_0=0$ and see problems 3.14 and 3.15). Note the deceptively small increase at prompt critical.

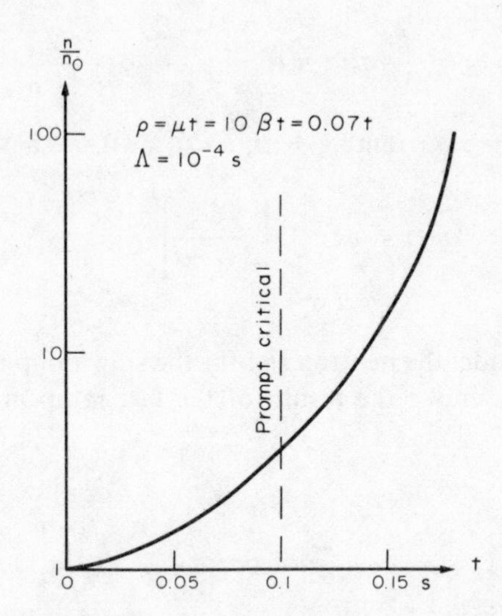

FIG. 3.3. Fast ramp approximation from critical.

PROMPT JUMP APPROXIMATION

This approximation exploits stiffness from the opposite end.† If the transient is slow, the prompt multiplying system has the opportunity of rapidly balancing itself and responding to further changes in the precursor decay rate as a source. This was evident in the analysis of the step reactivity change where, in a few Λ-values, the neutron level had "jumped" to a new value. This quasi-static approximation, therefore, takes dn/dt to be zero, reducing the *neutron* equation to an algebraic one. Again, this approach is useful in the *I*-group model also. We have here

$$0 = \frac{\rho - \beta}{\Lambda} n + \lambda c + s$$

so that

$$\frac{dc}{dt} = \frac{\lambda\rho}{\beta - \rho} c + \frac{\beta}{\beta - \rho} s; \quad c(0) = c_0 \tag{3.29}$$

Equation (3.29) shows that the approximation will fail as ρ approaches β and therefore should be used only below prompt critical. The formal solution for general reactivity variation and constant source is

†This also leads to a singular perturbation method based on powers of the reproduction time Λ.

$$c(t) = \left[c_0 + s\int_0^t \frac{\beta}{\beta-\rho}\exp\left[\int_0^{t'} \frac{\lambda\rho}{\beta-\rho}\,dt''\right]dt\right]\exp\left[\int_0^t \frac{(\lambda\rho}{\beta-\rho}\,dt'\right] \tag{3.30}$$

with the further relationships, assuming an initial steady state:

$$c_0 = \frac{\beta}{\Lambda\lambda}n_0; \quad n(t) = \frac{\Lambda\lambda c}{\beta-\rho} + \frac{s\Lambda}{\beta-\rho} \tag{3.31}$$

For the special case of the *slow* ramp, $\rho=\mu t$, $\mu<0$, $s=0$, we have

$$n(t) = n_0\, e^{-\lambda t}\left[\frac{\beta}{\beta-\mu t}\right]^{1+(\lambda\beta/\mu)} \tag{3.32}$$

as exemplified in Fig. 3.4.

In the prompt jump model the neutron rise for the slow ramp is unbounded as prompt criticality is approached, unlike the result for the fast ramp in the constant precursor

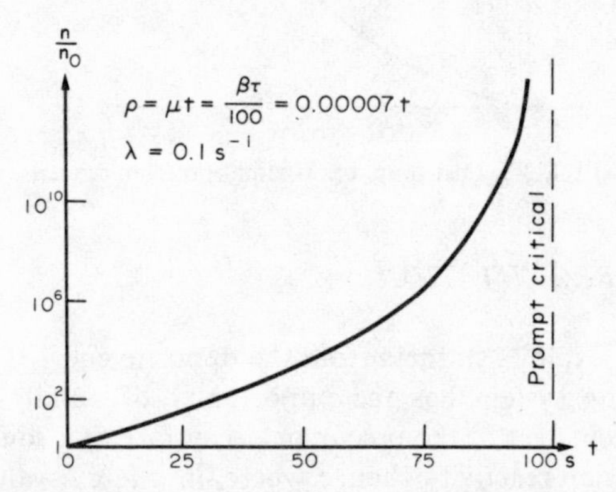

FIG. 3.4. Slow ramp approximation from critical.

approximation, pointing out the danger of the fast reactivity excursion of leading to a dangerous situation without adequate warning from the observable neutron behaviour.

Note also the many decades of flux level that may have to be covered from a clean start up to operation at power, placing considerable demands on detectors and associated instrumentation.

STEP REACTIVITY IN ONE-PRECURSOR GROUP

Returning to the original two equations of the model, we readily specialise the I-group result for a step change of reactivity starting in steady state. The approximate roots (for ρ not around β) may be found first by assuming an asymptotic root $p_1=[\rho-\beta]/\Lambda$ and then from the theory of algebraic equations, noting that the product of the roots p_0p_1 in this case is given by the coefficient $-\lambda\rho/\Lambda$, so that

$$p_0 \sim \frac{\lambda\rho}{\beta - \rho}; \quad p_1 \sim \frac{\rho - \beta}{\Lambda} \tag{3.33}$$

and with these values in the expression for the residues we have approximately

$$n(t) = n_0 \left[\frac{\beta}{\beta - \rho} \exp(p_0 t) - \frac{\rho}{\beta - \rho} \exp(p_1 t) \right] \tag{3.34}$$

$$c(t) = c_0 \left[\exp(p_0 t) + \frac{\Lambda\lambda\rho}{[\beta - \rho]^2} \exp(p_1 t) \right] \tag{3.35}$$

The prompt jump of the *neutron* level is immediately apparent. In the precursor equation the coefficient of the transient term is very small. Results are shown in Fig. 3.5 for some example cases.

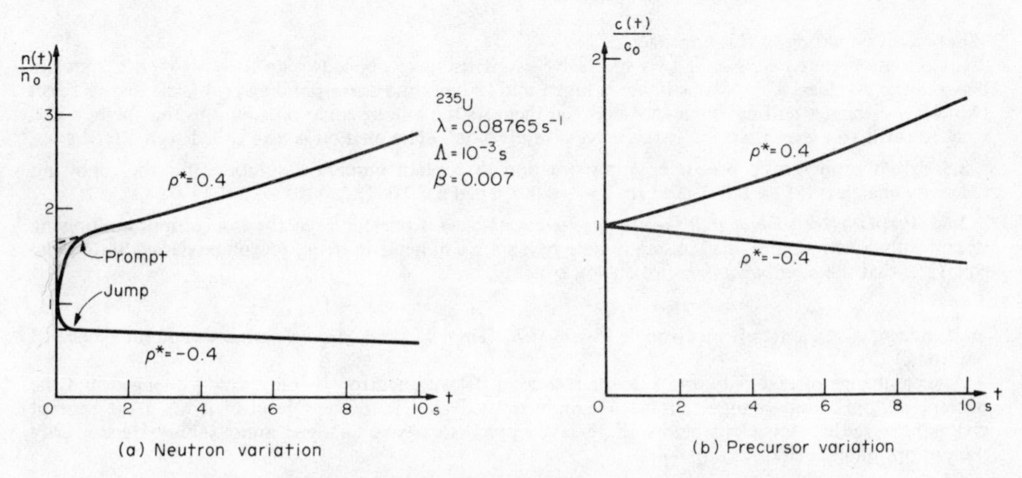

FIG. 3.5. Neutron and precursor changes to a step reactivity change.

Further approximate solutions are discussed in the problems to this chapter and in the context of analogue computing, Chapter 9. Problem 3.10 develops an exact solution of the ramp problem which is useful for checking the accuracy of an approximate method. But in limiting the reactivity variation studies to a few specified time dependent functions, we have only touched on one aspect of the behaviour of a reactor since this is to ignore the link with the neutron level itself through the various energy balances, etc. It is appropriate, therefore, to turn in the next chapter to the description of the process dynamics of a reactor and its associated plant in the light of such additional terms, making suitable use of further approximations to determine the characteristics.

Problems for Chapter 3

3.1. *Equations parametric in neutron lifetime and excess multiplication.* Take eqns. (3.1) and (3.2) and substitute for ρ and Λ in terms of k_{ex} and l. Solve the equations for constant properties in the impulse source problem and discuss the nature of the "inhour" equation for k_{ex} as a function of p.

3.2. *Sjorstrand pulsed source experiment*. Consider a sub-critical system at constant properties subject to an impulse source. Determine the number of neutrons produced in all time following the impulse by the *prompt* behaviour (i.e. $\int \exp(p_I t)dt$) and by the remaining exponential terms. Thus a detector system whose counts are proportional to the neutron level will determine a ratio of prompt to remaining counts that in turn will yield a measure of the reactivity (in kinetic units). Sketch the form of solution and the ultimate result of repeating the pulse regularly on a cycle that is long compared to Λ and short compared to $\beta\tau$.

3.3. *One-precursor group step reactivity problem.* Justify eqns. (3.34) and (3.35). How close must ρ be to β to have a 1% error using eqns. (3.33)?

3.4. *Numerical solution of the inhour equation.* Exact numerical solution of the inhour equation $f(p)=0$ is not always trivial especially if we are dealing with several isotopes and six groups per isotope, plus a few delayed photo-neutrons. However, suppose we are dealing with a ^{235}U system with the data given in Chapter 2 (and a common effectiveness ratio γ) with $\Lambda/\gamma=10^{-4}$ s. Write a digital program to find the seven roots given ρ. Use this to compute the $n(t)$ behaviour for a step change in reactivity and compare with the one-precursor group approximation, eqn. (3.34). (*Hint*. You will need a suitable iteration algorithm to "home" on the correct values.) Consider the following:

(a) *Newton–Raphson* $\quad f(p_{i'}) = 0 \simeq f(p_{i'} + \epsilon) + \epsilon \left.\dfrac{df}{dp}\right|_{p=p_i'+\epsilon}$

What might go wrong in this approach?

(b) *Successive interpolation.* If two successive estimates have opposite sign for $f(p)$ then a root lies between the p-values. The next estimate is taken midway and the corresponding $f(p)$ evaluated to select two closer estimates still having a root between them. Is the inhour equation suited to this method and what test will you use to terminate the successive division? (The method is also called *regula falsi.*)

3.5. Use the one-group precursor approximation to contain numerical solutions for the following reactivity changes: (a) $\rho=0.001$, and (b) $\rho=-0.003$ when $\beta=0.007$, $\Lambda=10^{-4}$ s and $\lambda=0.08$ s^{-1}.

3.6. *Alternative development of the prompt jump coefficient*. Consider a reactor as a prompt multiplying system (sub-critical) maintained in steady state by a source of neutrons from precursor decay. Show from eqn. (3.3) that the *multiplication* M of such a system

$$M = 1 + k_{\text{eff}} + k_{\text{eff}}^2 + k_{\text{eff}}^3 + \cdots = -1/k_{\text{ex}}$$

and, indeed, gives a steady neutron level $n=Ml S$. (Here k_{ex} is a *prompt* value discounting delayed neutrons.)

After a change of reactivity and k_{ex}, the source of delayed neutrons is only slowly changed, but the prompt multiplication changes almost instantaneously. What is the new level of n when the prompt chains have settled down but before the delayed precursors have changed appreciably? Hence verify the prompt jump coefficient $\beta/(\beta-\rho)$.

3.7. In the one-precursor group model the inhour equation may be written as the quadratic

$$p^2 + p\left[\lambda + \frac{\beta-\rho}{\Lambda}\right] - \frac{\lambda\rho}{\Lambda} = 0$$

For small values of $|\rho|$ develop the solution to this quadratic in powers of ρ. Making the usual approximation that $\Lambda\lambda/\beta \ll 1$, show that

$$p_0 \simeq \frac{\lambda\rho}{\beta-\rho}; \quad p_1 \simeq \frac{\rho-\beta}{\Lambda}$$

(and note that this result is true also for very large magnitudes of ρ).

3.8. Develop $f(p)$ as a *standard* polynomial in the form

$$p^n + a_1 p^{n-1} + a_2 p^{n-2} + \cdots$$

for the special case $\rho=0$. Since the roots are known to be real, distinct and in sequence of magnitude,

$$|p_0| < |p_1| < \cdots < |p_I|$$

we may use algebraic theory to approximate them as

$$-p_I \simeq a_1; \; p_{I-1} \simeq a_2/a_1; \quad \cdots - p_0 \simeq a_n/a_{n-1}$$

For a ^{235}U reactor with $\Lambda=10^{-4}$ s, show that this approximation yields the roots to about first figure accuracy.

3.9. *Graeffe method.* Develop the approximation of problem 3.8 into the Graeffe method. Here we note that if the p are solutions of

$$\prod_{i'} (p - p_{i'}) = 0$$

then also

$$\prod_{i'} (p - p_{i'}) \times \prod_{i'} (p + p_{i'}) = \prod_{i'} (p^2 - p_{i'}^2) = 0$$

This new equation in p^2 has roots $p_{i'}^2$ which are necessarily better spaced than the $p_{i'}$. Find the coefficients of this equation in standard form by multiplying the standard forms of the two product terms together, noting the change of sign for odd power terms in the equation with roots $-p_{i'}$.

Show that the application of this approximation twice in succession to the data of the last problem yields solutions as accurate as the data warrant.

3.10. *Exact ramp solution.* We may adapt Smets' exact solution for the ramp problem to a one-precursor group model. The result is helpful when an approximation method is to be tested against an exact solution. For $\rho = \mu t$ show that the neutron level satisfies the second order differential equation

$$\frac{d^2n}{dt^2} + \left[\lambda - \frac{\rho - \beta}{\Lambda}\right]\frac{dn}{dt} = \left[\frac{\lambda\rho}{\Lambda} + \frac{\mu}{\Lambda}\right] n$$

Show that each of the following integrals is (separately) a solution:

$$I_1(t) = \int_{-\infty}^{-\lambda} g(p,t)dp; \quad I_2(t) = \int_{-\lambda}^{\infty} g(p,t)dp$$

where

$$g(p,t) = [p + \lambda]^{\beta/\mu} \exp\left(-\frac{\Lambda p^2}{2\mu} - \frac{\beta}{\mu}p + pt\right)$$

so that a general solution to the problem in terms of two arbitrary constants is $n(t) = C_1 I_1(t) + C_2 I_2(t)$. Show how to determine C_1 and C_2 to match the initial conditions of $n(0) = n_0$ and $\dot{n}(0) = 0$ (for initial steady state). The results, of course, are exact in a theoretical sense but still require integration by quadratures to obtain specific values.

3.11. *Numerical solutions using difference equations.* This problem explores some aspects of using digital computers via difference approximations for the differential equation $dy/dt = f(y,t)$. Two approaches to the approximation are possible, the first based on a Taylor series expansion and the second on a difference operator approach. Approximations in various orders are possible in both.

(a) *Runge–Kutta.* First and second order approximations are:

(i) $y_n(t_n) = y_{n-1}(t_{n-1}) + hf_1$

where

$$f_1 = f(y_{n-1}, t_{n-1}) \quad \text{and} \quad t_n = t_{n-1} + h$$

(ii) $y_n = y_{n-1} + \frac{1}{3}h\, f_1 + 2f_2$

where

$$f_2 = f(y_{n-1} + \tfrac{1}{2}hf_1, t_{n-1} + \tfrac{1}{2}h)$$

The second order approximation is seen to be based on values at the last point *and* mid-way between points.

(b) *Predictor–Corrector.* First and second order approximations are:

(i) Identical to Runge–Kutta.

(ii) $\tilde{y}_n = 2y_{n-1} - y_{n-2}$ predictor

followed by

$y_n = y_{n-1} + hf(\tilde{y}_n, t_n)$ corrector

Apply these various approximations to a known situation such as

$$\frac{dy}{dt} = \lambda y$$

and explore the effect of a choice of step size h on the accuracy of the solution in terms of λh. Is there a marked upper limit to h?

Keeping in mind these results, apply the approximations to the simultaneous solution of the "stiff" equations in the one-group precursor model. Apply first to a constant property case (where the solution is known) and then to more general reactivity variations.

(*Note.* There are several methods of modifying the method to increase the step size that is acceptable and further reading in the references is indicated. See problem 3.13, for example.)

3.12. A reactor transient is to take the form of an increase at constant period from an initial steady rate terminated by a levelling off to a constant value after an interval T. Find a formal expression for the reactivity function to secure such a change in one-group and I-group precursor models.

Note that it is relatively easy to find $\rho(t)$ for *arbitrary* $n(t)$ but not vice versa.[9] Discuss the basis of this idea for a *reactivity meter*, an on-line device to display $\rho(t)$.

3.13. *Integral equation form.* Show that in the one-precursor group model

$$c(t) = c(t_0)\exp\{-\lambda[t-t_0]\} + \frac{\beta}{\Lambda}\int_{t_0}^{t} \exp\{-\lambda[t-t']\}n(t')dt'$$

Is it reasonable to suppose that the term in c_0 vanishes as $t_0 \to \infty$? If so

$$\frac{dn}{dt} = \frac{\rho - \beta}{\Lambda} n + \frac{1}{\Lambda}\int_0^\infty D(\tau)n(-\tau)d\tau + s$$

where the *delayed neutron kernel* is given by $D(\tau) = \beta\lambda\exp(-\lambda\tau)$. This device of integrating the precursor equations analytically has several uses. In this problem, use the integral equation form to prove the solutions to Smets' problem, 3.10.

Develop the kernel and integral equation in I-group theory.

3.14. *Difference equation from integral form.* By integrating the precursor equations as above we may develop a difference approximation to be applied to one equation only instead of $I+1$ simultaneous differential equations. The scheme is to approximate $n(t-\tau)$ in the kernel to the same accuracy as we choose for the difference approximation scheme. If we use the second order corrector scheme we might approximate $n(t-\tau)$ as varying linearly over a time step. Show that in this case

$$c(t) \simeq c_0 e^{-\lambda t} + \frac{\beta}{\Lambda\lambda}\left\{\left[n_0 - \frac{n_1 - n_0}{h\lambda}\right][1 - e^{-\lambda t}] + [n_1 - n_0]\frac{\lambda t}{\lambda h}\right\}$$

so that at step t_j

$$c_j = c_{j-1}e^{-\lambda h} + \frac{\beta}{\Lambda\lambda}\left\{\left[n_{j-1} - \frac{n_j - n_{j-1}}{\lambda h}\right][1 - e^{-\lambda h}] + n_j - n_{j-1}\right\}$$

The desired second order corrector applied to the neutron equation is

$$n_j = n_{j-1} + \left[\frac{\rho - \beta}{\Lambda} n_j + c_j\right] h$$

Show that this leads to the expression

$$\left\{1 - \frac{\rho h}{\Lambda} + \frac{\beta}{\Lambda\lambda}[1 - e^{-\lambda h}]\right\} n_j = h\lambda\left[c_{j-1} - \frac{\beta}{\Lambda\lambda} n_{j-1}\right] e^{-\lambda h} + \left\{1 + \frac{\beta}{\Lambda\lambda}(1 - e^{-\lambda h})\right\} n_{j-1}$$

which may now be solved explicitly for n_j and hence yields c_j.

Develop the corresponding expression for the I-group model. Compare the accuracy and computing time required with the evaluation of the step reactivity equation solution using the inhour equation and residues.

3.15. *Fast ramp accident starting from delayed critical.* Specialise eqn. (3.27) to the case that $\rho = \mu t$ with μ greater than, say, 10β, i.e. the fast ramp reactivity accident starting from delayed critical. Develop an integration scheme suitable for digital computation (e.g. Simpson's rule) and reproduce Fig. 3.3. Note how the neutron level is still relatively low as the reactivity reaches prompt critical so that the apparent reactor period is deceptively low.

3.16. Express the formal solution of problem 3.15 as

$$\frac{n(t)}{n_0} = \exp(x^2 - 2\epsilon x) + \epsilon\sqrt{\pi}\exp(x - \epsilon)^2[\mathrm{erf}(x - \epsilon) + \mathrm{erf}(\epsilon)]$$

where the error function

$$\mathrm{erf}(x) = \int_0^x e^{-y^2}dy \quad \text{and} \quad \epsilon = \beta\sqrt{2\mu\Lambda},\ x = \sqrt{\frac{\mu}{2\Lambda}}\,t$$

If $\epsilon > 1$ (i.e. fast ramp but not too fast) use the asymptotic expansion of erf(x) to show that as the reactor passes prompt critical the neutron level has risen by the factor $\epsilon\sqrt{\pi}$, (*Hint.*: At prompt critical, $x = t$.)

Reference for Chapter 3

1. D. L. Hetrick, *Dynamics of Nuclear Reactors*, Univ. Chicago Press, Chicago, 1971.
2. J. Lewins, Use of the generation time, *Nucl. Sci. and Engr.* **7**, 122 (1960).
3. Z. Akcasu, G. S. Lellouche and L. M. Shotkin, *Mathematical Methods in Nuclear Reactor Dynamics*, Academic, New York, 1971.
4. R. T. Fenner, *Finite Element Methods for Engineers*, Macmillan, New York, 1975.
5. H. B. Smets, Exact solutions of the reactor kinetics equations, *Bull. Acad. Royale Belgique* **3**, 256–271 (1959).
6. M. Werner, Solutions of the space and time dependent equations of nuclear reactor theory, *Adv. in Nucl. Sci. and Tech.* **10**, Plenum, New York, 1977.
7. A. Radkowsky, (ed.) *Naval Reactors Physics Handbook*, **1**, USAEC, Washington, 1964.
8. W. M. Stacey Jr., *Space–Time Nuclear Reactor Kinetics*, Academic, New York, 1969.
9. R. L. Murray, C. R. Bingham and C. F. Martin, Kinetics Analysis by an Inverse Method. *Nucl. Sci. and Engr.* **18** 461 (1964)

CHAPTER 4

Linear Reactor Process Dynamics with Feedback

Introduction

Solutions obtained in Chapter 3 were limited to cases where the reactivity was a known function of time. This is not a realistic situation for a reactor at power as it ignores mechanisms leading to variations of the reactivity dependent on the state of the reactor, including the neutron population and power level we are ostensibly seeking. These mechanisms include:

(a) *Inherent feedback*. Neutron levels affect, for example, heat and xenon production. Consequent temperature changes and variations of "poisons" affect the reactivity.
(b) *External plant feedback*. The reactor is coupled to heat exchangers, turbines, generators and ultimately to the load centres of an electric grid. Changes in demand or running conditions affect reactor conditions.
(c) *Control effects*. To provide for the desired behaviour in the face of the first two effects, a control system will be added. While some controlling actions, such as startup, are essentially time specified, others arise from the use of automatic feedback controllers designed to modify the system response and thus interact with reactivity and the neutron level.

Thus the *neutronics* equations developed in Chapter 2 and studied initially in Chapter 3 are but one part of the description of the state of a reactor; the neutron level $n(t)$ is only one of several significant state variables. In this chapter we treat the process dynamics of the reactor itself, principally the inherent feedback effects, using simple forms for associated plant and control effects and the classical or frequency approach. Such an approach is natural when the system is linear in its state variables; this is not wholly the case in the present context, and it will be necessary to make certain approximations in the development as we go along. The results will be a description of the dynamics of the system for normal operating levels and small departures therefrom, which will allow the designer to make statements about important features such as stability and speed of response to (small) changes in demand. We concentrate in this chapter on how the internal elements can be described in the frequency space and upon the techniques of combination, leaving until Chapter 5 the description of particular reactor systems with their associated plant.

To provide some framework for the discussion it is helpful to have in mind an overview of the time scale of the processes occurring, as shown in Table 4.1. It is a saving grace that the same stiffness that makes for difficulties in solving the coupled equations describing the dynamics of a reactor does lend itself to a grouping of phenomena on

TABLE 4.1.
Characteristic times of nuclear reactor dynamics

Fuel burn-up Transuranic isotope production	years
Samarium production	months
Xenon build-up and shutdown	days
Diurnal load variations Xenon flux tilting	tens of hours
Coolant transit times through system	tens of seconds
Coolant transit times through core Precursor dominated effects	seconds
Heat transfer, fuel to coolant	tenths of seconds
Prompt neutron-dominated effects	hundredths of seconds

different time scales that then permits us to ignore many of the complications outside a particular band of interest. Experience is needed, of course, to know when such decoupling is acceptable.

Campbell[1] has a useful description of how plant in general may have process dynamics studied in the frequency (or transform) space, while Schultz[2] remains a valuable application of classical control ideas to nuclear reactors. We turn now to the derivation of the transfer functions for the faster processes listed in Table 4.1, finishing with a summary description of control elements.

Neutronics Transfer Function

Our view point has changed from that of Chapter 3. We expect now that the reactivity will be dependent on the neutron level itself through other processes, and so we regard ρ as one of the state variables. In this light we reconsider the neutronics equations

$$\left.\begin{aligned} \frac{dn}{dt} &= \frac{\rho - \beta}{\Lambda} n + \sum_i \lambda_i c_i + s; \quad n(0) = n_0 \\ \frac{dc_i}{dt} &= \frac{\beta_i}{\Lambda} n - \lambda_i c_i; \qquad\quad c_i(0) = c_{i0} \end{aligned}\right\} \tag{4.1}$$

If ρ depends on a state variable such as temperature, then the term in $\rho n/\Lambda$ in eqns. (4.1) implies we are now dealing with a nonlinear system. This introduces substantial difficulties in any attempt to obtain solutions. The method we adopt is based on the thought that for normal operations of the system the departure from a steady operation level n_0 and c_{i0} will be *relatively* small and that we may hope to obtain a linear approximation for these *departures* even if the over-all equations are nonlinear.

On this basis, put $\delta n(t) = n(t) - n_0$ and $\delta c_i(t) = c_i(t) - c_{i0}$ and also $\delta\rho(t) = \rho(t) - \rho_0$. That is, we develop the original neutronics equations in terms of perturbations about the unperturbed state n_0, c_{i0}, ρ_0. These latter satisfy the unperturbed equations

$$\left.\begin{aligned} 0 &= \frac{\rho_0 - \beta}{\Lambda} n_0 + \sum_i \lambda_i c_{i0} + s \\ 0 &= \frac{\beta_i}{\Lambda} n_0 - \lambda_i c_{i0} \end{aligned}\right\} \tag{4.2}$$

which naturally reduce to $0 = \rho_0 n_0/\Lambda + s$. It is assumed that the only coefficient of eqns. (4.1) to be varied is indeed the reactivity ρ and, as discussed in Chapter 2, this is a satisfactory assumption. Of course we may well start with a source-free situation, in which case ρ_0, the unperturbed reactivity in the steady state, is zero; we now assume this to be the case as is appropriate for a reactor already at power.

After substitution of the perturbation forms and use of the unperturbed equations we have

$$\left.\begin{aligned} \frac{d}{dt}\delta n &= \frac{\delta\rho}{\Lambda} n_0 + \frac{\delta\rho}{\Lambda}\delta n - \frac{\beta}{\Lambda}\delta n + \sum_i \lambda_i \delta c_i \\ \frac{d}{dt}\delta c_i &= \frac{\beta_i}{\Lambda}\delta n - \lambda_i \delta c_i \end{aligned}\right\} \tag{4.3}$$

equations where the "nonlinear" term takes the form $\delta\rho\delta n/\Lambda$. If the variation $\delta n(t)$ about n_0 is relatively small, we may be able to ignore this final product, a philosophy that can be discussed further when we consider stability of the reactor. If we may, then

$$\left.\begin{aligned} \frac{d\delta n}{dt} &= \frac{\delta\rho}{\Lambda} n_0 - \frac{\beta}{\Lambda}\delta n + \sum_i \lambda_i \delta c_i \\ \frac{d\delta c_i}{dt} &= \frac{\beta_i}{\Lambda}\delta n - \lambda_i \delta c_i \end{aligned}\right\} \tag{4.4}$$

which is linear in (δn, $\delta\rho$). Taking Laplace transforms is now fruitful and noting that $\delta n(0) = \delta c_i(0) = 0$,

$$\left.\begin{aligned} \left[p + \frac{\beta}{\Lambda}\right]\delta\bar{n}(p) &= \frac{n_0}{\Lambda}\delta\bar{\rho} + \sum_i \lambda_i \delta\bar{c}_i \\ [p + \lambda_i]\delta\bar{c}_i(p) &= \frac{\beta_i}{\Lambda}\delta\bar{n} \end{aligned}\right\} \tag{4.5}$$

Further substitution yields the transform $G(p)$

$$G(p) \equiv \frac{\delta\bar{n}}{\delta\bar{\rho}} = \frac{n_0}{\Lambda p + \sum_i \frac{p\beta_i}{p + \lambda_i}} = \frac{n_0}{\Lambda} \frac{1}{f(p)} \tag{4.6}$$

where $f(p)$ is just the function discussed at length in Chapter 3, for the special case envisaged here, that $\rho_0 = 0$.

The "linearised" eqns. (4.4) are seen to be of the *form* of the original equations for the complete neutron and precursor populations except that the role of a source is now played by the term $\delta\rho n_0/\Lambda$. We might have expected, therefore, that the transform would be of the same form as the source transfer of Chapter 3 with the additional factor n_0/Λ. This factor is evidently neutron level dependent and it is more convenient perhaps to use the expression normalised to n_0 so that we employ $G_n(p)$ and the corresponding transfer function $G_n(j\omega)$, which may be interpreted as the *fractional* transform of the neutron/power output for a reactivity input:

$$G_n(p) \equiv \frac{\delta\bar{n}/n_0}{\delta\bar{\rho}} = \frac{1}{\Lambda f(p)} \tag{4.7}$$

The corresponding transfer function $G_n(j\omega)$ may be more conveniently computed in some cases if G_n is rearranged as a rational polynomial

$$G_n(p) = \frac{[p + \lambda_1][p + \lambda_2] \ldots [p + \lambda_I]}{\Lambda p [p - p_1] \ldots [p - p_I]} \tag{4.8}$$

where $p_0, p_1, \ldots, p_I$ are the roots of the inhour equation (3.13) lying on the $\rho = 0$ axis of Fig. 3.1. In particular, $p_0 = 0$ for the assumed $\rho_0 = 0$ and is therefore absent from eqn. (4.8).

Figure 4.1 plots this low power transfer function just derived for ^{235}U and a range of neutron reproduction times. The remarks of Chapter 2 concerning effective values apply, particularly therefore the factorising of a common effectiveness fraction $\gamma_i \rightarrow \beta/\bar{\beta}$. Certain details are worth emphasising. Because of the location of the roots of $f(p) = 0$ the transfer function in $j\omega$ has a series of alternate lead and lag terms so that the phase lies between zero and a $\frac{1}{2}\pi$ lag. The phase approaches $\frac{1}{2}\pi$ at low frequencies and similarly above the highest break frequency, $p_I \simeq \beta/\Lambda$. It is only at these high frequencies that the transfer function is sensitive to the value of the reproduction time Λ. When the roots p_i are known, the transfer function can readily be approximated using the break frequency method. Correspondingly, it can be approximated by a one- or two-group equivalent of eqn. (4.8) matched graphically to the accurate transfer function in any desired range of interest.

APPROXIMATIONS TO THE NEUTRONICS TRANSFER FUNCTION

The full six-group $G_n(p)$ is obviously sometimes more than can be handled conveniently. Some useful approximations in increasing effectiveness are:

(a) Effective lifetime model

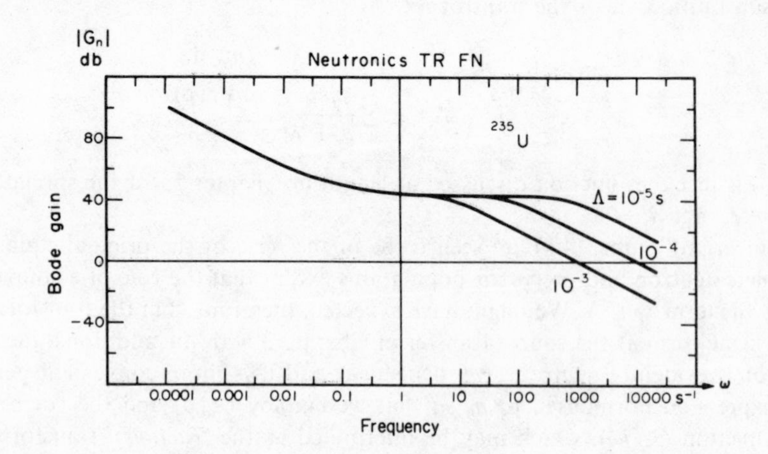

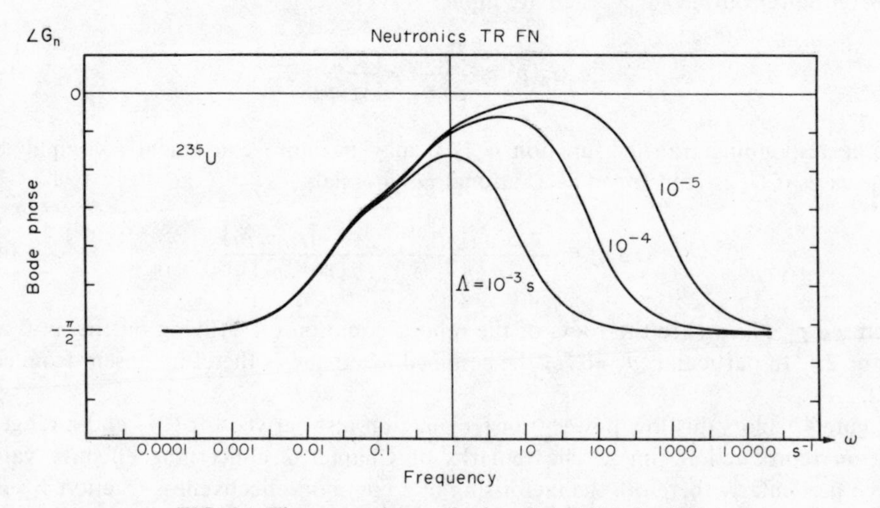

FIG. 4.1. The neutronic transfer function $G_n(j\omega)$ for ^{235}U.

$$G_n(p) \rightarrow \frac{1}{[\beta\tau + \Lambda]p} \simeq \frac{\lambda}{\beta p} \tag{4.9}$$

(b) Prompt jump model

$$G_n(p) \rightarrow \frac{p + \lambda}{\beta p} \tag{4.10}$$

(c), One precursor group model

$$G_n(p) \rightarrow \frac{p + \lambda}{\Lambda p[p + \lambda + \beta/\Lambda]} \simeq \frac{p + \lambda}{\Lambda p[p + \beta/\Lambda]} \tag{4.11}$$

Approximation (a) is of limited value. Approximation (b) arises from (c) on letting Λ tend to zero.

The transfer function $G_n(j\omega)$ has a direct interpretation as the response of a reactor at low power (where feedback effects are not present) to a sinusoidal variation of reactivity of unit amplitude, measured as the *variation* of neutron population in amplitude and phase. Thus the transfer function is experimentally determinable from the oscillation of a control rod or other suitable absorber. Simultaneous measurements are made of the neutron level against phase/amplitude of the oscillation. In such an experiment the complete transfer function would be found including any other effects not described directly in the neutrons equations, but if the reactor itself is at low power and for frequencies greater than, say, hr^{-1} (but not so great as to invalidate diffusion theory and the "lumped" representation by virtue of neutron waves taking appreciable time to cross the reactor), then only the neutronics effects will be significant. Certainly some care is needed in the location of control rod oscillator and neutron detectors if spatial effects are to be eliminated, but the classical work at the CP2 reactor [(9)] fully justifies the use of this standard frequency analysis method of classical control theory for nuclear reactors.

STABILITY OF CONTROLLED SYSTEM AT LOW POWER

Figure 4.2 is the Nyquist diagram representation of the same information on the transfer function $G_n(j\omega)$ in another form. To illustrate questions of stability treated via the Nyquist diagram let us suppose that (at low power) the neutron level was to be

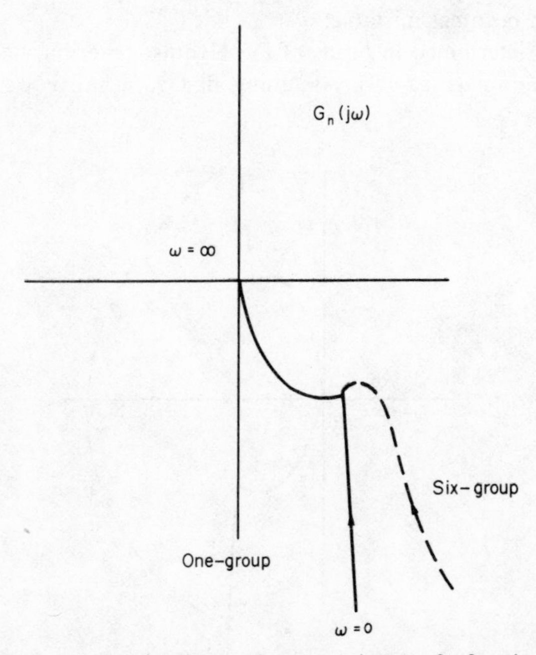

FIG. 4.2. Nyquist diagram for neutronics transfer function.

controlled by a feedback system with an idealised characteristic of having an arbitrary gain K and zero phase lag/lead at all frequencies as in the block diagram of Fig. 4.3. Would this system be stable?

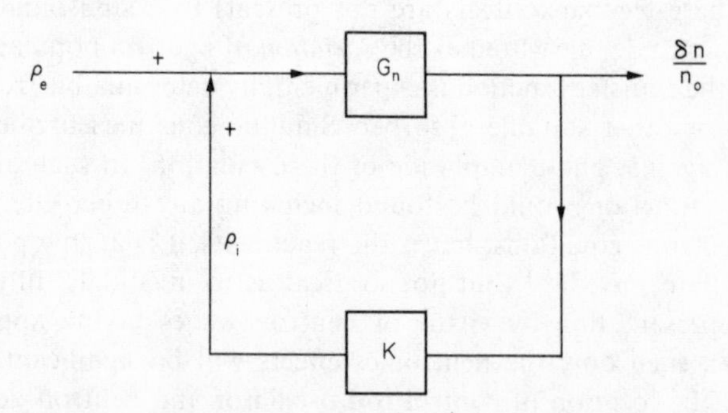

FIG. 4.3. Block diagram for low power stability with idealised feedback.

The open loop transfer function is KG_n and the closed loop transfer function is $KG_n/[1-KG_n]$, where we have absorbed the power level into the definition of the feedback gain K. The control system envisages a measurement of the actual neutron level n, a comparison with the demand level n_0 and the provision of a reactivity control signal ρ_i proportional to the difference δn via say the movement of a control rod. If we consider a small error and the known direction of the effect of reactivity on n, it is elementary to see that stability will require the correction reactivity to be of opposite sign to the error signal δn, i.e. that K will have to be negative for static stability. There remains conceptually the possibility of *dynamic* instability, that at too high a gain $|K|$ the feedback system becomes unstable.

Stability may be determined in terms of the Nyquist criterion, however, making use of the Nyquist diagram (Fig. 4.4). Assuming first K negative we are to investigate

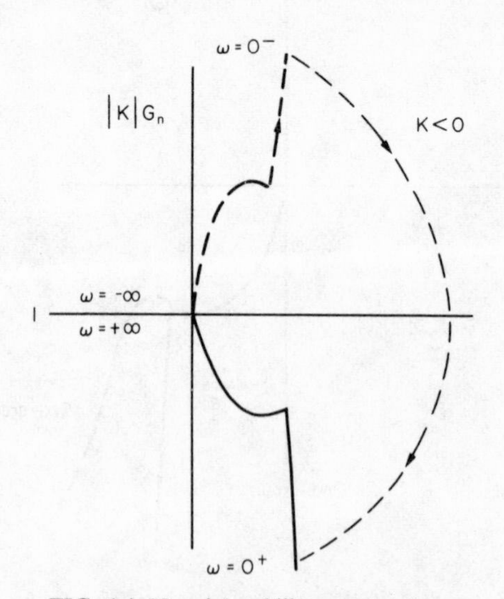

FIG. 4.4. Nyquist stability construction.

$1+|K|G_n$ around the origin or equivalently, $|K|G_n$ around the point $-1, 0$. Because of the behaviour of G_n as $\omega \to 0$, however, it is necessary to "close" the Nyquist diagram and we note that as the variable p tends to zero,

$$\mathscr{L}_{p \to 0}\, G_n(p) = \frac{\lambda}{\beta p} = \frac{\lambda}{\beta}\frac{1}{v}\, e^{-j\theta}$$

where we have substituted the polar form for $p = re^{j\theta}$. The counter-clockwise semicircle in the p-plane around the origin (see Fig. 1.9) leads to $1/p$ traversing a *clockwise* semicircle of large radius, $1/v\, e^{-j\theta}$. The completed Nyquist diagram is seen *not* to enclose the test point for *any* magnitude of gain. Since the open loop system has no poles in the right half p-plane we have shown that the closed system is stable at all gains for negative feedback control of the idealised sort postulated.

In the sense of the linearisation of the neutronics equations it is indeed true that the open loop system has no poles within the right half-plane and the claim is therefore substantiated. Note, however, that the open loop is unstable (the transfer function tends to ∞ as ω tends to zero) so that in a somewhat special sense the closed system is only conditionally stable.

If the gain is positive, we are investigating $1-|K|G_n$ around $-1, 0$ and the effect of the change of sign is to require a double reflection of $|K|G_n$ in real and imaginary axis. The "closure" construction is still a clockwise semicircle of large radius and it can be left to the reader to sketch the resulting figure and see that it surrounds the test point; the system with positive feedback is unstable for all values of the gain.

The example is of limited practical interest and serves chiefly to introduce the Nyquist criterion to a nuclear reactor problem because we have ignored other feedback mechanisms and we have assumed no lead or lag associated with K. The control mechanism would have to be astonishingly rapid to justify this in relation to the natural frequencies in G_n. However, there are occasions in dealing with slower natural phenomena such as xenon poisoning, where it is more reasonable to assume that the feedback has a much shorter time scale than the effect (xenon) being studied and this form of *power coefficient* can then be employed.

Void Effects

When material is removed from a local region of a reactor to leave a void (or perhaps to leave a tenuous region of steam, etc.) we may expect a change of reactivity. The effect may be particularly prominent in liquid moderated reactors where the liquid coolant has an appreciable moderating effect and voidage is more likely to occur. In this case fuel is not being removed, but clearly the removal of moderator will affect absorption, moderation and allow streaming of neutrons through the "void". It is generally desirable to design such reactors undermoderated for elementary safety; if the power goes up and the coolant expands or bubbles form to force the coolant out, it is desirable that the reactivity should fall to offset the rise in power. However, there may be local variations in reactivity effects so that the size and even the sign of the reactivity effect of a void varies along a coolant channel.

This section discusses voids from the viewpoint of the boiling water reactor (BWR) where they are significant. In many other cases the effect of voidage can be adequately related to temperature effects alone and are thus covered in a subsequent section on temperature effects. But in the BWR boiling can occur at constant temperature so that voids have to be considered in their own right; the physics of the process are also complicated by the role of pressure in a BWR.

There can be some attempt, at least at zero power, to measure the reactivity effect of voids experimentally by introducing artificial voids such as polystyrene blocks, etc. It is likely, of course, that the reactivity effect varies with the volume of void, but for convenience we shall treat the ratio of reactivity change to void volume for small volumes of void, writing a void coefficient of reactivity α_v such that

$$\delta\rho \simeq \alpha_v \delta V = \frac{\partial \rho}{\partial V}\,\delta V \tag{4.12}$$

where δV is the change in void volume.† For practical purposes, steam can be considered a true "void" in comparison with the liquid coolant. Not only will α_v depend on the location of the void in any detailed description but it will also vary with the power level at which the reactor is run via the temperature and pressure conditions. If, however, we are studying small transients around a set power we can make the simplification of a constant α_v.

Some of the physical origin of void effects include:

(a) growth and collapse of vapour bubbles in the nucleate boiling region, together with time delays for bubble formation;
(b) radiation-induced bubbles where the radiation provides a nucleus for growth;
(c) the sweeping of bubbles in the coolant flow into regions of different temperature or pressure and different reactivity coefficients;
(d) the increased drag in two-phase flow where voids have lowered the density leading, from continuity of matter, to an acceleration in coolant flow and thus compounding pressure effects;
(e) as a consequence of (d) the possibility of instability arising as the less effective coolant becomes hotter and thus produces more voids—interaction between channels.

It can be seen that this is a complex situation. Early natural circulation BWR do show tendencies to instability. In a simplified picture the main phenomena treated are the growth of bubbles in the boiling region of a fuel channel and the shift of the boundary between boiling and non-boiling with power levels. Other consideration would be given to interaction between channels and hydrodynamic instability arising from the shift between different boiling regimes. Fortunately the influence of the void coefficient is so strong in BWRs as to make it possible to neglect more direct temperature effects in a first approximation. (See problems.)

Figure 4.5 gives the block diagram for such a natural circulation BWR (modelled

†It is sometimes convenient to define a non-dimensional or fractional void coefficient based upon $\delta V/V$.

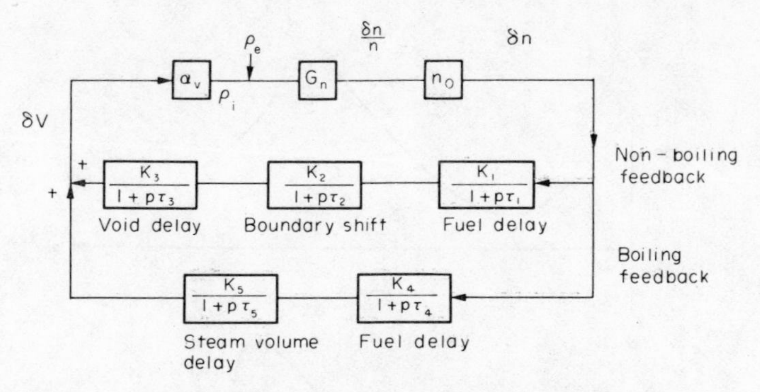

FIG. 4.5. Natural circulation BWR block diagram.

after the EBWR[5]) and Table 4.2 gives the meaning and appropriate values of the various gains and time constants.

TABLE 4.2.
Parameters for a natural circulation BWR at 20 *MW based on one channel analysis*

Void coefficient α_v		-0.4×10^{-3} per mm³
Non-boiling region		
K_1	non-boiling power/total power	0.35
τ_1	fuel plate time constant	0.45 s
K_2	interface shift/non-boiling power	-30 mm/MW
τ_2	half water transit time	0.30 s
K_3	steam volume/interface shift	-0.25 mm³/m
τ_3	half steam transit time	0.20 s
Boiling region		
K_4	boiling power/total power	0.65
τ_4	fuel plate time constant	0.40 s
K_5	steam volume/boiling power	6.0 m³/MW
τ_5	power-void time constant	0.10 s

Two loops are shown—for the boiling region and the non-boiling region; the latter incorporates the description of the change of interface between the regions with changing power. The number of simple delay terms and the comparability of the time constants suggests the possibility of dynamic instability as the accumulated phase lag of the open loop approaches π. Such instabilities have indeed been observed as tending to occur at a sufficiently high power. Figure 4.6 shows a comparison of measured and predicted *closed* loop transfer functions, $G(j\omega)=\delta n/\delta\rho$ at 20 MW, and the amplitude resonance around 10 radians/s is evident. (Note that Fig. 4.6 represents the *closed* loop since it is not feasible to break the loop and measure the open loop transfer function; the open loop information would be calculated if it was proposed to use the Nyquist criterion to predict instability.)

Effects not accounted for in the foregoing model of the EBWR include:

direct fuel temperature effects
water temperature effect in non-boiling region
changes in feed flow (forced circulation)

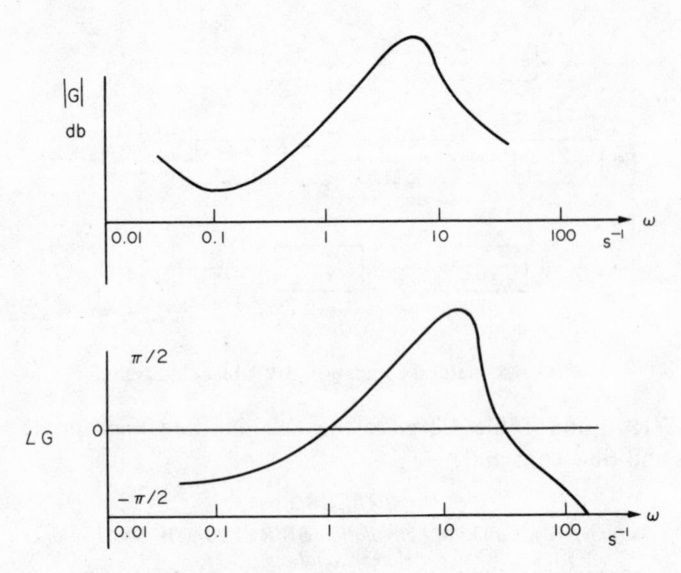

FIG. 4.6. EBWR closed loop transfer function at 20 MW.

changes in steam load/losses affecting pressure
water acceleration, steam flashing

since these tend to be insignificant in the EBWR. Increasing power in the EBWR leads to the onset of instability around 50 MW. This behaviour in natural circulation BWRs is one of the reasons for the modification of the BWR cycle for current power reactors to include substantial forced circulation and will be illustrated in Chapter 5. Reference 13 gives further analysis of thermal and hydraulic stability in a power BWR.

Reactor Thermal Transients

Temperature effects in reactors can be conceptually divided between inherent and external feedback, the latter arising as changes in demand affect the turbine and the way in which coolant is drawn off or its state on return to the reactor. It is significant that there is a certain amount of thermal inertia so that there are time delays in establishing new temperatures.

The effect of temperature on reactivity is again a matter for detailed reactor physics calculations supported, to the degree possible, by experimental verification. These effects may be quite detailed and for our purpose we make the gross simplifying assumption that we can speak of a temperature coefficient of reactivity α_T being $\partial\rho/\partial T$, so that we may write due to temperature effects that

$$\delta\rho_T = \alpha_T \delta T \tag{4.13}$$

Of course, even such a lumped reactivity effect is likely to be dependent on the temperature itself and hence the power level, but for our present purpose, discussing small changes around a set power level, it is reasonable to adopt a constant coefficient.

Temperature effects arise from several different mechanisms, however, and it may well be necessary to include more than one type of temperature coefficient. These mechanisms might be categorised as follows:

(a) FUEL COEFFICIENT

If its temperature rises the fuel element can be expected to expand and thus affect the density and cross-sections appearing in the neutron balance. There may also be over-all geometric effects dependent on the design, e.g. bowing into a different flux or importance. Another contribution to the fuel coefficient arises from Doppler broadening of resonant cross-sections in the higher temperature fuel. The sign as well as the magnitude of these coefficients depends on the detailed nuclear processes. Doppler coefficients are likely to be a significant parameter in the safety of large fast reactors, and it has to be said that reactor physics calculations are not yet as precise in this area as is desirable. As the isotopic composition of the fuel changes, with burn-up of ^{235}U and production of plutonium, the fuel coefficient may vary; as the fuel is by and large the fastest reacting contributor to temperature effects, it is important for safety that the fuel coefficient should be negative throughout the core life.

(b) MODERATOR COEFFICIENT

An increase in moderator temperature increases the effective temperature of the neutrons in a thermal reactor causing fission. This in turn leads to a change in the neutron balance and reactivity because of the non-$1/v$ nature of so many of the cross-sections of fissionable materials (and poisons). Again the sign as well as magnitude of the coefficient will depend upon the isotopic composition of the charge. In a Magnox reactor it is found that whereas the moderator coefficient is negative at the beginning of core life in fresh unirradiated fuel, it may well become positive in an irradiated fuel. Fortunately this effect is accompanied by a somewhat longer time constant than for the fuel coefficient since the energy has largely to pass outwards through the fuel into the moderator and in the large graphite masses of the Magnox and AGR systems the moderator time constants can be an order of magnitude slower.

(c) COOLANT COEFFICIENT

The direct effect of the coolant temperature on reactivity is usually small and perhaps negligible. There is, however, an important role for the coolant in the energy balance affecting the temperature elsewhere in the reactor and thus the coolant temperature plays a significant part in the feedback equations.

ELEMENTARY MODEL

The simplest model will be to "lump" the reactor as a single thermal structure with a characteristic temperature T and heat capacity m, say. This core temperature will be linked to the coolant energy removal mechanism involving coolant inlet temperature T_i and outlet temperature T_o, say, together with a coolant flow rate F and its heat capacity c, say. For simplicity we take the coolant as incompressible so that pressure as a further thermodynamic state variable can be ignored. We neglect details of coolant expansion but suppose that external variations of the flow rate F and the inlet temperature T_i can be imposed upon the reactor.

Figure 4.7 illustrates the simple model proposed. We next suppose that the mean temperature of the reactor, i.e. T, is the average of the coolant outlet and inlet temperature, thus enabling us to substitute for T_o at any stage via the relation $T=\frac{1}{2}[T_o+T_i]$.

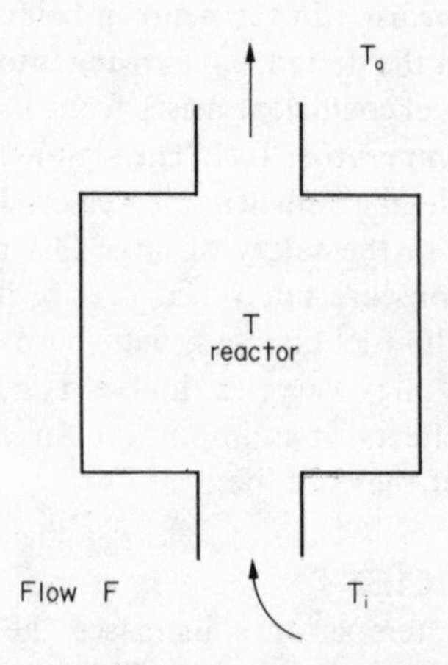

FIG. 4.7. Elementary temperature model.

An energy balance for the reactor allowing for the energy transported by the coolant gives

$$m\frac{dT}{dt} = \kappa n - cF\,[T_o - T_i] \tag{4.14}$$

with κ a suitable constant connecting power to neutron level. This may be taken in a simple model as $30/\Lambda$ picowatts per neutron if Λ, the reproduction time, is in seconds. If the steady state power is known, we may work backwards to determine from this the steady state neutron population n_0. Similarly, cF_0 can be determined from a knowledge of the steady state temperature rise at a known power: $cF_0 = \kappa n_0/[T_o - T_i]_0$.

To deal with the product term $cF[T_o - T_i]$, we assume m and c constant and study the change δT due to changes δn and δT_i and δF imposed upon the reactor. With the usual linearisation process and substitution for T_o,

$$m\frac{d\delta T}{dt} = \kappa\delta n - c\triangle T_c\delta F - 2cF_0[\delta T - \delta T_i] \tag{4.15}$$

where $\triangle T_c$ is $[T_o - T_i]$ in the reference state and F_0 the coolant flow in the reference state. This form, with constant coefficients, is now suitable for treatment by Laplace transformation and we may identify the following functions:

$$G_{T'} = \frac{\delta\bar{T}}{\delta\bar{n}} = \frac{\kappa}{2cF_o}\frac{1}{1+p\tau};\quad G_f = \frac{\delta\bar{T}}{\delta\bar{F}} = \frac{\Delta T_c}{2F_o}\frac{1}{1+p\tau};\quad G_i = \frac{\delta\bar{T}}{\delta\bar{T}_i} = \frac{1}{1+p\tau} \tag{4.16}$$

where τ, the thermal time constant,† is $m/2cF_0$. The corresponding block diagram is given in Fig. 4.8.

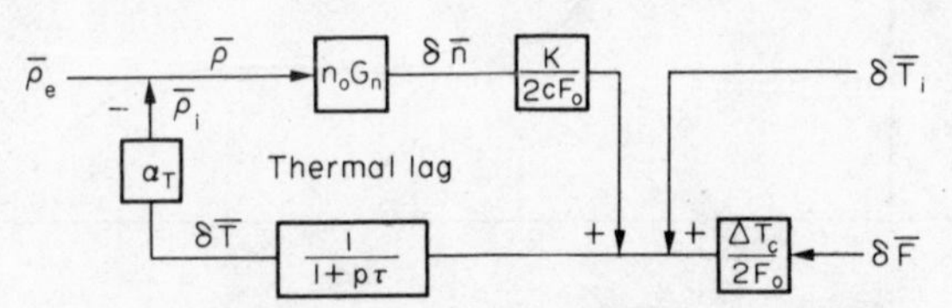

FIG. 4.8. Elementary temperature effect block diagram.

This time constant may be determined experimentally in certain reactor types by circulating the coolant at zero fission power. In a Magnox reactor, for example, CO_2 circulation requires an appreciable fraction of nominal power, and in these circumstances the reactor will heat up (and the sign of $[T_o - T_i]$ will be reversed). If the heat exchangers are isolated and their heat capacity is neglected, eqn. (4.14) can be used to determine cF/m from the temperature information.

TRANSFER FUNCTION WITH INTERNAL FEEDBACK

Consider for the time being the inherent internal feedback effects only with a view to determining whether the system is stable. The closed loop transform function corresponding to the open loop $G_n n_0 G_T$ is

$$G = \frac{n_0 G_n}{1 - n_0 G_n \alpha_T G_T} = \frac{n_0[p + \lambda][1 + p\tau]}{p\Lambda\left[p + \dfrac{\beta}{\Lambda} + \lambda\right][1 + p\tau] - [p + \lambda]\dfrac{n_0\alpha_T\kappa}{2cF_0}} \tag{4.17}$$

in a one-precursor group model. The corresponding transfer function $G(j\omega)$ has obvious limits for large and small ω:

$$\underset{\omega\to\infty}{\mathscr{L}} G(j\omega) = \underset{\omega\to\infty}{\mathscr{L}} n_0 G_n(j\omega); \qquad \underset{\omega\to 0}{\mathscr{L}} G(j\omega) = \underset{\omega\to 0}{\mathscr{L}} -\frac{1}{\alpha_T G_T} \tag{4.18}$$

$$= n_0/j\omega\Lambda \qquad\qquad = -2\,cF_0/\alpha_T\kappa$$

So for large frequencies the feedback is negligible on the basis of characteristic thermal times τ, long compared to the neutronics process. For small frequencies, however, the magnitude of the forward transfer function is curtailed by the feedback if $\alpha_T < 0$ and correspondingly the phase is brought up to zero at low frequencies. Indeed, as seen in Fig. 4.9, the temperature effect may lead to a portion of the closed loop transfer function having a phase *advance*. Note that G is a neutron level transform function for $\delta\bar{n}$ and *not* a relative level function for $\delta\bar{n}/n_0$.

The p-plane may be developed graphically from the root locus method in order to study the effect of sign and magnitude of α on stability from the expression

†τ will be used for thermal time constants in the remainder of this chapter and we shall not have need to refer to $1/\lambda$ of Chapter 3.

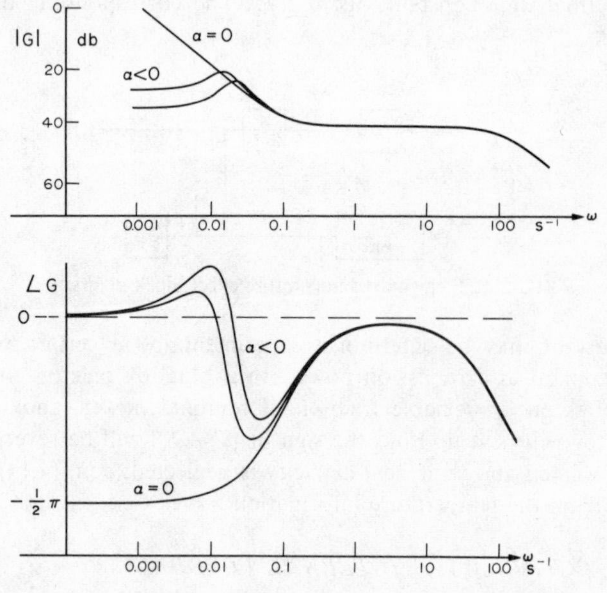

FIG. 4.9. Closed loop transfer function $G(i\omega)$ for typical temperature effect.

$$\frac{p\left[p+\dfrac{\beta}{\Lambda}+\lambda\right][p+1/\tau]}{p+\lambda}=\frac{n_0\alpha_T\kappa}{2cF_0\tau}\equiv\mu \tag{4.19}$$

Considering typical values

$$\frac{1}{\tau}<\lambda<\frac{\beta}{\Lambda}+\lambda \tag{4.20}$$

and Fig. 4.10 shows the two principal cases, using methods summarised in Chapter 1. In deriving these results, we assume $1/\tau>\lambda$; the case for $1/\tau<\lambda$ can be left as an exercise.

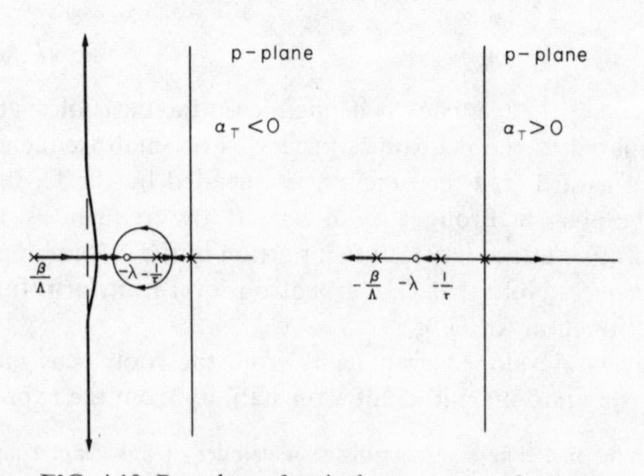

FIG. 4.10. Root locus for single temperature feedback.

The two asymptotes at "infinity" are at $\pm\frac{1}{2}\pi$ for $\alpha_T<0$ and on the real axis for $\alpha_T>0$. The intersection is at approximately $-\frac{1}{2}\beta/\Lambda$. This value is also close to one of the departure points p_c from the real axis; the other values of p_c are between the right hand pair of poles and again close to the zero. Thus for positive feedback there are no complex roots, but there is always a positive root in the right half-plane indicating instability. For negative feedback of sufficient magnitude there are oscillatory components, but all p-values lie in the left half-plane; the system is stable. This transition to an oscillatory component will also arise if n_0 is increased. The frequency of onset of a pure oscillatory term is readily shown to be given by substituting $p=j\omega$:

$$\left.\begin{aligned} &\text{Real:} && -\omega^2\left[\frac{\beta}{\Lambda}+\lambda+\frac{1}{\tau}\right]=\mu\lambda \\ &\text{Imaginary:} && -\omega^2+\frac{\beta}{\Lambda\tau}=\mu \end{aligned}\right\} \tag{4.21}$$

whence $\omega^2<0$ for assumed pure imaginary oscillations, showing there is no such behaviour. The general nature of the solution is the same as long as $1/\tau$ is small compared to λ. We can reasonably conclude that the same stability characteristics are valid in the I-group precursor model; this may be left as an exercise for demonstration via the Nyquist diagram. In particular, the limits of $G(j\omega)$ given in eqn. (4.18) are valid in the more general model.

Figure 4.11 gives the Nyquist stability construction for this single temperature feed-

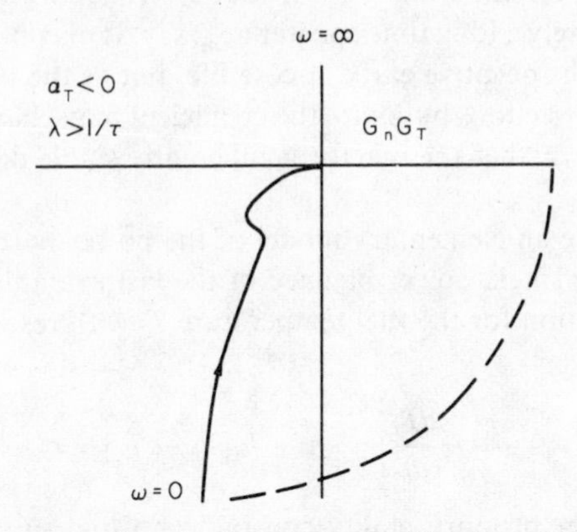

FIG. 4.11. Nyquist diagram for one group with temperature feedback.

back model for $\alpha_T<0$ and using the one-precursors group model. Stability is determined by the behaviour of $G_n n_0 G_T$ (together with closure) in respect of rotations around the point $-1, 0$. The system is inherently stable.

Other models for the neutronics, as given in eqns. (4.9) and (4.10) may be employed graphically in the same way, and while there is no change in the conclusion to be drawn with respect to stability and the sign of the temperature coefficient of reactivity, it is instructive to observe the difference at high frequencies of these approximations (Fig. 4.12).

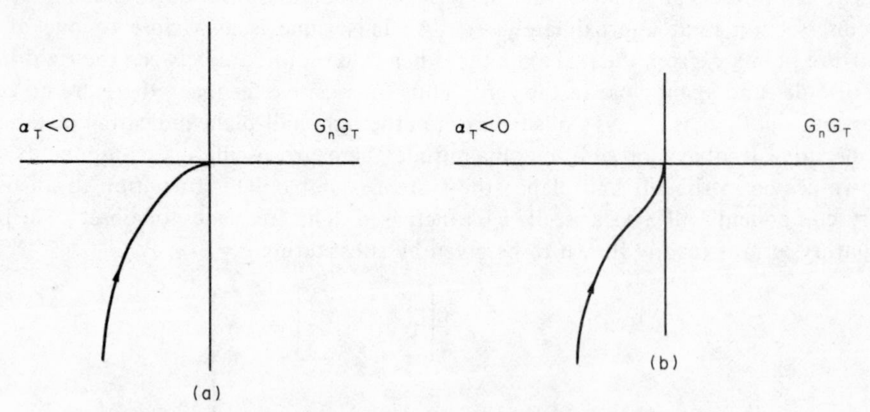

FIG. 4.12. Nyquist diagrams for (a) effective lifetime, and (b) prompt jump.

TWO TEMPERATURE COEFFICIENTS

Several mechanisms for reactivity change exist which require representation in the form of two feedback terms rather than one. Thus we may have the fuel/moderator situation. Increase in fuel temperature occurs relatively quickly after a neutron transient and it is reasonable to assume that the design provides for a negative corresponding reactivity coefficient, α_f say, and with τ_f as the time constant for changes of fuel temperature. The moderator for the Magnox or AGR systems, however, have large thermal inertia and correspondingly a long time constant τ_m, say. It may be that the corresponding reactivity coefficient α_m is negative early in core life, but as the plutonium isotopes with non-$1/v$ resonance cross-sections build up, the coefficient may change sign. It is important therefore to determine whether the reactor is inherently stable despite a positive, slower acting feedback.

To study this, we take an elementary model of the power balances for fuel and moderator analogous to the single power balance of the last example. The model envisages *fuel* leading to the equation for the fuel temperature T_f with respect to a (mean) coolant temperature T_c:

$$m_f \frac{dT_f}{dt} = \kappa n - h_f[T_f - T_c] \tag{4.22}$$

with κ the former power proportionality constant and h_f a suitable heat transfer term to relate the energy removal to the temperature difference between fuel and coolant. For small changes in T_f with T_c constant we assume h_f (like κ and m_f, the fuel heat capacity) to be constant also so that

$$\frac{d\,\delta T_f}{dt} = \frac{\kappa}{m_f}\,\delta n - \frac{h_f}{m_f}\,\delta T_f \tag{4.23}$$

Writing $\tau_f = m_f/h_f$, the fuel temperature time constant, and expressing h_f itself in terms of the steady state values of eqn. (4.22) as $h_f = \kappa n_0/[T_f - T_c]_0 = \kappa n_0/\triangle T_f^*$ say, gives after Laplace transformation

$$\delta \bar{T}_f = \frac{\Delta T_f^*}{1 + p\tau_f} \frac{\delta \bar{n}}{n_0} \tag{4.24}$$

To deal with the moderator, we assume an equation for the same *form* and suppose that the parameters can be fitted adequately from observed behaviour. Thus

$$\delta \bar{T}_m = \frac{\Delta T_m^*}{1 + p\tau_m} \frac{\delta \bar{n}}{n_0} \tag{4.25}$$

where T_m is the moderator temperature, τ_m the moderator time constant and $\triangle^* T_m$ is conceptually $[T_m - T_c]_0$ in the reference state and is a measure again of the coupling of the moderator to the thermal system.

The moderator equation has not been derived as such, but the model is found to be empirically satisfactory. In general, moderators receive some 10% of the energy of fission via γ-radiation, but changes in temperature arise predominantly from the coupling to fuel coolant. The degree of coupling is given in relative terms by $\triangle T_m^* / \triangle T_f^*$ and depends upon the reactor type. This ratio may be relatively large in an LWR (intimate mixing of moderator and coolant); modest in a Magnox or AGR; small in a SGHWR (where pressure tubes are effective insulators of the moderator). To take a Magnox station as an example, Table 4.3 indicates some typical values. Note in passing that if

TABLE 4.3.
Magnox temperature parameters

	Fuel	Moderator
Coefficient of reactivity α	-2×10^{-5} [K^{-1}]	-15 to 3×10^{-5} [K^{-1}][a]
Time constant τ	3 [s]	80 [s]
Coupling temperature ΔT^*	250 [K]	130 [K]
Feedback transform G_T	$\frac{\Delta T_f^*}{1+p\tau_f}$ [K]	$\frac{\Delta T_m^*}{1+p\tau_m}$ [K]

[a] Range from beginning to end of core life/fuel charge life.

we change the reference power level (or n_0) we might expect a change in $\triangle T_f^*$ and $\triangle T_m^*$ but leaving their ratio constant to a first approximation.

Figure 4.13 indicates the block diagram for this form of two temperature feedback

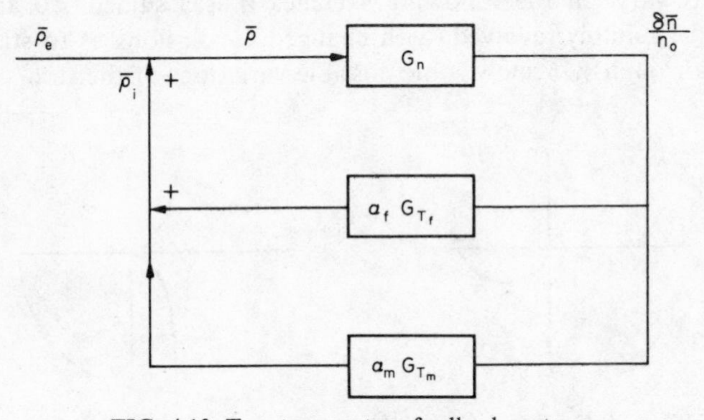

FIG. 4.13. Two temperature feedback systems.

showing the internal or inherent feedback only. The equations given are of rather simple or decoupled form in that T_f, say, does not appear in the moderator balance or vice versa. However (see problems) a more general system of two coupled equations may be manipulated into this normal or decoupled form, so the following analysis has wider applicability than just to two temperature feedback paths if the constants are suitably interpreted.

The two feedback paths can be combined into a single feedback term $H(p)$ as

$$H(p) = \frac{\alpha_f \triangle T_f^*}{1 + p\tau_f} + \frac{\alpha_m \Delta T_m^*}{1 + p\tau_m} \tag{4.26}$$

A value $p=\omega=0$ corresponds to behaviour without oscillation, i.e. static stability/instability. The feedback at $\omega=0$ is

$$H(0) = \alpha_f \triangle T_f^* + \alpha_m \triangle T_m^* \tag{4.27}$$

and this itself illustrates the role of the reference temperature increments $\triangle T^*$ and their coupling to the reactivity coefficients. What is significant is not only the coefficient α but also the magnitude of the weighting $\triangle T^*$. We can also anticipate that static stability will require $H(0) < 0$ as could be argued from simple physical grounds.

We are dealing with a system of third order complexity with two parameters of interest, α_f and α_m, so that the results will take a little effort to unravel. The problems suggest various simplifications which have the virtue of displaying easily some of the characteristics of the solution we are looking for (though even here we are to employ a simple model of the neutronics transfer function). As a preliminary it is useful to write $G_n H$ in standard form and explore some of the possible behaviour we should anticipate. We can put

$$H(p) = A \frac{p + \gamma}{[p + 1/\tau_m][p + 1/\tau_f]} \tag{4.28}$$

where $$A = \frac{\alpha_f \Delta T_f^*}{\tau_f} + \frac{\alpha_m \Delta T_m^*}{\tau_m}; \quad \gamma = \frac{H(0)}{A\tau_m\tau_f}$$

We may generally suppose $\tau_f < \tau_m$ but must note that the sign as well as the magnitude of A and γ may vary. In the following sketches A is assumed <0 and the Nyquist diagrams must be suitably revolved (with changed conclusions as to stability) if $A > 0$.

Figures 4.14 through 4.16 show some possible variations of the zero, $-\gamma$, in $H(j\omega)$ and

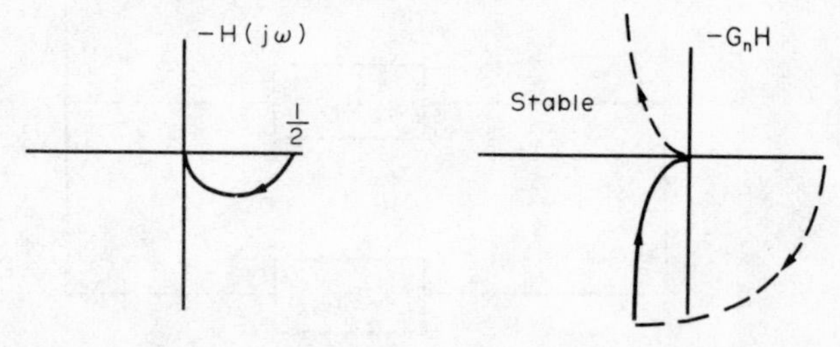

FIG. 4.14. Feedback and stability: $-H(p) = [p+5]/[p+1][p+10]$.

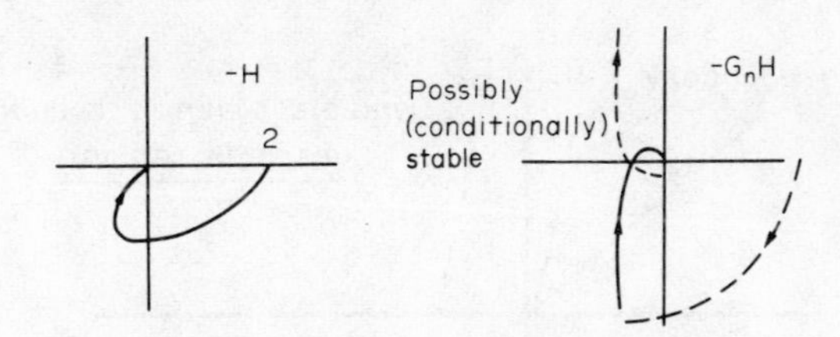

FIG. 4.15. Feedback and stability: $-H(p) = [p+10]/[p+1][p+5]$.

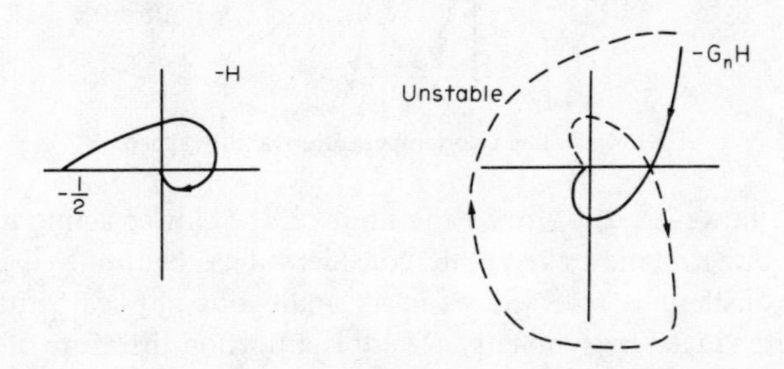

FIG. 4.16. Feedback and stability: $-H(p) = [p-5]/[p+1][p+10]$.

$-G_n(j\omega)H(j\omega)$, the latter being studied around the point $-1, 0$ for Nyquist stability deductions.

It is instructive to sketch further variations of A and γ.

From such sketches we may derive an expectation that when $H(0)>0$ instability is ensured, but, nevertheless, there are possibilities for stability if $\alpha_m>0$ as long as $\alpha_f<0$. To make this precise we employ Routh's criteria (see Chapter 1 for the array). We find from consideration of $1-G_nH$ that we are to study the third order equation

$$p\left[p+\frac{1}{\tau_f}\right]\left[p+\frac{1}{\tau_m}\right]-\frac{\lambda}{\beta}\left\{\frac{\alpha_f\triangle T_f^*}{\tau_f}\left[p+\frac{1}{\tau_m}\right]+\frac{\alpha_m\triangle T_m^*}{\tau_m}\left[p+\frac{1}{\tau_f}\right]\right\}=0 \quad (4.29)$$

and that since one of the three criteria is met explicitly we require:

(1) static stability: $H(0) < 0$,

(2) dynamic stability: $\alpha_f\triangle T^*\tau_m^2 + \alpha_m\triangle T^*\tau_f^2 - \frac{\beta}{\Lambda}[\tau_m+\tau_f] < 0$,

i.e. within the region admitting static stability with no non-oscillating but divergent solution, there may, nevertheless, arise a pair of oscillating and diverging solutions.

These conditions can be represented graphically in Fig. 4.17 in the $\alpha_f\triangle T_f^*$, $\alpha_m\triangle T_m^*$ space. Also shown in this figure is the half region parallel to the dynamic stability region in which any uniform scaling of $\triangle T_f^*$ and $\triangle T_m^*$, which would arise to a first approximation by increasing the reference power level, leaves the system stable.

Thus the third quadrant, with α_f and α_m both negative, is absolutely stable. Stable operating points can be found in the second and fourth quadrants. Given the assumed $\tau_f<\tau_m$, the limitation on α_m for the usual $\alpha_f<0$ is governed by the consideration of

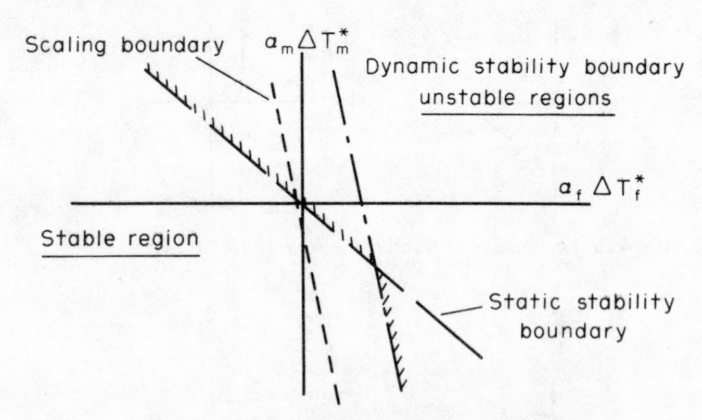

FIG. 4.17. The two-temperature stability space.

static stability. If, however, $\alpha_f > 0$, then the limit on the slower acting moderator effect might arise from either static or dynamic considerations. Naturally if the relative size of the two time constants is reversed, as in an analogous physical situation, the arguments can be interchanged by symmetry. There is a portion therefore of the second and the fourth quadrants, as well as the whole of the third quadrant, in which over-all power increase (leaving T_m^*/T_f^*) constant do not change the stability considerations.

So far we have neglected the transport of energy out of the core. This consideration leads us to the dynamics of the whole plant, and the representation of the energy transport is illustrated in the derivation of the boiler or heat exchanger transfer function, to which we now turn.

Reactor Plant Dynamics

Important elements in the plant associated with the reactor for our purposes may include:

(a) heat exchanger or boiler;
(b) turbine together with its admission throttle;
(c) piping and circulating pumps;
(d) generating equipment or final mechanical or electrical load.

In this book it is not possible to cover details of design of the several types of boilers and other plant in use or the methods to analyse them and categorise their behaviour (see, for example, refs. 3, 4, 8 and 12). But before developing one simple model in an elementary representation, some general introduction may be helpful.

The simplest type of heat exchanger will consist of a number of tubes lying within a shell. Straight tubes have some disadvantages over thermal expansion joints and maintenance, so U-tubes in a shell are more common. There is evidently a choice as to whether the primary coolant passes through the inside of the tubes or the outside in the shell.

Since the purpose of the heat exchanger in this context is limited to raising steam (including perhaps preheating as well as evaporation and perhaps superheating and reheating intermediate pressure steam) it is, indeed, preferable to call it a boiler. In any case, it would be desirable that the boiler provides a reservoir of steam, in the form of hot water able to evaporate under a slight drop in pressure, to meet fluctuations in the demand for steam that could not be met by the slower acting heat source in the reactor itself, and thus smooth out fluctuations or meet rapid changes in demand. In this case it would be natural to have the water in bulk form in the shell with the primary coolant in the smaller volume tubes.

This is not generally satisfactory, however, when the primary coolant is gaseous. A more important consideration is the work needed to circulate the compressible gas coolant. If this was passed through the narrow tube side, the resulting drop in pressure would make too big a demand on pumping power.

In any case, the energy stored can be quantified in terms of the enthalpy needed to raise the normal contents of the boiler above the feed water inlet temperature and is conventionally expressed in terms of full-power seconds (fps).

A simple tube and shell may have only a few fps; to increase the storage a drum or La Mont boiler may be used with a larger quantity of water in the shell and a steam layer over it with the water recirculated within the boiler (or some variation). Commonly the recirculation flow rate is several times the steam draw off rate. Such boilers in nuclear reactor practice may give some tens of fps (though they are generally small in energy storage capacity compared to the La Mont boilers used in fossil-fuelled stations). One disadvantage of the La Mont boiler will be a corresponding increase in the natural time constant of the boiler, measured simply as the ratio of heat capacity to coolant flow rate, a measure that indicates the speed with which the primary side temperatures say react to changes in the secondary side conditions. Typical time constants may have the order of seconds.

In some reactors a different approach has been adopted with the water/steam passing through tubes and known as a "once-through" arrangement. This may be adopted when space is limited in a closely integrated design and also as a more suitable way for providing superheat since appropriate sections of tubes can be given appropriate steels. That is, the cheapest mild steels will be used where corrosion, etc., problems are low and the increasingly expensive stainless steels and austenitic steels (up to a factor of ten in cost) for higher duties in the evaporator and superheat sections respectively. Other disadvantages of stainless steel are the activation of corrosion products swept into the reactor (chromium) and the poor heat conductivity.

It follows that the temperature variations in these tube sections must be constrained, and although the energy storage is already perhaps only a quarter of a comparable drum type boiler, it would be unwise to allow a transient to make use of the energy stored in the metal of the superheater section by letting evaporation take place there. These stricter limits on transients when the higher thermal efficiency of superheated steam is sought are a factor to be considered by the control engineer. The drawback is partly offset by the shorter time constant of once-through boilers that makes them more responsive to natural load-following effects.

We assume that we have been provided with a heat exchanger design including some knowledge of the nominal temperature drop both across the primary coolant and between the mean primary temperature and the mean secondary temperature. For our purposes this latter may well be identified with the saturation temperature of the steam being produced while a simple approximation for the mean primary temperature is to take the average of the primary coolant inlet and outlet temperatures. The temperature drop from primary to secondary is conventionally represented as being proportional to the power passing through a heat transfer coefficient proportionality constant. It is in fact unsound to suppose that the heat transfer coefficient is a constant, though we shall make this assumption and merely indicate the need in more refined work to make allowance for the effect of changing temperature differences and mass flow rates on the apparent heat transfer coefficient.

In the analysis of process dynamics of mass flows, there are two significant complications:

(a) *Transport delays.* The time of transit through the piping, both internal to the boiler and external connections, denoted as θ, will lead to a delay in temperature variations as they enter the system being felt at the exit. If it is assumed that there is no mixing within the pipes, i.e. slug flow, and in the absence of energy additions or losses (well insulated) the transport delay will lead, in the Laplace transform treatment, to exponential terms $e^{-p\theta}$.

(b) *Mixing.* On the other hand, the flow may be mixed. If it is well mixed, i.e. flow into a reservoir rather than down a pipe, we can anticipate that there would result the simple lag transform of an ideal well-mixed reservoir. Real situations are going to be a mixture of the two effects, and a judicious balancing of the two types of time constants is called for.

BOILER TRANSFER FUNCTION

Figure 4.18 shows the elementary "lumped" model used to derive a boiler transfer function on the basis of a single ordinary differential to describe the rate of change of steam temperature as a balance between heat from the primary side and energy removed by flow of steam from the boiler.

It is assumed that:

(a) The primary temperature T_p is the average of the boiler primary inlet and outlet temperatures T_{bi} and T_{bo}. These, of course, will be connected to the reactor outlet and inlet temperatures respectively by the appropriate exponentials in the Laplace transform space representing transport delays. We have $T_p = \frac{1}{2}[T_{bi} + T_{bo}]$.

(b) The temperature of the secondary side is essentially that of the saturated steam being produced T_s which is uniquely related to its saturated pressure p_s by the saturation curve $p_s(T_s)$.

(c) The steam water mass is constant by virtue of make up or feed water but enthalpy is transported out with the steam flow.

(d) Although the turbine throttle is unchanged, nevertheless the rate of steam flow from the boiler will vary with p_s and, indeed, to a good approximation the flow rate is proportional *at a given* throttle to p_s since the turbine back pressure is negligible. In turn, the enthalpy rise of saturated steam above the feed water is essentially constant so the power output of the boiler is proportional to p_s.

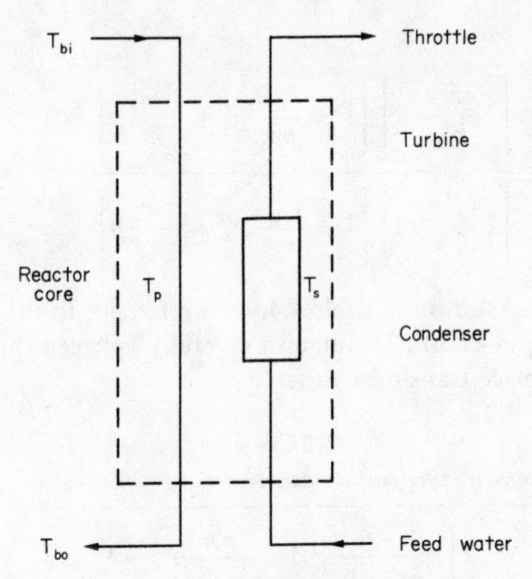

FIG. 4.18. Simplified boiler diagram.

There is, therefore, a back effect upon the primary coolant since a change in temperature affects the heat conducted into the secondary side while the heat removed from the secondary side itself depends on the resulting secondary temperature T_s as it governs p_s.

Let cF be the primary coolant flow rate times heat capacity (W/K), $c_s m_s$ the effective heat capacity to the secondary side (J/K) (to include steam, water and possibly metal), F_s the secondary steam flow (kg/s), n the "heat transfer coefficient" (J/K), and $H = H_s - H_f$ the increase in enthalpy from feed water to steam (J/kg).

A balance of boiler power BP from primary to secondary gives

$$BP = cF[T_{bi} - T_{bo}] = h[T_p - T_s] = F_s H \tag{4.30}$$

which will also enable us later to eliminate h in favour of power and temperature rises, assumed known for the nominal design. An energy balance for the secondary with allowance for heat capacity yields

$$c_s m_s \frac{dT_s}{dt} = h[T_p - T_s] - F_s H \tag{4.31}$$

The immediate purpose is to find how T_{bo} varies with T_{bi} allowing for the change of steam flow from the boiler, and to this end we suppose the *primary* coolant flow is constant. We proceed by taking small variations, assuming h constant and Laplace transform, noting that we have assumed $\delta F_s/F_s = \delta p_s/p_s$ since the throttle is fixed.

$$\left[1 + p\frac{c_s m_s}{h}\right]\delta\overline{T}_s = \delta\overline{T}_p - \left[\frac{F_s H}{h p_s}\right]\delta\overline{p}_s = \delta\overline{T}_p - \left[\frac{F_s H}{h p_s}\right]p_s'\delta_s\overline{T} \tag{4.32}$$

where $p_s' = dp_s/dT_s$ is the slope of the saturation curve for water at the appropriate value of p_s. So

$$\left[\frac{F_s H}{h p_s}p_s' + 1 + \frac{c_s m_s}{h}p\right]\delta\overline{T}_s = \delta\overline{T}_p = \tfrac{1}{2}[\delta\overline{T}_{bi} + \delta\overline{T}_{bo}] \tag{4.33}$$

and using eqn. (4.30)

$$G_b(p) \equiv \frac{\delta \overline{T}_{bo}}{\delta \overline{T}_{bi}} = \frac{\left[2\frac{cF}{h} - 1\right]\left[1 + \frac{F_s H}{hp_s} p'_s + \frac{c_s m_s}{h} p\right] + 1}{\left[2\frac{cF}{h} + 1\right]\left[1 + \frac{F_s H}{hp_s} p'_s + \frac{c_s m_s}{h} p\right] - 1} \equiv K \frac{1 + \tau^- p}{1 + \tau^+ p} \quad (4.34)$$

where the parameters are shown in Table 4.4, with reference to the steady state values, particularly the boiler power BP. Distinguish carefully between the secondary (steam) pressure p_s and the Laplace transform variable p.

TABLE 4.4.

Boiler transfer function parameters and typical values

K	$\dfrac{2\left[\dfrac{T_p - T_s}{T_{bi} - T_{bo}}\right]\left[1 + \dfrac{p_s/p'_s}{T_p - T_s}\right] - 1}{2\left[\dfrac{T_p - T_s}{T_{bi} - T_{bo}}\right]\left[1 + \dfrac{p_s/p'_s}{T_p - T_s}\right] + 1}$	0.7
τ^+	$\dfrac{c_s m_s/BP}{1/(T_{bi} - T_s) + \dfrac{p'_s}{p_s}}$	30 s
τ^-	$\dfrac{c_s m_s/BP}{1/(T_{bo} - T_s) + \dfrac{p'_s}{p_s}}$	10 s
$K\dfrac{\tau^-}{\tau^+}$	$\dfrac{2\left[\dfrac{T_p - T_s}{T_{bi} - T_{bo}}\right] - 1}{2\left[\dfrac{T_p - T_s}{T_{bi} - T_{bo}}\right] + 1}$	0.25

Consider first of all the gain K. The ratio $[T_p - T_s]/[T_{bi} - T_{bo}]$ is essentially a design parameter. For the sample reactor of Chapter 5 we might have values $T_p - T_s = 32$ K and $T_{bi} - T_{bo} = 40$ K at full power; that chapter also gives the logarithmic derivative of saturated pressure with saturated temperature (Fig. 5.4) from which a typical value of $p_s/p'_s = 80$ K. The quoted ratio, 32/40 in the example, is determined in the design. If the primary coolant flow rate F is constant and we make the more questionable assumption that the heat transfer coefficient h does not vary with power level, then this ratio remains unchanged at part power. However, the term in p_s/p'_s as a ratio to $T_p - T_s$ is evidently power dependent so that K varies at part power. It may be seen that $0 \leqslant K \leqslant 1$ with $K \rightarrow 1$ at low power, though a typical full power value may be $K = 0.7$.

Secondly, consider the time constants τ^+ and τ^-. An arbitrary time constant might be defined in the form $c_s m_s \triangle T/BP$ with $\triangle T$, a temperature difference to be determined. It may be seen that $T_p - T_s$ plays this role to a first approximation. However, the drop in temperature across the primary from inlet to outlet, which affects the behaviour by providing an additional reservoir of heat, leads to a splitting into two time constants. A second correction arises from the role of the saturation line between steam temperature and pressure. Indeed, at very high power (beyond normal design range) the time constants tend to a common limit, $\tau \to c_s m_s p_s / BPp'_s$, and, similarly, if the primary circulation rate F is low so that $T_{bi} - T_{bo}$ is high with a fixed ratio to $T_p - T_s$.

Since $T_{bi} > T_s$, it can be seen that τ^+ is necessarily positive, but exceptionally τ^- may be negative. (However, this would be outside normal design limits and the simple model would be suspect.) Unlike many transfer functions, $G_b(j\omega)$ provides for a non-vanishing attentuation $K\tau^-/\tau^+$ at very high frequencies (the last entry in Table 4.4) with a typical value of 0.25 since typically $\tau^-/\tau^+ = \frac{1}{3}$. $K\tau^-/\tau^+$ is independent of power in this model, so that τ^+/τ^- varies as K with part power.

Figure 4.19 is the Bode diagram for typical values, sketched from the break frequencies.

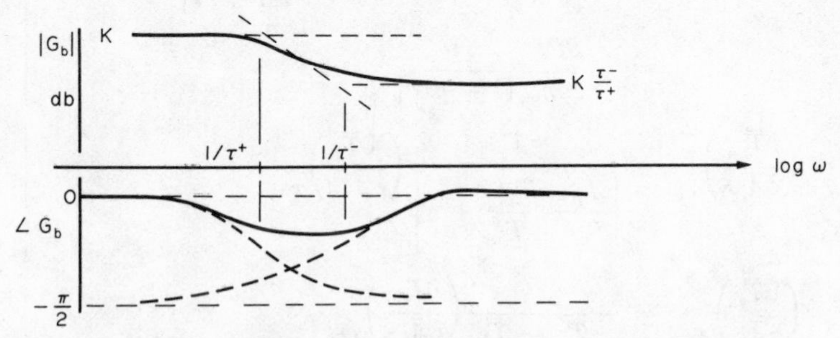

FIG. 4.19. Boiler transfer function bode diagram (primary outlet to inlet).

At low frequencies, the attenuation is the gain K and there is zero lag. At high frequencies, the attenuation is $K\tau^-/\tau^+$ and again no phase lag. At intermediate frequencies there is a small phase lag which might affect primary loop stability. Atypical cases are left for the problems.

A very similar derivation can be used in discussing the corresponding heat transfer through a body of water or metal for the passage of the primary coolant through the reactor itself.

THROTTLE VARIATIONS

We have derived a connection between T_{bo} and T_{bi}. There are, however, further effects when the throttle opening is changed, varying the flow of steam and the power from the boiler to the turbine, with back effects on the primary. It is a situation with two inputs, δT_{bi} and δA (the throttle opening), and two outputs, δT_{bo} and either δT_s or δp_s, as we wish, or ultimately with an expression for power flow due to variations of both inputs. There are, therefore, four partial derivative relations of which we have so far derived $(\partial T_{bo}/\partial T_{bi})_A$, thus indicating which input is kept constant.

We suppose that the steam flow F_s is proportional to the pressure p_s at a constant throttle setting through a proportionality A. When the throttle is opened or closed we can refer to a change δA though clearly this is merely a statement of the change of flow *at a given pressure*; of course, $\delta A/A$ is only a definition, the fractional change of steam through the throttle at constant pressure. When the pressure also varies we have

$$\frac{\delta F_s}{F_s} = \frac{\delta A}{A} + \frac{\delta p_s}{p_s} = \frac{\delta A}{A} + \frac{p'_s}{p_s}\,\delta T_s \tag{4.35}$$

again involving the saturation curve for water. As before we assume power flow is proportional to steam flow through the approximately constant enthalpy rise of water to steam despite saturated pressure variations.

Using the same model, procedure and equations as before results in the following relations:

$$\left.\begin{aligned}
\left(\frac{\partial \bar{T}_s}{\partial \bar{A}/A}\right)_{T_{bi}} &= -\frac{1}{\left[\dfrac{1}{[T_p - T_s] + \frac{1}{2}[T_{bi} - T_{bo}]} + \dfrac{p'_s}{p_s}\right]}\,\frac{1}{[1 + p\tau^+]}\\
&= -\frac{BP\tau^+}{c_s m_s}\,\frac{1}{[1 + p\tau^+]}\\
\left(\frac{\partial \bar{T}_{bo}}{\partial \bar{A}/A}\right)_{T_{bi}} &= \frac{1}{\left[\dfrac{T_p - T_s}{T_{bi} - T_{bo}} + \frac{1}{2}\right]}\left(\frac{\partial \bar{T}_s}{\partial \bar{A}/A}\right)_{Tbi}\\
\left(\frac{\partial \bar{T}_s}{\partial \bar{A}_{bi}}\right)_{A} &= -\frac{1}{T_{bi} - T_{bo}}\left(\frac{\partial \bar{T}_{bo}}{\partial \bar{A}/A}\right)_{Tbi}
\end{aligned}\right\} \tag{4.36}$$

Figure 4.20 shows the schematic derived so far, which is perhaps as complicated as we might wish to get in a preliminary analysis.

LIMITATIONS OF MODEL

This elementary account of the role of the boiler in an indirect cycle omits several points, including:

(a) pressure-smoothing devices;
(b) detailed treatment of piping delays versus mixing;
(c) variations in thermodynamic properties and heat transfer relations;
(d) complications of superheating and multi-pass flows; feedheating and reheating;
(e) primary side pumping power (significant for gaseous coolants);
(f) water–steam acceleration terms (significant for boiling water reactors).

In addition to the phenomena indicated above whose neglect might invalidate a particular application, it is clear that the derivation has been on a "lumped" basis and since real systems are composed of distributed items, the model is correspondingly limited. The account of mixing or transport in piping, or the splitting of the heat exchanger into two parts (primary balance and secondary balance) are attempts, of course, to provide a more detailed model. As a general guide to whether such an approach is acceptable, it may be said that as a representation of the over-all dynamics of the reactor

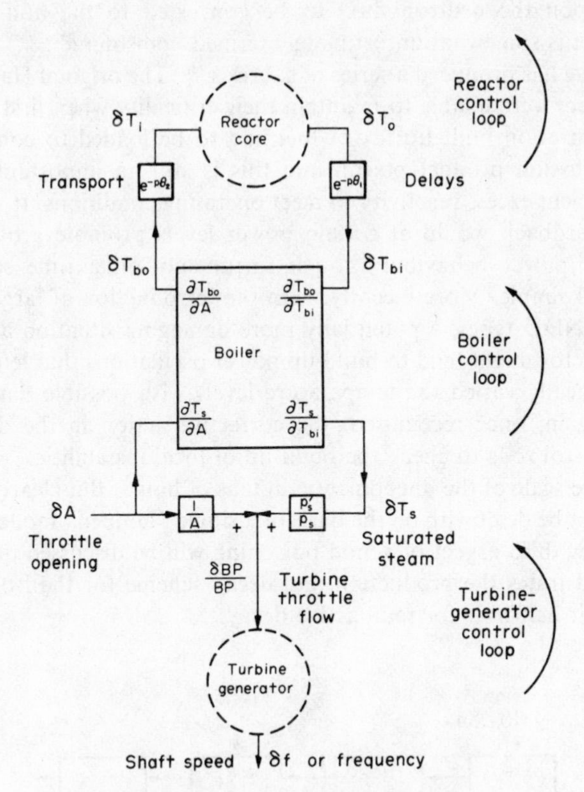

FIG. 4.20. Boiler transfer function schematic.

plant this approach is satisfactory as long as we do not seek to draw conclusion about stability and transient behaviour on a time scale shorter than the time constants derived in the simple model. If therefore, for example,. we are talking about time constants of a few seconds for mixing and transport delays, then this puts a limit on the range of phenomena representable. We might expect that special problems of BWRs with instability possibilities on a shorter time scale will not be elucidated by such a "coarse" model.

Xenon Poisoning

The ^{135}Xe isotope has the largest thermal neutron capture cross-section known, of the order of 300 pm^2 (3 megabarns)† associated with the formation of a "magic" nuclear closed shell ^{136}Xe. Since ^{135}Xe is produced in fission it has a significant effect upon reactor dynamics, governed by the characteristic time for ^{135}I to decay to produce the troublesome ^{135}Xe and the time for the ^{135}Xe either to decay to a non-absorbing isotope

†The 2200 m/s cross-section is reported as 272 pm^2 but correction factors for the non-$1/v$ absorption would raise the effective value.

or (dependent upon the neutron flux) to be converted to the non-absorbing ^{136}Xe isotope. The effect is somewhat unfortunately termed "poisoning".

Xenon poisoning has produced a series of suprises.[10] The original Hanford plutonium production reactors were unable to maintain their criticality when first run at power as the xenon concentration built up; extra fuel had to be loaded to compensate for the build-up of this fission product poison and this is still an important element in the provision of sufficient excess reactivity to meet operating conditions. It was then realised that the xenon feedback would at certain power levels promote gross instabilities in the neutron level/power behaviour though fortunately on a time scale that makes correcting control simple. More recently, with the introduction of large reactors of the Magnox and CANDU types, a potentially more damaging situation arose where local regions of the reactor might tend to build up power oscillations that left the total power unchanged but locally varied the temperature levels with possible damage to the fuel elements, etc. Again, once recognised, the corrective action in the design phase—to provide local control rods to check the build-up of local instabilities—is relatively easy in view of the time scale of the phenomenon in tens of hours. But clearly internal spatial instabilities cannot be dealt with on the basis of a single "lumped" model for the neutron processes, and this third aspect of xenon poisoning will be discussed qualitatively only.

Figure 4.21 illustrates the production and decay scheme for the isotopes of interest together with their half-lives for radioactive decay.

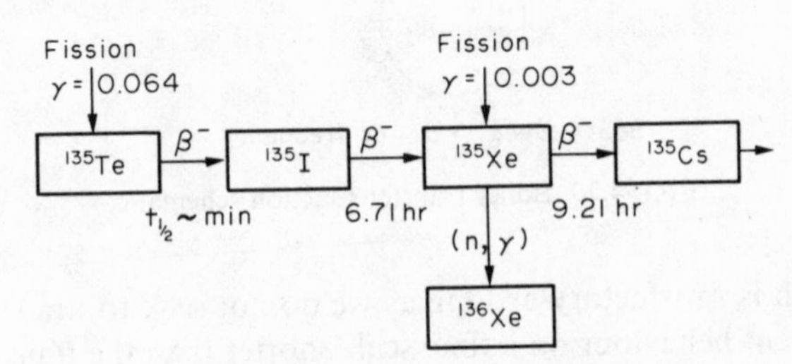

FIG. 4.21. Decay scheme for ^{135} Xe isotopes.[5]

Whilst there is a small direct production of the isotope ^{135}Xe from fission, the greater part of the production arises indirectly from decay of the fission fragment chain. Because the preliminary steps in the decay chain are rapid, we may for practical purposes suppose that the production starts with ^{135}I which then decays into ^{135}Xe. Qualitatively, the hold-up in iodine means that the reactor can be started up and run at power for some hours before the "poisoning" effect is felt; then, in the absence of a compensating reactivity increase, the reactor power falls. If the reactor power is suddenly brought to zero after an operating period, however, xenon is still being produced and is now being removed only by its own decay process. There is, therefore, a transient build-up of xenon which may, at moderate to high flux levels, far exceed the steady operating level of ^{135}Xe. Thus to restart the reactor before this xenon poisoning has decayed would call for even greater excess reactivity than that required to continue steady operation.

This in turn implies a safety problem; the extra reactivity has to be provided in the first instance, and as the xenon is burnt out in the process of the restart operation this excess reactivity is once again present and needs to be contained by suitable control rod insertion.

These gross changes are of operating interest, especially perhaps in marine work where it would be embarrassing to lose the ability to restart the reactor after a temporary shutdown. These problems are studied by means of analogue computer solutions in Chapter 9 (although analytical solutions suitable for digital evaluation are also readily found). Here we shall concentrate on the consequences for stability of small variations in power level when xenon feedback is taken into account. For simplicity we shall drop the explicit reference to the isotope number 135.

XENON–IODINE EQUATIONS

Local balances of production and removal can be drawn up in terms of the neutron flux or density. We assume that the flux is uniform throughout the core of the reactor. This is not perhaps too bad an assumption in a reactor where flux flattening is likely to be part of the design process for other reasons; if not, it would be appropriate to use some form of adjoint weighting in passing from local balances to lumped reactivity effects, which we omit here for simplicity. Using X and I to represent the time dependent concentrations of the xenon and ^{135}I isotopes, we have

$$\frac{dI}{dt} = \gamma\Sigma^f\phi - \lambda_i I \tag{4.37}$$

$$\frac{dX}{dt} = \lambda_i I - \lambda_x X - \phi\sigma X \tag{4.38}$$

where γ is the yield of iodine, atoms per fission, Σ^f the fission cross-section and $\phi\sigma$ the flux times xenon absorption cross-section.

The small amount of direct yield of xenon has been neglected in these equations. It is shown in a problem that some of the conclusions about the possibility of instability at low fluxes are in fact ruled out by consideration of the direct yield and we shall indicate the limits of the present analysis as we go along. The assumption is made here on the grounds of simplicity and the results are conservative since a more detailed analysis at low flux is more favourable. The microscopic cross-section σ times the xenon concentration X is, of course, a macroscopic cross-section for neutron absorption and hence reactivity loss. Note also that the loss of one iodine atom by decay is accompanied by the production of one xenon atom and that the removal process for xenon is a combination of decay and neutron capture. In view of the magnitude quoted these two removal processes are of equal significance at a neutron flux, say, 7×10^{16} neutrons/m^{2}s in a typical power range.

The build-up of xenon leads to a reactivity loss or reactivity poisoning. The poisoning P itself is defined as the ratio $\sigma X/\Sigma^f$, and it is readily seen that in the steady state

$$P(\phi,\infty) \equiv P_{ss} = \frac{\gamma\sigma\phi}{\lambda_x + \sigma\phi} \tag{4.39}$$

so that at very high fluxes, greater than 10^{18} per m^{2}s, say, $P_{ss} \rightarrow \gamma$. The reactivity loss in a simple model is given approximately by

$$\rho_x = -\frac{\sigma X}{\nu\Sigma^f} = -\frac{P}{\nu} \tag{4.40}$$

so that in the steady state a power reactor may well have a reactivity deficiency of some 3–4% due to xenon build-up. Correspondingly define $\rho_{ss} = -P_{ss}/\nu$ as the steady state reactivity effect, with a limit of $-\gamma/\nu$ for very high fluxes (note negative ρ_{ss}).

The equations are "nonlinear" by virtue of the term in $\phi\sigma X$, so that for process dynamics we go through the usual linearisation procedure, considering small departures from the steady reference state. On neglecting terms in $\delta\phi\sigma\delta X$,

$$\left.\begin{aligned} \frac{d}{dt}\delta I &= \gamma\Sigma^f\delta\phi - \lambda_i\delta I \\ \frac{d}{dt}\delta X &= \lambda_i\delta I - \lambda_x\delta X - \phi_0\sigma\delta X - X_0\sigma\delta\phi \end{aligned}\right\} \tag{4.41}$$

Noting that $X_0 = \gamma\Sigma^f\phi_0/[\lambda_x + \sigma\phi_0]$ and taking Laplace transforms we have the (relative) xenon transform function

$$G_x(p) \equiv \frac{\delta\bar{\rho}_x}{\delta\bar{n}/n_0} = -\sigma\phi_0\rho_{ss}\frac{p - \lambda_i\lambda_x/\sigma\phi_0}{[p + \lambda_i][p + \lambda_x + \sigma\phi_0]} \tag{4.42}$$

and transfer function

$$G_x(j\omega) = \rho_{ss}\frac{\lambda_x}{\lambda_x + \sigma\phi_0}\,\frac{1 - j\omega\nu\phi_0/\lambda_i\lambda_x}{\left[1 + \dfrac{j\omega}{\lambda_i}\right]\left[1 + \dfrac{j\omega}{\lambda_x + \sigma\phi_0}\right]} \tag{4.43}$$

where we have made use of the relation $\delta n/n_0 = \delta\phi/\phi_0$. Equation (4.42) is preferred as the transform in p while eqn. (4.43) may be more helpful in plotting the transfer function $p \to j\omega$.

The transform contains two poles in the left half-plane and one zero in the right half-plane. At fluxes above some 10^{16} neutrons/m^{2}s, the $\lambda_x + \sigma\phi_0$ pole is of larger magnitude than the λ_i pole but the zero $\lambda_i\lambda_x/\sigma\phi_0$ is smaller. This component of the transfer function for high fluxes shows an increase in gain to the break frequency λ_i therefore followed by a decrease. At low fluxes, however, the amplitude decreases with frequency. The gain also depends on the term $\rho_{ss}/[\lambda_x + \sigma\phi_0]$ with its maximum at $\phi_0 = \lambda_x/\sigma = \phi^*$.

At low frequencies there is a reactivity *loss* for a power increase due to the xenon poisoning, i.e. a phase of π. At higher frequencies, however, the successive lags in the system reduce the phase to zero where it may promote marginal instability (positive feedback convention) and ultimately to $-\frac{1}{2}\pi$ at high frequencies, above the $\lambda_x + \sigma\phi_0$ break point. If there had been only a *direct* yield we could have anticipated a minimum phase of $\frac{1}{2}\pi$ due to the single lag term and therefore no possibility of dynamic instability. This supports the claim that an allowance for the direct yield would mitigate the effects of xenon poisoning, and we proceed on this assumption in order to simplify the development, leaving a fuller analysis to the problems.

Figure 4.22 gives $G_x(j\omega)$ for the data of Fig. 4.19 at various fluxes from $0.01\phi^*$ upward

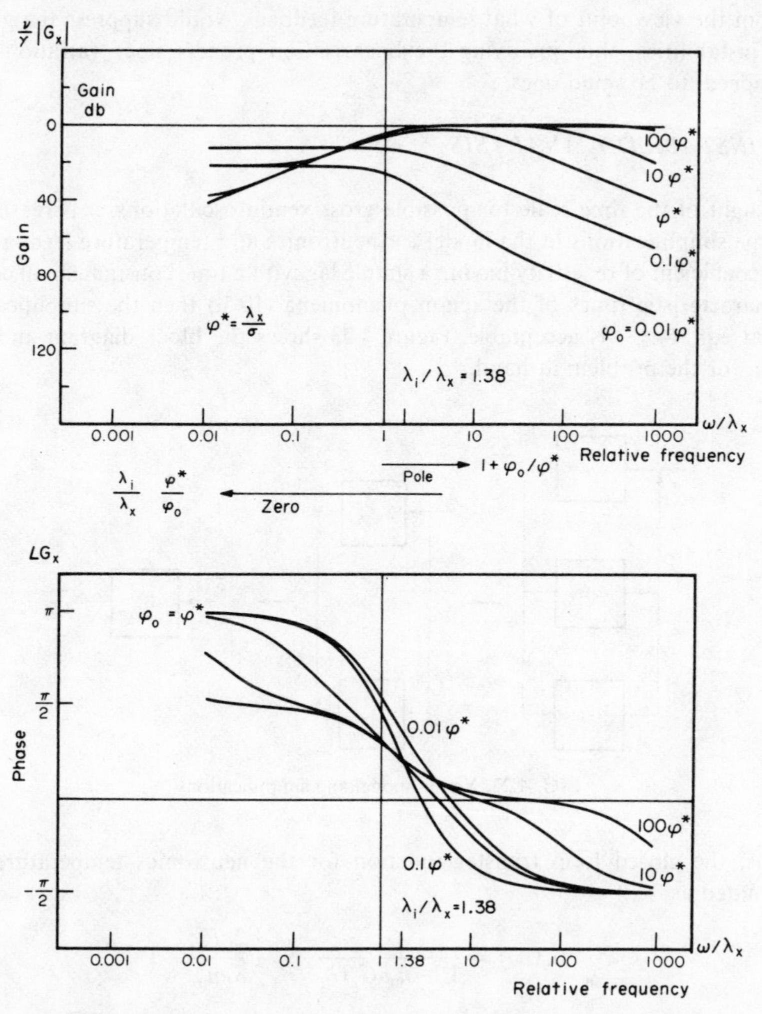

FIG. 4.22. Xenon relative transfer function (neglecting direct yield).

(ϕ^* about 7 10^{16} neutrons/m^{2}s). Typical power ranges are from 1 to $10\phi^*$ and it is seen that in this range the gain of the fractional transfer function $(\partial\rho)/(\partial\phi/\phi_0)$ is close to its maximum, γ/ν, in the frequency range of interest where the phase is zero.

Several general deductions can be made. At high frequencies the gain falls off and dynamic instability in practice will not occur. At low fluxes the gain is too small to support serious oscillations. At high fluxes, the *relative* gain is high but the reactivity change for an absolute power level change falls off with increasing ϕ_0. It is only for fluxes around λ_x/σ and for frequencies around λ_i that we observe a sufficient amplitude and a phase lag around zero such that dynamic instability is to be anticipated (positive feedback convention). This general conclusion remains true when the small *direct* yield of xenon is included (see problems).

We now pursue the possibility of gross xenon instability at moderate and higher fluxes from the viewpoint of what temperature feedback would suppress the growth of possible instabilities, thus justifying the linearisation process since variations will be forced, indeed, to be small ones.

GROSS INSTABILITY ANALYSIS

In the light of the time scale for possible gross xenon oscillations, it is reasonable to make some simplifications in the model for neutronics and temperature feedback. It we take one coefficient of reactivity having a simple lag with a time constant small compared to the characteristic times of the xenon phenomena (10 h) then the simplification developed at eqn. (4.18) is acceptable. Figure 4.23 shows the block diagram and its simplification for the problem at hand.

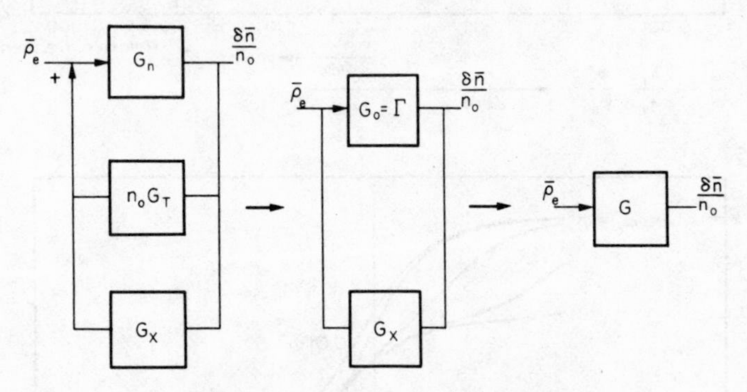

FIG. 4.23. Xenon model and simplifications.

That is, the closed loop transfer function for the neutronics–temperature path is approximated as

$$\mathscr{L}_{\omega\to 0}\; G_0(j\omega) = \frac{G_n}{1 - n_0\alpha G_n G_T} \to -\frac{2cF_0}{\alpha\kappa n_0} \equiv \Gamma \tag{4.44}$$

defining a constant Γ which will be positive for the negative temperature feedback coefficient envisaged. Then the closed loop transform for the complete system is

$$G(p) = \frac{\Gamma}{1 - \Gamma G_x} = \frac{\Gamma[p + \lambda_i][p + \lambda_x + \nu\phi_0]}{[p + \lambda_i][p + \lambda_x + \nu\phi_0] + \Gamma\nu\phi_0\rho_{ss}[p - \lambda_i\lambda_x/\nu\phi_0]} \tag{4.45}$$

Thus the behaviour is governed by the roots of the denominator polynomial in eqn. (4.45). In terms of our simple model

$$\Gamma\frac{P}{\nu} = \frac{2cF_0}{\alpha\kappa n_0}\rho_{ss} = \frac{2}{\alpha[T_o - T_i]}\rho_{ss} \equiv \mu \tag{4.46}$$

defining the parameter μ suitable for a root locus method, where $[T_o - T_i]$ is the steady state rise in coolant temperature. Note that $\mu > 0$ for $\alpha < 0$ and μ is large for $|\alpha|$ small.

The coefficients of the denominator polynomial are generally functions of $\sigma\phi_0$ but we may regard μ as providing the size of temperature feedback for a given ϕ_0 value. The poles of the *open* loop transfer function are $-\lambda_i$ and $-[\lambda_x+\sigma\phi_0]$ and the zero is $\lambda_i\lambda_x/\sigma\phi_0$, leading to the root locus expression

$$\frac{[p+\lambda_i][p+\lambda_x+\sigma\phi_0]}{\left[p-\dfrac{\lambda_i\lambda_x}{\sigma\phi_0}\right]}=\mu \tag{4.47}$$

and Fig. 4.24, *interpreted in terms of increasing* $|\alpha|$.

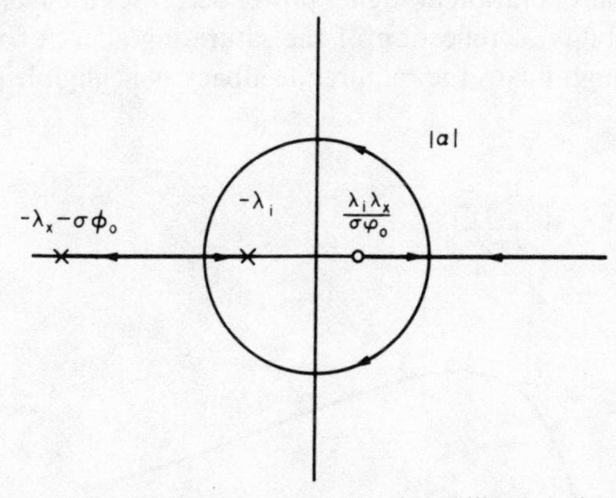

FIG. 4.24. Root locus for gross xenon instability (no direct yield).

Without temperature (or some other) feedback mechanism, the xenon effect leads to an unstable behaviour with a real, increasing root. With some feedback control the divergent root becomes oscillatory. At very low ϕ_0 these incipient transients seem to be very fast acting, but the situation is not really that bad because of the low amplitude at low ϕ_0 (and see problem 4.18 for the mitigating effect of the direct yield of xenon).

Figure 4.24 shows that sufficiently large $|\alpha|$ will secure the transfer of the roots into the left half-plane and suppress growing unstable cross-oscillations. Our immediate interest is in this dynamic stability limit and the necessary μ-value to suppress the instability. This may be determined by substituting $p=j\omega$ in eqn. (4.48) and equating real and imaginary parts. From the latter the critical value is given by

$$\mu_c=\lambda_i+\lambda_x+\sigma\phi_0 \tag{4.48}$$

and from the former, at the point of dynamic instability, the oscillatory frequency is given by

$$\omega_c^2=\lambda_i\left[2\lambda_x+\sigma\phi_0+\lambda_x\frac{\lambda_i+\lambda_x}{\sigma\phi_0}\right] \tag{4.49}$$

Equation (4.48) suggests that there is a minimum value of μ_c, i.e. $\lambda_x \times \lambda_i$ at low fluxes, and hence a maximum magnitude $|\alpha|$ required in any design.

Since the direct yield is significant at low fluxes and itself suppresses oscillations, a better value for μ_c minimum is probably $3\lambda_x$ rather than $\lambda_x+\lambda_i$. At high flux, μ_c increases and the required $|\alpha|_c$ becomes negligible. If we substitute for the particular model of this chapter for μ:

$$\alpha_c \left[\frac{T_o - T_i}{\sigma\phi_0}\right] \frac{1}{2} \frac{\nu}{\gamma} = - \frac{\sigma\phi_0}{[\lambda_x + \sigma\phi_0][\lambda_i + \lambda_x + \sigma\phi_0]} \tag{4.50}$$

In comparing a given design at different power levels, some parameters will be kept constant. In this expression it is reasonable to suppose that the coolant flow rate is the same at the different proposed power levels so that with higher power, both T_o-T_i, the coolant temperature rise, and the flux ϕ_0 will increase by the same ratio. We see in this model therefore that operation at higher power decreases the magnitude of feedback required to limit instability, a reflection of the saturating effect of xenon poisoning at higher fluxes; at very high fluxes the required feedback is negligible (see Fig. 4.25).

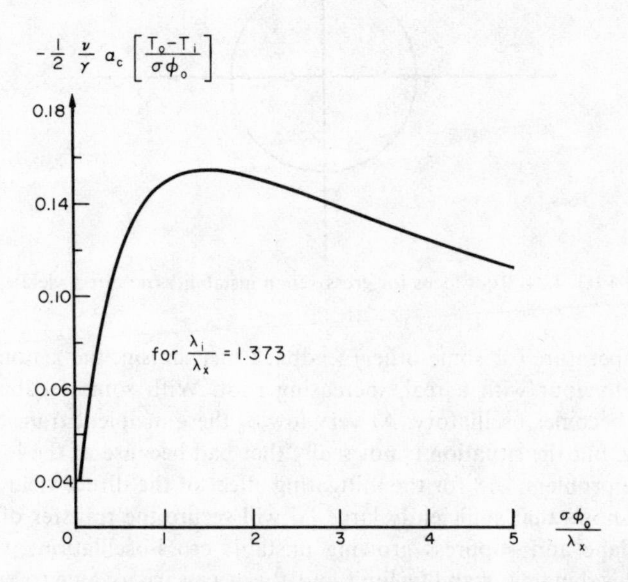

FIG. 4.25. Stabilising temperature feedback required at different flux levels to suppress gross xenon instability (direct yield neglected).

From eqn. (4.49) we may find the minimum ω^2 as $2\lambda_i\lambda_x[1+\sqrt{+1\lambda_i/\lambda_x}]$ or about 1 cycle per day. The practical range of fluxes above the levels where gross xenon oscillations are significant is limited to $\sigma\phi_0$ values of about 100 λ_x so that even at the highest fluxes the gross instability would cycle only a few times a day.

It should perhaps be added that if the feedback term is too small to suppress the potential oscillation, nonlinearities of the system will certainly limit the temperature fluctuations, but not perhaps before they reach damaging proportions. Fortunately the slow rate of oscillations makes it simple to control the gross power fluctuation by either manual or automatic control.

The discussion of xenon effects, both transfer function and root locus construction, has been carried out in terms of neglecting the *direct* yield of xenon in fission. The general conclusions are not worsened by an allowance for the small direct yield; the problems encourage the reader to explore this complication for himself.

XENON SPATIAL OSCILLATIONS

The interesting phenomenon of *spatial* as opposed to gross xenon oscillations will be covered qualitatively more briefly perhaps than it deserves, but a spatial analysis would take us outside the limitations of "lumped" kinetics that brevity demands. Qualitatively, however, it is possible in a large reactor for local variations to build up as follows: a local increase of neutron density burns out the existing poison and locally raises the net neutron production, thus increasing the power locally before the increase in fission products leads, through the iodine decay hold-up, to a compensating increase in xenon density. If the total power is constant, a local increase must be accompanied by a local decrease in power elsewhere. Here the reduced flux no longer burns out as many xenon atoms so that the local neutron balance is unfavourable and few neutrons are produced. Ultimately the process catches up with itself after the decay of the iodine atoms makes itself felt and the corrective action swings the system into an oscillation in the opposite direction. Left to itself, the oscillations can build up in magnitude (although non-linearities again distort their nature) and produce locally damaging temperature excursions even though the *gross* power is kept constant.

The nature of these oscillations can be seen from Fig. 4.26 as a higher "mode" superimposed upon the fundamental flux shape, analogous to overtones in music.

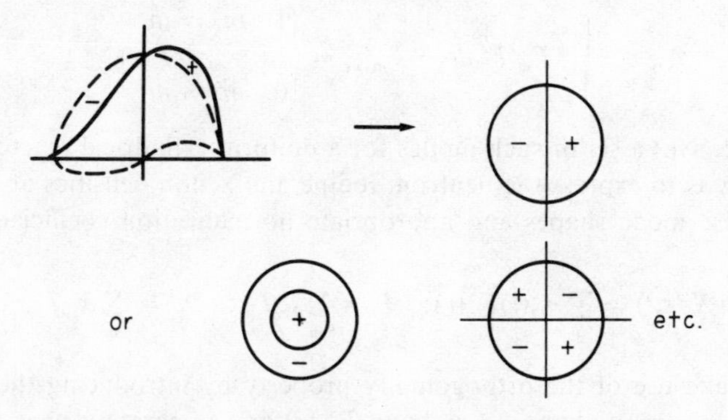

FIG. 4.26. Conceptual development of xenon spatial oscillations.

MODAL ANALYSIS OF FLUCTUATIONS†

Spatial instability may be explored quantitatively and approximately using a *nodal* approximation in which the reactor is lumped in regions and a difference expression developed to represent the transfer of neutrons between regions (see problem 4.22). Alternatively, a *modal* approximation can be used in which the spatial shapes are

†This section may be omitted on a first reading.

represented as a sum of mode shapes. The methods of mathematical physics based on separating the time and space dependent equations to provide a set of normal or orthogonal modes in which the solution can be expressed is well known and provides a convenient set of expansion modes. In the present context with the nonlinear terms involving products of neutron and xenon density, some severe approximations will be needed. Nevertheless, we suppose that the neutron equation in the absence of both precursors and xenon effects provides a suitable set of shape dependent functions, i.e. that in the time and space dependent equation $\partial N/\partial t=[P-R]N$ we write $N(r,t)=n(t)F(r)$ with n dependent on time only and F dependent on position only (and energy in a generalisation). Diffusion theory, for example, lends itself to such a separation, and for simplicity we assume one group diffusion theory. If the coefficients of the diffusion equation are time independent, then the equation may indeed be separated though it is found that there are many solutions which may be enumerated $n_m(t)F_m(r)$. The space dependent shapes $F_m(r)$ incorporate the boundary conditions for the neutron density at the edges of the reactor. One such shape, the lowest mode F_0, is unique in being everywhere of the same sign within the reactor region, but the others, $F_m(r)$; $m>0$, display zeros with regions of positive and regions of negative value. It follows that the curvature, and therefore the neutron leakage as a proportion of other events, is larger in the higher modes since neutrons are leaking between regions of different sign as well as out of the reactor in such a mode; the modal reactivity is more negative in higher modes. Only the lowest mode, of course, can have independent existence. In the time dependent linear model the higher modes provide for a description of the transient shape which must ultimately decay to the lowest mode shape. These natural modes have a useful orthogonality property which, with a suitable normalisation, means we may write for the integrals over the reactor volume

$$\int_V F_m{}'F_m dr = \delta_{m'm} = \begin{cases} 1 & m' = m \\ 0 & m' \neq m \end{cases} \tag{4.51}$$

Problem 4.22 gives a set of such modes for a uniform cylindrical reactor.

The idea now is to express the neutron, iodine and xenon densities at power N' and X' through these mode shapes and appropriate normalization coefficients, dependent on time:

$$N'(r,t) = \sum_m n_m(t)F_m(r); \quad I' = \sum_m i_m F_m; \quad X' = \sum_m x_m F_m \tag{4.52}$$

and then to make use of the orthogonality property by introducing these expansions into the equations, multiplying by a shape $F_{m'}(r)$ and integrating over the volume of the reactor. Cross-modes in m', m from the linear terms will then vanish exactly. If this is done to the equations

$$\frac{\partial N'}{\partial t} = [P - R - X'\sigma v]N'; \quad \frac{\partial I'}{\partial t} = \gamma\Sigma^f vN' - \lambda_i I'; \quad \frac{\partial X'}{\partial t} = \lambda_i I' - \lambda_x X' - X'\sigma vN'$$

(and this may be generalised to include delayed neutrons if we wish) we still have nonlinear terms of the form of $\int_V F_{m'}F_{m''}F_m dr$ to deal with. The approximation is now made that only for $m''=0$ will such integrals be appreciable and that higher modes will not tend to induce each other. The integrals over the (xenon-free) production and removal operators $P-R$ will be expressible as a reactivity and reproduction time, though the reactivity will vary noticeably with each mode and must be enumerated accordingly. For $m'=0$ the retained terms give

$$\frac{dn_0}{dt} = \frac{\rho_0}{\Lambda} n_0 - v\sigma \int_V F_0^3 dr x_0 n_0; \quad \frac{di_0}{dt} = \Sigma^f v n_0 - \lambda_i i_0;$$
$$\frac{dx_0}{dt} = \lambda_i i_0 - \lambda_x x_0 - \sigma v \int_V F_0^3 dr x_0 n_0 \tag{4.53}$$

The spatial integral is seen to have the nature of an effectiveness factor for the xenon cross-section and the equations essentially that of the lumped model treated earlier; the effectiveness correction should be included but is normally close to unity.

In higher modes, neglecting cross-terms except for those containing F_0,

$$\left.\begin{aligned} \frac{dn_m}{dt} &= \frac{\rho_m}{\Lambda} n_m - \sigma v \int_V F_0 F_m^2 dr[x_0 n_m + x_m n_0]; \quad \frac{di_m}{dt} = \gamma \Sigma^f v n_m - \lambda_i i_m; \\ \frac{dx_m}{dt} &= \lambda_i i_m - \lambda_x x_m - \sigma v \int_V F_0 F_m^2 dr[x_0 n_m + x_m n_0] \end{aligned}\right\} \tag{4.54}$$

Here again the integral in $F_0 F_m$ provides an effectiveness factor which we may assume incorporated into the microscopic cross-section σ.

These (truncated) higher mode equations are, then, of the same form as the lowest mode equations or the lumped model equations after linearisation though now including a mode reactivity which in turn depends on the solution $F_m(r)$ of the low power equations.

Since the equations at power do not separate exactly, and in view of the discarding of cross-terms, there are several approximations involved in deriving these equations but they are not dissimilar in this from the linearisation of the lumped model. Of course much depends on the availability of the modal shapes which can be found analytically only for the most simple geometries (see problem 4.21). If the real reactor configuration varies appreciably from such a simple geometry, the approximations become doubtful; the alternative of numerical computation of the modes of a more complicated geometry is hardly worth while since we might as well solve the original equations digitally and directly.

Subject to these qualifications, the preceding lumped analysis may be taken over and used for the higher mode equations, allowing, however, for the role of a negative modal reactivity. This term, caused by the increased leakage probability within a mode, always tends to stabilise the higher modes, requiring a smaller negative temperature coefficient, for example, to limit the growth of the higher xenon modes than in the lowest mode. If this temperature stabilisation is not available, however, we have the possibility of control rods being available to hold the gross power, in the lowest mode, but not perhaps the possible higher modes because the control rods have been positioned too close to the zeros of higher modes or because no instrumentation is available to sense the growth of such modes.

Thus eqn. (4.43) can be used for the transfer function for xenon reactivity effect for higher modes, incorporating an effective σ and keeping in mind that instability cannot occur until the xenon reactivity effect exceeds the modal reactivity decrease.

Oscillations will be more likely to occur in large reactors, such as the Magnox and CANDU systems where the regions of the reactor are close to self-sustaining criticality rather than LWRs where the system is compact and tightly coupled (more negative modal reactivities). Like all xenon phenomena depending on the intrusion of the resonance cross-section of ^{135}Xe into the thermal flux region, there will be little problem in fast reactors.

In passing it should be mentioned, however, that spatial oscillations may potentially occur for other reasons. The positive moderator temperature coefficient of a Magnox reactor is one such situation and this would not rule out a fast reactor.

As far as xenon is concerned, the cure is simple in view of the time scale of the problem—provide comprehensive regional instrumentation to detect the fluctuations in neutron density or temperature level and provide a pattern of control able to halt incipient growth in the likely lower modes.

The remaining inherent dynamical processes in a reactor that might be considered are further "poison" build-up, particularly samarium poisoning, and the changes in isotopic composition of the fuel over long exposure as the fissile inventory is burnt out and to some extent replaced by breeding from fertile material. The slow time scale of these processes, however, somewhat divorces them from the immediate control problems; they can better be studied directly as time dependent differential equations, which we do in Chapter 9.

In the remainder of this chapter we discuss the process dynamics of the control elements that may be used to produce, in normal operation, the required stability against potential divergencies and the response to changes in demand in a satisfactory manner.

Reactor Control Elements

Local control of the reactor to supplement its inherent process behaviour is exercised through some device affecting the reactivity. Here we discuss reactor control only from the process viewpoint. In other chapters we discuss control from the viewpoint of safety (Chapter 7) and startup/shutdown (Chapters 3 and 9). In Chapter 8 we take a look at controlling such a total plant or system optimally.

Automatic control devices may also be applied elsewhere in the plant, e.g. the throttle of the turbine or the rate of circulation of coolant in the primary circuit of reactor and heat exchanger.

However, we must be clear about the various levels of control needed, which may be summarised as follows:

(a) To have available at all times in the range of expected operation, startup and all conceivable malfunction situations, a sufficient reactivity control margin to guarantee the shutting down of the fission process when required and in the time scale appropriate to safety.

(b) To provide excess reactivity, held back under control, to meet such requirements as:
 (i) reactivity loss on heating up from a cold start;
 (ii) reactivity loss in xenon and other poisoning;
 (iii) reactivity loss (occasionally gain) in fuel burn-up and isotope production.

(c) To provide immediate control reactivity in such cases as:
 (i) startup and shutdown;
 (ii) localised adjustments to control temperature distributions and the balance between fuel elements channels;
 (iii) process control of fluctuations in load demand.

In this section we are discussing essentially the last only of these requirements. While at present the economics of nuclear power plants often dictate that they shall be used at full power to meet base load demand, this will not always be true (e.g. marine plant) and there may be local fluctuations inevitably felt on the plant. Thus the user may well dictate that the plant should be capable of responding to, say, a 10% demand change in a matter of one or two minutes. Of course, the plant must also be protected against more severe changes or outages.

Table 4.5 illustrates possible control margin requirements for a Magnox reactor. It is

TABLE 4.5.
Typical Magnox control margins

Control requirement	Reactivity
Bulk control	
Cold to hot margin	0.015
Xenon poisoning	0.020
Error allowance	0.005
Safety rods	0.010
Operational control	0.010
Long term fuel changes	0.020
Total margin	0.080

seen that the change of state from cold start to hot operating entails a substantial margin of excess reactivity, to be held down in the cold, shutdown state.

This design of thermal reactor has a relatively modest xenon poisoning and no great capacity for xenon override in the case of an unexpected shutdown.

CONTROL METHODS

Thermal reactors are generally controlled by the addition or removal of neutron-absorbing materials since at these neutron energies materials such as cadmium, boron, hafnium, etc., have large capture probabilities. In addition to the control *rod* form (a mechanical insertion of absorber) such poisons can also be introduced in a soluble form if there is a liquid coolant. Boron and its compounds are found suitable and can be both added and subtracted from the coolant by a suitable process plant. Soluble poisons have a big advantage over rod form in being distributed and not causing local flux disturbances which tend to promote localised temperature variations, but the adjustment of level is slow compared to rod movement. Thus it would be usual to supplement them for process control with weak control rods. Variations of coolant circulation and coolant pressure are also useful control media in light water reactors. These again are incapable of providing the necessary localised control to make comparative adjustments to temperature distributions, etc. Some water reactors can vary gross reactivity through the level of a liquid moderator (e.g. SGHWR) or absorber, "liquid control rods".

Fast reactors have a problem in providing satisfactory control, for in the high energy spectrum the conventional resonance absorption cross-sections are small and moderator effects in the coolant are absent. It becomes important in the first instance to provide for a good negative fuel coefficient through the Doppler effect. Attempts have been made to control reactivity by removing fuel (so-called driver rods in which fuel is removed as the control absorber is inserted) or by varying the effectiveness of reflectors. The methods, however, cause mechanical difficulties in providing adequate cooling in a power reactor so that semi-conventional rods are used though with particular attention to their operation in a sodium environment. Fortunately there are smaller demands to control reactivity in a fast reactor in the sense that reactivity varies less in xenon and temperature effects on going to power and over the life of the core than in thermal reactors.

The special needs of adjusting reactivity against fuel burn-up leads in some designs to the use of burnable poisons which can be adjusted approximately to decrease in effectiveness at the same rate as the fuel charge loses reactivity. These are evidently slow-acting devices. Again, since soluble poisons are slow to be adjusted, we shall continue the analysis in terms of the conventional control rod motion.

Starting at the right of Fig. 4.27, the block diagram for control rod effect, it may be

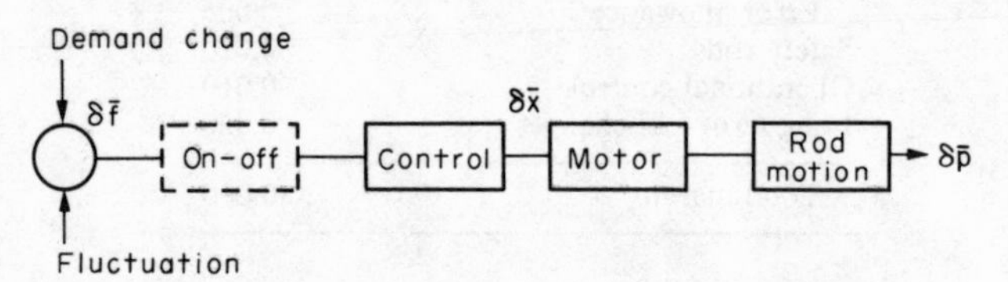

FIG. 4.27. Block diagram for control rod effects.

supposed that the tip of the rod is moved from a low flux to a higher flux region and more neutrons are absorbed, thus increasing the reactivity loss associated with the rod. It is shown in standard reactor physics texts (e.g. ref. 10) when adjoint weighting is allowed for, that to a good approximation the change of reactivity per unit length of travel can be taken as a constant, $-K_c$ reactivity/m, say.

The control rod motor design will vary widely to achieve standards of reliability and convenience of operation in what may be difficult ambient conditions. Very often, however, such a motor can be represented as a second order damped oscillatory system with a transform function.

$$\frac{\partial \bar{\rho}}{\partial \bar{x}} = G_m(p) = \frac{K_m}{p^2 + 2\zeta\omega_n p + \omega_n^2} \tag{4.55}$$

where ζ is the damping factor, ω_n is the natural undamped frequency, K_m is the static gain and x the output signal from the controller (Fig. 4.27).

The natural frequency is probably determined by the choice of motor but the damping factor may be adjustable within a range. The main parameter of adjustment is the gain K_m.

The shaping function or controller may be of two main types—continuous or discontinuous. For the former we attempt corrective action in proportion to a signal derived from observed performance and demand variation. While apparently attractive, this has some defects: (i) delays in the circuit may lead to dynamic instability and "hunting" of the system without achieving the objective, and (ii) the control motor, etc., will be subjected to continuous wear since even the smallest discrepancy will call it into action.

Discontinuous control envisages a band of discrepancy between performance and demand in which no corrective action is taken, with control rod movement switched in when the discrepancy exceeds some set limit. This cuts down wear and may well achieve a better performance.

A common "shaping" circuit for continuous control (which can be adapted to some extent in discontinuous control) is the three-term controller mixing different amounts of integral and differential control with a proportional feedback control. The gain setting of proportional control determines the droop or degree of residual discrepancy against demand variation; it may not be set too high because of dynamic instability. Due to the nature of the neutronics transfer function (with its term in $1/p$) there is an inherent integral performance and an integral term, designed to ultimately eliminate residual error, may be superfluous on the neutron or reactor side of control. A differential term, designed to speed up the response process and anticipate the development of a demand transient is usually desirable. A simple idealised controller might be chosen as

$$G_c(p) = K[1 + \lambda p] \tag{4.56}$$

therefore (and Chapter 9, Analogue Computing, discusses practical realisation of such an element). More complicated circuits to link demand felt on the generator with control of the reactor may be used, now probably based on a numerical model of the plant realised through a digital computer. Discontinuous control evidently introduces non-linearities in which control effect is not proportional to the net demand signal. The *describing function* technique has been developed to allow classical control in the frequency space to be used for systems including such elements.

THE DESCRIBING FUNCTION

The essence of the describing function technique is to subject the element to an oscillating input of known amplitude and to represent the resulting oscillatory output by a Fourier series. Due to the nonlinearity, the output will consist not only of a fundamental term of the frequency of the input but also of a series of harmonics of higher frequencies. The amplitude and phase of the output terms can be expected to depend nonlinearly on the amplitude of the input signal, and in these two respects the nonlinear system differs from the linear system. It is then usually assumed that the remainder of the system attenuates higher frequency signals (low pass filter) so that only the fundamental term in the Fourier series is retained. Naturally the nonlinearity has introduced complications compared to the linear system, but for simple elements the input–amplitude dependent transfer function that results from the above technique is readily handled.

The method can be illustrated with the nonlinear switch as shown in Fig. 4.28 which

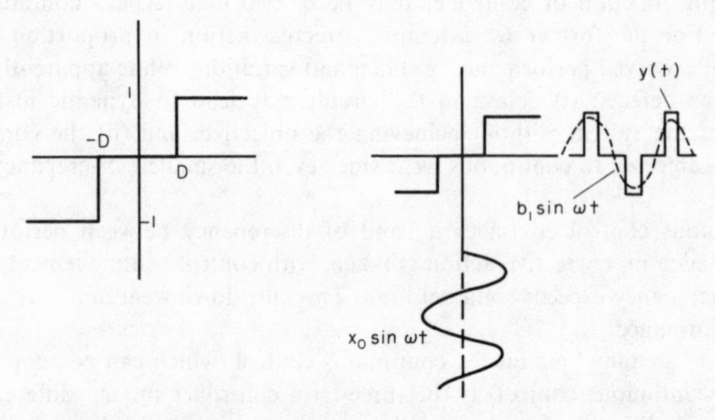

FIG. 4.28. Two-way switch with dead band for describing function.

is designed to move a control rod in or out in response to a demand signal, only when the demand exceeds some dead band D, say.

The output $y(t)$ for an oscillatory input $x(t)=x_0 \sin \omega t$ is readily determined as a function of the ratio x_0/D. The Fourier representation of $y(t)$ is evidently a sine series with no cosine component (e.g. there is zero mean value) so

$$y(t) \simeq \sum_n b_n \sin n\omega t \tag{4.57}$$

and the lowest mode component has an amplitude b_1 obtained from the orthogonal integral

$$b_1 \int_0^\pi \sin^2\omega t \, d\omega t = \int_0^\pi y \sin \omega t \, d\omega t = \int_\theta^\pi \sin \omega t \, d\omega t \tag{4.58}$$

where $\theta = \sin^{-1} D/x_0$. Hence

$$b_1 = \frac{4}{\pi}\sqrt{1-[D/x_0]^2}$$

for $x_0 > D$, and is zero for $x_0 < D$ of course.

We may now represent the phase and amplitude of the input–output relationship though now as a more complicated function depending on the input amplitude x_0. However, in this case (Fig. 4.29) we have the simplification that the phase is always zero and the amplitude, though dependent on x_0/D, is independent of frequency ω.

Note that the gain is zero until such time as the input signal exceeds the deadband; it rises quickly to a peak and then falls off with increasing input signal since the output signal is limited to amplitude one.

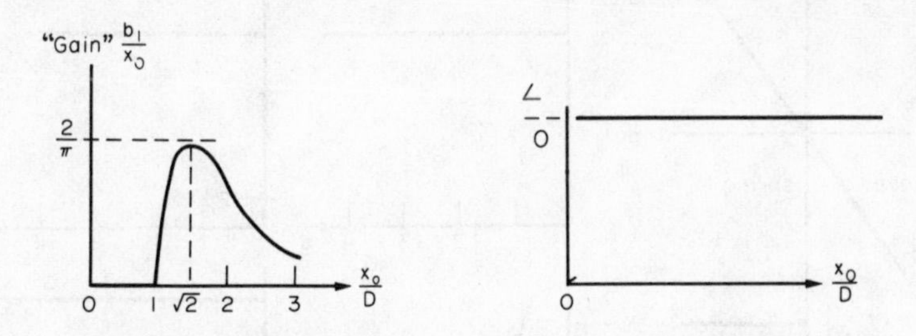

FIG. 4.29. Describing function for two-way switch with deadband D.

Figure 4.30 sketches similar results for a range of nonlinear elements. It may be noted in this figure that only in the case of hysteresis is the phase of the describing function other than zero. In other cases, therefore, the introduction of such elements in a loop permits conventional Nyquist criteria to be used with a simple rescaling of the diagram against the nonlinear element gain. It might also be noted in passing that the idea of a describing function may be employed to develop more accurate results for the *neutronic* output due to a sinusoidal variation of input reactivity than the linearised model provides for.

Load Dynamics

TURBINE CONTROL

We mention briefly a typical conventional controller for a turbine-generator as it might be specified for a power station interconnected to a national grid or utility system. The system output (Fig. 4.31) is taken to be the frequency f of the electricity supplied which is compared in the comparator with a set demand frequency f_0 to form the control or error signal δf. The error is used in a feedback loop to determine the throttle opening. The nature of the feedback controller is indicated in the diagram as being:

(a) a saturating proportional element to give an initial action proportional to δf;
(b) an integral controller to increase the action as long as the error signal persists;
(c) a saturating element to limit the control to the maximum throttle changes (i.e. open and closed) designed to relate to the maximum droop in the frequency f_1 to f_2 about the set frequency f_0 (48–52 Hz in the UK system).

Evidently the two nonlinear elements will require the describing function treatment.

In view of the high speed of steam expanding through the turbine we may assume that this reacts instantaneously compared with some of the other phenomena such as the boiler transients and neglect in this book a detailed analysis. In any case, the transients are likely to be comparable in time scale with the mechanisms of the turbine governor so that reference should be made to specialist material if this is considered important.

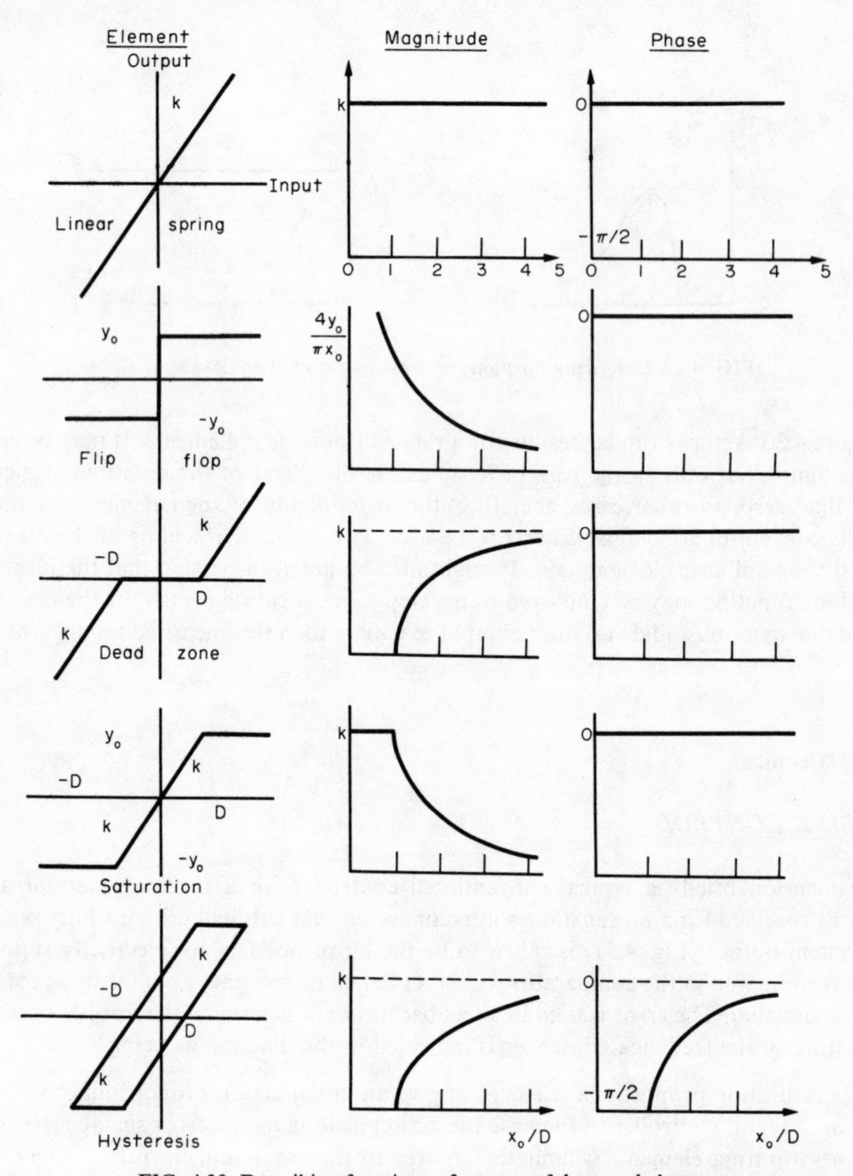

FIG. 4.30. Describing functions of some useful control elements.

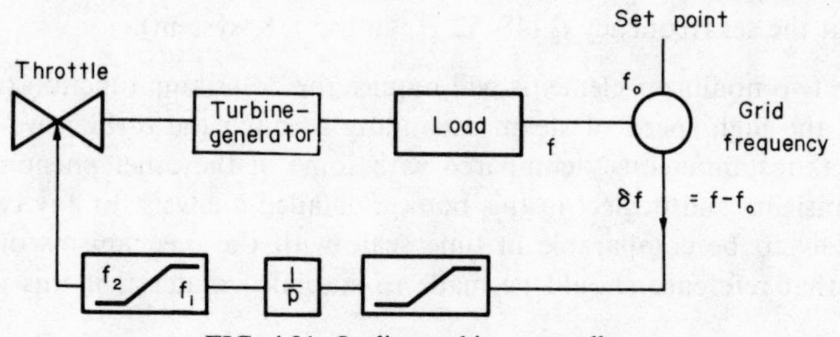

FIG. 4.31. Outline turbine controller.

ELECTRICAL LOAD DYNAMICS

For applications to electrical generators, typically the reactor produces steam for a turbine–generator set. The generator supplies the electrical load.

Since there is no effective storage of electrical energy as such, difference between electrical demand and supply—caused by load variations—is felt as a change of speed of the turbine-generator.

The energy stored in a rotary system may be expressed as $\frac{1}{2}I\Omega^2$, where I is the amount of inertia and Ω the rotational speed. Thus a small difference δP in supply and demand leads to a change of speeds $\delta\Omega$ given by a "linearisation" as

$$\delta P = \frac{dE}{dt}\,\delta t = \frac{d}{dt}\,[I\Omega\delta\Omega] = I\Omega\,\frac{d}{dt}\,\delta\Omega \tag{4.59}$$

It is convenient to consider not the shaft speed Ω but the system frequency f, e.g. 50 Hz in the UK/European system and 60 Hz in the USA, the connection being $2\pi n$, where n is the number of pairs of electrical poles in the system generating the frequency f. (Thus f may be common throughout an electrical grid but the shaft speed in a particular station depends, *inter alia*, on its number of pairs of poles.) Making this substitution and taking Laplace transforms,

$$\frac{\delta \bar{f}}{\delta \bar{P}} = \frac{K_f}{p} \tag{4.60}$$

where $K_f = 1/2\pi nI\Omega$ is an (approximate) constant for the generating set. The change in frequency is then a suitable source parameter for control action.

Of course the power input P in this analysis is mechanical shaft power and is related to the thermal power delivered by the steam to the turbine via the thermal efficiency of the turbine. If we may assume this to be constant, somewhere between 8% for older nuclear reactors and 40% for the most recent gas cooled reactors, it becomes possible to relate the thermal and mechanical power. More detailed models allow for changes of thermodynamic efficiency η with steam conditions, variations in feed water, etc.

Conclusion

In this chapter we have developed the process dynamics of a reactor and its associated plant together with elements of its control system. Despite the relatively simple treatment, these elements should be sufficient for synthesis in a design study or should illustrate the techniques by which equations may be set up to model important phenomena, and from these equations transforms and transfer functions can be derived suitable for small variations and the application of classical linear or frequency space analysis.

By talking in terms of the classical frequency space analysis we are able to make deductions about system stability directly. Where we go on to synthesise a system with adequate transient response, we may use the standard devices of gain and phase margin. Alternatively, if the frequency components are realised by using an analogue computer, as discussed in Chapter 9 (especially Table 9.1) then, of course, the (linearised) transient response is directly available.

Material on the frequency space analysis of reactor systems, together with a description of typical power plants, is continued in Chapter 5.

Problems for Chapter 4

NEUTRONICS TRANSFER FUNCTIONS

4.1. If delayed neutrons are neglected, the equation

$$\frac{dn}{dt} = \frac{\rho}{\Lambda} n; \quad n(0) = n_0$$

admits an *exact* solution to the case with $\rho = \epsilon \sin \omega t$.

Find the response. Show that in this model the time-averaged $n(t)$ is zero, i.e. constant mean power.

4.2. When there is a spread of time constants, i.e. more than one (coupled) kinetic equation, the result of problem 4.1 is no longer valid. An approximate solution for a sinusoidal and other reactivity variations can be developed consistently and represented in the form of a describing function using a series expansion[16]. Suppose $\rho(t) = \epsilon \sin \omega t$ with ϵ a "strength" factor. In a one-group model, find the equations satisfied by $m(t)$ and $g(t)$ if we assume

$$n(t) = \exp \int_0^t m(t')dt'; \quad c(t) = g(t)n(t)$$

Now assume the following series whose leading terms are constants:

$$m(t) = m_0 + \epsilon m_1(t) + \epsilon^2 m_2(t) + \ldots$$
$$g(t) = g_0 + \epsilon g_1(t) + \epsilon^2 g_2(t) + \ldots$$

The series are assumed uniformly convergent. Find the solutions m_0 and g_0 valid for $\epsilon \to 0$. Find $m_1(t)$, g_1 by substitution in the equations for m and g and neglect of terms in ϵ^2 or higher. Show by expansion of the exponential form of solution that leading terms can be considered equivalent to the transfer function solution for $\rho = \epsilon \sin \omega t$.

4.3. Construct a transfer function for natural uranium from the data of Chapter 2 for effective reproduction times $\Lambda = 10^{-4}$, 10^{-5}, 10^{-6} s (use of a digital computer is assumed). Using the break frequency method, fit a two-group precursor model to the data graphically in the range of 0.1 to 100 radians/s and determine suitable values of λ_1, λ_2, p_1 and p_2. Find the frequency at which the phase difference $< G_n(j\omega)$ is a minimum and show in a one-model this is approximately $\sqrt{\beta\lambda/\Lambda}$.

TEMPERATURE FEEDBACK MODELS

4.4. Justify the construction of Fig. 4.10 for the values

$$\beta/\Lambda = 10^2 \text{ s}^{-1}; \quad \lambda = 0.1 \text{ s}^{-1}; \quad \tau = 100 \text{ s}$$

Find the equivalent result if $\tau = 1$ s.

Sketch an equivalent to Fig. 4.10 in a two-precursor group model.

4.5. Consider the neutronics transform function $n_0 G_n(p)$, eqn. (4.6), in circumstances where a "perfect" proportional controller with negative feedback was available (see diagram) having a gain K. Show for

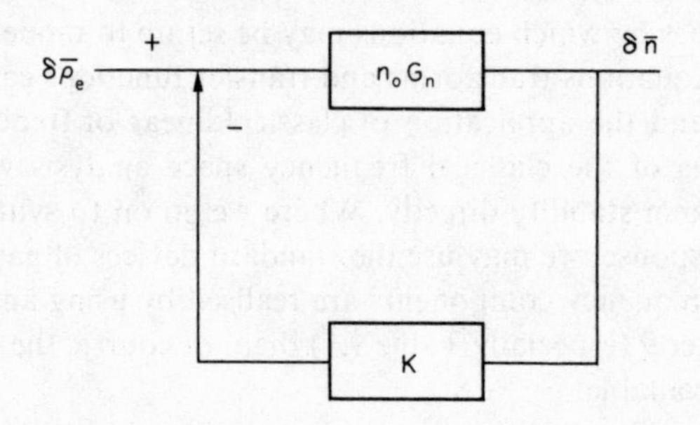

the Nyquist diagram that no dynamic instability is anticipated at any gain setting. Sketch a root locus in one- or six-group approximations.

If the controller loop incorporated (a) a simple lag, or (b) a quadratic lag, would you modify the first conclusion?

4.6. *Two coefficient model.* Suppose (artificially) that the fuel time constant τ_f goes to zero in eqns. (4.22) et seq. Show that the dynamic stability condition reduces to

$$\frac{1}{\tau_m} - \frac{\alpha_f \Delta T_f^*}{\beta} \lambda < 0$$

Consider the static condition $H(0) < 0$; would this now admit $\alpha_f < 0$? Show that if further $\lambda \to 0$, the only condition to be met is $H(0) < 0$. Can this model now admit oscillatory solutions? Sketch Nyquist diagrams for a range of two-temperature feedback functions $H(p)$ combined with two- and six-group precursor models for $G_n(p)$.

4.7. Sketch root locus diagrams for the two-temperature coefficient models from eqn. (4.29) on the basis of fixed $\alpha_f < 0$ and using α_m as the parameter.

4.8. Investigate the stability of a boiling water/steam generating heavy water reactor under the following assumptions:

(i) the steam pressure is maintained constant by throttle control;

(ii) use an effective lifetime model in the form

$$\frac{d}{dt}\delta n = \frac{\rho}{\Lambda} n_0; \quad \frac{\delta n}{n_0} = \frac{\delta P}{P}$$

(iii) take account of voidage and fuel temperature on reactivity

$$\rho = \alpha_f \delta T_f + \alpha_v \delta V$$

(iv) represent the fuel temperature delay as

$$\tau_f \frac{d}{dt}\delta T_f = \Delta T_f^* \frac{\delta P}{P} - \delta T_f$$

(v) represent the bubble growth leading to voids as

$$\tau_v \frac{d}{dt}\delta v = K\frac{\delta P}{P} - \delta V$$

with τ_v a time constant for bubble growth and $\kappa > 0$, an effectiveness factor dependent on the power distribution along the coolant channel. Hence show that if $\alpha_f < 0$, stability requires

$$\alpha_v < -\frac{\alpha_f T_f^*}{KV_0}$$

thus indicating that a strong negative fuel coefficient admits a relaxation of the negative void coefficient requirements for stability.

4.9. A LWR at 6 MPa (60 bar) pressure can be operated in *either* a boiling region at saturated temperature *or* non-boiling, sub-cooled at 220°C. Taking typical values of void and temperature effect to be

$$\alpha_v \equiv \left(\frac{\partial \rho}{\partial V}\right)_T = -12 \text{ kg/m}^3$$

$$\alpha_T \equiv \left(\frac{\partial \rho}{\partial T}\right)_V = -1.5\ 10^{-4} \text{ per } K$$

(and note V now refers to specific volume, m³/kg), determine the two contributions to reactivity change in either region due to an enthalpy increase of 30 kJ/kg. Thus show the effect in the boiling region is substantially greater than in the non-boiling region, with corresponding implications for stability in BWRs. Compare the contribution of temperature and volume effects in the non-boiling region.

PLANT DYNAMICS

4.10. In an experiment to determine the time constant of a 500 MW(thl) Magnox reactor, the reactor is shut down and the heat exchanger isolated (no secondary flow). The temperature is raised from cold by circulating the CO_2 employing 50 MW(e) circulating pumps. The stagnant temperature was observed to increase at 3 K/s. What value do you ascribe to τ?

4.11. A PWR has a pressuriser, a vessel part filled with steam connected to the reactor core to take up pressure transients, to establish the working pressure and to allow for expansion at high temperatures

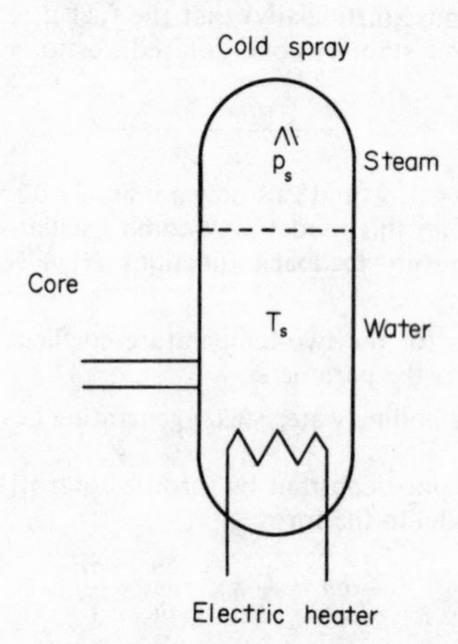

(thus developing the desirable negative void coefficient). A simplified pressuriser is shown.

Keeping in mind the dependence of pressure p_s, temperature T_s *and* density (ρ), develop an expression for

$$G_p(j\omega) = \frac{\delta \bar{p}_t}{\delta \bar{T}_s}$$

BOILER TRANSFER FUNCTION

4.12. Derive a primary coolant transfer function through the core along the lines of the model used for the boiler transfer function. Sketch Bode diagrams, using a suitable data such as given in Chapter 5.

4.13. Extend the boiler transfer function to allow for:

(a) variations of the primary coolant flow rate;
(b) a two-section energy balance, appropriate perhaps for a once-through boiler, in which masses are ascribed to the primary as well as the secondary side and time delays arise for temperature changes in both sections.

4.14. The boiler transfer function has two imputs ($\delta \dot{T}_{bi}$ and $\delta A/A$) and two outputs (δT_{bo} and either δF_s or δT_s, etc.). Derive a time constant without the final correction term by considering a boiler with no primary loop but an instantaneous power source from an electric heater, say. Express the original transfer function as a 2×2 matrix; does the result lie immediately in the form established in problem 1.10? Can you modify the result to take this form and so express the transfer function in the form of a loop with two elements in it? Draw a block diagram.

4.15. For what range of parameters is τ^- of the boiler transfer function negative? Is this reasonable? Draw the Bode diagram. Is this range physically reasonable? Using the data of the sample reactor, Chapter 5, evaluate K at full and part loads. To what extent is K sensitive to the change of the slope p_s'/p_s? How will you estimate the power of the reactor in the example (not given directly in Table 5.3)? Prove the claim that $K \to 1$ for low power operation. Sketch Nyquist diagrams for $\tau^- < 0$ and $\tau^- > 0$. Show that

$$\tau^+/\tau^- = [T_{bi} - T_s][T_{bo} - T_s + p_s/p_s']/[T_{bo} - T_s][T_{bi} - T_s + p_s/p_s']$$

4.16. Derive the primary coolant transfer function $(\partial T_{io}/\partial T_{ci})_{P,F}$ for the reactor side of a PWR. State the assumptions inherent in the equations

$$T_c = \tfrac{1}{2}[T_{co} + T_{ci}] \quad \text{mean coolant temperature}$$

$$c_f m_f \frac{dT_f}{dt} = P - h[T_f - T_c] \quad \text{fuel energy balance}$$

$$cm_c \frac{dT_c}{dt} = h[T_f - T_c] - cF[T_{co} - T_{ci}] \quad \text{coolant energy balance}$$

and show that the result has the form

$$\left(\frac{\partial \bar{T}_{co}}{\partial \bar{T}_{ci}}\right)_{\rho, F} = \frac{2\dfrac{T_f - T_c}{T_{co} - T_{ci}} - 1 + p\tau_c + \dfrac{1}{1 + p\tau_f}}{2\dfrac{T_f - T_c}{T_{co} - T_{ci}} + 1 + p\tau_c - \dfrac{1}{1 + p\tau_f}}$$

where τ_i and τ_f are suitably defined time constants.

4.17. Develop the transfer functions of a primary loop allowing for:
(a) boiler transfer function of one simple lag;
(b) two exponential transport delay terms for connecting piping;
(c) reactor coolant transfer function based on one fuel mass *and* mass of primary coolant in channel. Take τ_f=0.1 s; τ_{pc}=1 s; τ^-=10 s; τ^+=15 s; $\theta_1=\theta_2$=1.5 s. Is it possible for this loop to be unstable if K_b=0.5 for any reactivity coefficient $\alpha<0$? (*Hint*: Plot a Nyquist diagram.)

XENON OSCILLATIONS

4.18. The equations including a direct yield of xenon from fission become

$$\frac{dI}{dt} = \gamma_i \Sigma^f \phi - \lambda_i I$$

$$\frac{dX}{dt} = \lambda_i I + \gamma_x \Sigma^f \phi - \lambda_x X - \phi \sigma X$$

Find the new form of the transfer functions and show that it takes the form

$$\frac{p - z}{[p - p_1][p - p_2]}$$

Show that z may be positive or negative dependent on the reference value $\sigma\phi_0$. What effect does the low value of $\sigma\phi_0$ have on the phase of the transfer function, and hence show that for fluxes below about 3×10^{15} neutrons/m² s there is no onset of gross xenon oscillation? What frequency do oscillations start at as $\sigma\phi_0$ increases? Sketch the transfer function by the break frequency method (if a digital computer is available, draw exact transfer function) and Nyquist diagrams for fluxes 10^{12} through 10^{18} neutrons/m² s in decade intervals. Carry out a root locus analysis with direct yield included and generalise Fig. 4.25.

CONTROL ELEMENTS

4.19. Derive the describing functions of Fig. 4.30.

4.20. *Soluble poison dynamics.* Suppose that a soluble boron compound can be removed from a primary system via a bypass process device at a rate proportional to the difference in demand and actual boron concentrations, assume immediate mixing of the bypass flow on return to the primary and thus show that the transfer function for the process is

$$\frac{\delta \bar{B}}{\delta \bar{B}_0} = \frac{1}{1 + p\tau}$$

where B is the actual concentration, B_0 the demand concentration and $\tau = V/a$, where V is the system volume and a the proportionality constant.

4.21. *Separable fluxes and spatial control.*† A reactor is designed as a uniform right circular cylinder, height H and radius R. Show that at low power a one-group diffusion model leads to fully separable solutions of the form

$$N_m = n_m(t) \cos\left((1 + 2u)\frac{\pi z}{H}\right) \cos(v\theta + \psi) J_0\left(\alpha \frac{r}{\omega R}\right)$$

†May be omitted on a first reading.

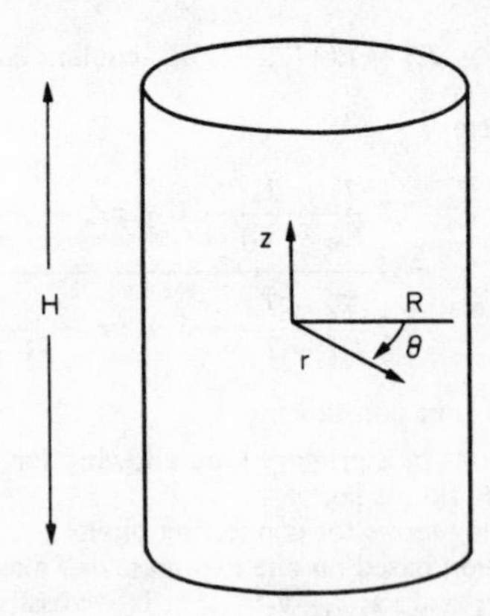

where α_w denotes the zeros of the Bessel function J_0 of zero order, first kind and ψ is an arbitrary polar phase, with $m, u, v, w=0, 1, 2, \cdots$ (m rises to 7 for all combinations to $u=v=w=1$).

Why is the fundamental mode with $m=u=v=w=0$ the independently realisable mode? Where are the zeros of the modes $m=1$ to 7? Where are the *largest* values of these modes (i.e. the antinodes)? Where, then, will you place axial control rods to suppress xenon oscillations in these modes allowing for (a) axial modes, and (b) the arbitrary phase? Sketch the first eight modes.

Show that the transfer function $G_n(j\omega)$ in higher modes becomes

$$G_n(j\omega) = \frac{n_0}{\Lambda f(j\omega)} = \frac{n_0}{\Lambda p - \rho_m + \sum p\beta_i/[p + \lambda_i]}$$

and that eqn. (4.18) generalises to

$$\mathop{\mathscr{L}}_{\omega \to 0} G(j\omega) = -1/\left[\frac{1}{2}\frac{\alpha_T \kappa}{cF_0} + \frac{\rho_m \beta}{n_0 \lambda \Lambda}\right]$$

If $H=2R=10$ m, $\eta=\nu\Sigma^f/\Sigma=1.9$ and $\Lambda=10^{-4}$ s, find the reactivity in the modes 100, 010 and 001 if the lowest mode reactivity (000) is zero, at a flux $\phi=10^{16}$ neutrons/m² s. (*Hint*: You will need to know the Laplacian operator ∇^2 in this geometry and to evaluate the additional leakage probability in the higher modes.)

If there is no temperature coefficient of reactivity, say whether you expect xenon instability in these modes and, if so, whether the divergence is oscillatory.

4.22. *Xenon spatial oscillation* (after Henry[18]).† Consider a reactor described as two regions in one-dimensional one-group diffusion theory, each region thickness H separated by a pure moderating regime of characteristic thickness L. If j_1, j_2 are the external currents of neutrons directed from region 1 to 2 and vice versa, then in steady state (and a prompt jump approximation neglecting delayed neutrons),

$$j_1 - D\frac{\phi_1 - \phi_2}{L} + [\Sigma_1 + X_1\sigma - \nu\Sigma_1^f]\phi_1 H = 0$$

$$j_2 - D\frac{\phi_2 - \phi_1}{L} + [\Sigma_2 + X_2\sigma - \nu\Sigma_2^f]\phi_2 H = 0$$

with D the diffusion coefficient, ϕ_1 and ϕ_2 the region fluxes and X_1 and X_2 the region xenon concentrations (from an integration over the volumes).

If the properties of the regions are the same, other than flux and xenon values, and if the over-all power level is constant and $j_1/\phi_1=j_2/\phi_2$. show that

$$x \equiv \frac{X_1 - X_2}{X_1 + X_2} = -\frac{D}{(X_1 + X_2)HL}\frac{f}{1 - f^2}, \quad \text{where } f \equiv \frac{\phi_1 - \phi_2}{\phi_1 + \phi_2}$$

Writing average value $\phi=\frac{1}{2}[\phi_1+\phi_2]$ [constant], show that

†May be omitted on a first reading.

$$\frac{d^2x}{dt^2} + \left[\lambda_i + \lambda_x + \sigma\phi - \Sigma^f \frac{LH}{2D} \frac{\sigma\phi}{\lambda_x + \sigma\phi}\right] \frac{dx}{dt} + \lambda_i \left[\lambda_x + \sigma\phi \left[1 + \Sigma^f \frac{LH}{2D} \frac{\lambda_x}{\lambda_x + \phi}\right]\right] x = 0$$

on making the linear approximation, that f^2 is small compared to 1.

Hence show that x may tend to oscillate and find the period on the verge of instability, comparing to the corresponding single region period also neglecting direct yield. This conceptual model can be developed into *nodal* approximations for the spatial behaviour using corresponding difference approximations to the diffusion model.

References for Chapter 4

1. D. P. Campbell, *Process Dynamics*, Wiley, New York, 1958.
2. M. Schultz, *Control of Nuclear Reactors and Power Plants* (2nd edn.), McGraw-Hill, New York, 1961.
3. *Naval Reactors Physics Handbook*, **1**, USAEC, 1964 (including a useful review of boiling stability analysis methods).
4. *Nuclear Power Plant Control and Instrumentation, IAEA, Vienna*, 1972 Conference, (1973).
5. L. J. Templin (ed.), *Reactor Physics Constants* (2nd edn.), UASEC, 1963.
6. R. E. Skinner and D. L. Hetrick, The transfer function of a water boiler reaction, *Nucl. Sci. and Engr*. **3**, 573 (1958).
7. D. C. Hetrick, *Dynamics of Nuclear Reactors*, Chicago, 1971.
8. H. Etherington (ed.), *Nuclear Engineering Handbook*, McGraw-Hill, 1958 (especially section 8).
9. J. M. Harrer, R. E. Boyer and D. Krucoff, *Measurement of CP2 Reactor Transfer Function*, ANL-4373, 1952.
10. A. M. Weinberg and E. P. Wigner, *The Physical Theory of Neutron Chain Reactors*, University of Chicago Press, Chicago, 1958.
11. C. Sastré, Reactor transfer functions, *Adv. in Nucl. Sci. and Tech.* **2**, 1 (1964).
12. G. T. Lewis jr., M. Zizza and P. de Rienzo, Heat exchanges in nuclear power plants, *Adv. in Nucl. Sci. and Tech.* **2**, 41 (1964).
13. G. M. Roy and E. S. Beckford, Performance characteristics of large boiling water reactors, *Adv. in Nucl. Sci. and Tech.* **1**, 179 (1962).
14. *Boiler Dynamics and Control in Nuclear Power Stations*, BNES Conference London, 1972.
15. A. Pollard, *Process Control*, Heinemann, London, 1971.
16. Z. Akcasu, General solution of the reactor kinetic equations, *Nucl. Sc. and Engn.* **3**, 456–67 (1958).
17. J. C. Tyror and R. I. Vaughan, An introduction to the neutron kinetics of nuclear power reactors, Pergamon, Oxford, 1970.
18. A. Henry, *Nuclear-Reactor Analysis*, MIT, Mass., 1975.
19. T. W. Kerlin, *Frequency Response Testing in Nuclear Reactors*, Academic, New York, 1974.

CHAPTER 5

Power Reactor Control Systems

Introduction

In this chapter types of power reactors and their associated plant are discussed from the view point of their control systems, selecting principally those reactors that are in current commercial use for the production of land-based electricity. A reactor will have four main control systems including a protection system to initiate safety devices, a radiation monitoring system and a process plant system for secondary items such as deionising make-up water. What we concern ourselves with is, how-ever, the over-all operating control system whose purpose is the automatic governing of the plant in its primary aim—to meet demands varying between, say, 20–100% of full power.

We omit detailed discussion of operation at very low power, which is essentially the startup/shutdown regime where control may be manual or automatic but certainly very different from general running. This is not to say that some manual or operator action may not be incorporated with changes of power setpoint in the power range. Also omitted is the treatment of slower-acting control devices such as offset xenon effects, since there is time indeed to use manual control if desired. The focus is on the time scale between seconds and minutes where control must be always available, stable and flexible to react to varying demands in a consistent and constrained way.

Control of the reactor and its plant such as turbines and coolant circulators, depends significantly on the reactor type: light water, gas cooled, etc. Nevertheless, there are some general considerations for reactor control that are worth establishing before being illustrated in the context of specific systems. These general considerations cover not only an understanding of the purpose of power reactors, i.e. the nature of the load to which they are being applied, but to aspects that may be called:

(a) part load specification;
(b) load coupling and decoupling;
(c) load following;
(d) system stability.

LOAD DEMAND

Power reactor applications to produce work are diverse; from an extreme, say, of a nuclear marine propulsion plant as an independent unit to a closely integrated central electricity supply (utility). For the former example, load demand arises essentially in the mind of the captain, and control will be required to maintain a steady output at the shaft as wanted and to vary this on demand from the bridge.

The closely integrated utility is a more complicated case. Electrical load commonly shows a number of variations in which only general trends can be anticipated; the trends are overlayed with fluctuations. General trends include the annual cycle—peak loading in winter for heat, light and power as well as (particularly in the United States) a second peak for air conditioning in the summer—and the diurnal cycle—day and evening peaks affected by the time of year, with lows in the early hours of the morning. The size of interconnected utilities make international time zones a part of the demand structure. Random fluctuations can be caused, for example, by the startup or shutdown of an electric train, about 1 MW, or an advertising break on television (several million electric kettles). There must therefore be a facility for varying the supply to meet sudden changes of demand. Figure 5.1 illustrates the experience of the UK Central Electricity Generating Board (CEGB). The steepness of the rise or fall in demand in a day should be noted

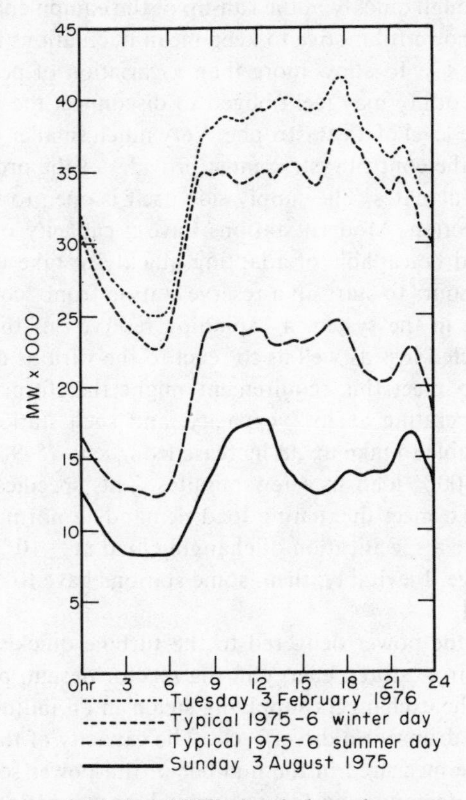

FIG. 5.1. CEGB electrical supply.

together with the requirement to have installed capacity more than five times the minimum demand.

There are two further significant characteristics of an electrical grid. First, electricity cannot effectively be stored. If, therefore, variations of demand create an imbalance between supply and demand, the difference is taken up by a change in the stored energy of the turbine-generators and any other synchronous motors, and hence in shaft speed as reflected in a change of the frequency of the supply system. Secondly, the strong electrical interconnection of the coupled motors and generators means that for our purposes all supply units act with a common inertia and common frequency change in this tightly synchronised situation.

There are many pressing reasons for keeping a constant supply frequency. Synchronous motors are used for clocks, domestically and industrially. Frequency deviation leads to a steadily increasing clock error which may be inconvenient or even dangerous. Electricity meters, on which the utility revenue depends, are thrown out of calibration by departures from standard frequencies. Other motors lose efficiency and therefore create economic problems. Overspeeding of the turbine-generator has obvious safety as well as economic implications in view of the damage that can be done and subsequent loss of availability from a blade torn loose; however, underspeeding may also be dangerous in bringing shaft speeds close to resonant frequencies below normal running speed that are usually passed through quickly in the run-up of the equipment to synchronous speed. There is, therefore, a powerful motive to keep mean fluctuations in the supply frequency down and in no case, say, to allow more than a variation of perhaps 2%. At that frequency deviation, the utility may feel obliged to disconnect the entire load with results that may approach the level of a catastrophe. Very much smaller mean square deviations are sought for which the control system must provide, of the order of 0.1%.

Not only demand fluctuates. The supply side itself is open to failures of a generating unit or an interconnection. Modern stations have a capacity of over 1000 MW(e) so that the system should be capable of adapting quickly to take up the loss of one such unit. It takes several hours to start up a reserve station from "cold" and therefore there is a necessity to have in the system a "spinning reserve" of this capacity, capable of taking up the unexpected loss as well as to react to the various demand fluctuations.

Spinning reserve to meet this requirement might therefore take the form of four 1000 MW stations operating at 75% capacity, and such stations are specified in the design stage as being able to take up an increase from, say, 75–90% load within 2 or 3 s, with the balance to 100% load in a few minutes. This specification meets the loss of supply requirement. To meet fluctuating load demands a normal station in the supply system is likely to have a specification of changing load at ±10% a minute over at least 50% of its power range. Inevitably, then, some stations have to be run at an apparently unecomonic part load.

An ability to vary the power delivered to the turbine quickly is bound up with the existence of a reservoir of stored energy in the reactor design, particularly therefore in the heat capacity of the exchangers supplying steam in an indirect cycle or the coolant in the core or storage drum for a direct cycle. The capacity of the reactor plant to meet such a demand can be measured in the number of full-power seconds of energy stored available to meet this demand, and this in turn will be a function of the heat exchanger or boiler design. Larger capacity is purchased at a cost. There will be a trade-off point between flexibility and capital cost; the control engineer may well find himself having to argue the dynamic merits against the more static viewpoint of the reactor physicist.

We shall not go into the scheduling of power stations to meet demand changes other than to state the obvious: that stations are brought into use according to the marginal cheapness of the electricity they supply. Nuclear stations are generally high capital/low fuel cost units and therefore are used for base load. Peak demand might be met by cheaper or older, lower efficiency, fuel expensive and likely fossil-fuelled units. However, the amount of nuclear electricity now provided in many utility systems is such as to exceed the minimum base load (summer night low) and therefore of necessity, nuclear stations must be built to provide some degree of flexibility. (The UK installed nuclear capacity for the CEGB 1975/6 was 6% sent out but the proportion of nuclear electricity generated in the year was 10%.)

General Control Considerations

CONTROL ACTION

Figure 5.2 is a general reactor plant schematic to illustrate some general features of any control system, particularly places where information or signals may be obtained

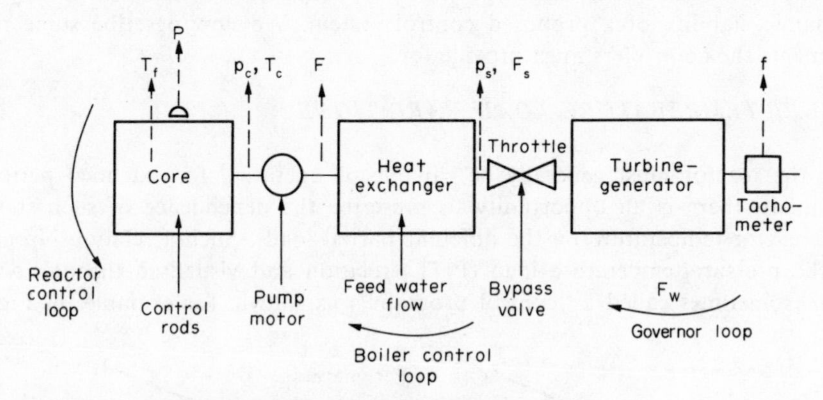

FIG. 5.2. A reactor power plant control schematic.

and places where control action might be applied; both are equally significant in a control system.

Principle signals available on which to base automatic control are:

(a) reactor power P: from neutron or flux measurements (less conveniently from combined temperature increments and coolant flow measurements);
(b) pressure p_c and temperature T_c measurements on the primary coolant at core exits and boiler entrance/exit; primary flow rates;
(c) turbine-generator shaft speed, f, perhaps measured or at least expressed as generated frequency;
(d) secondary pressure, particularly at the boiler exit p_s and at the high and intermediate entrances to the turbines; secondary water level in boiler.

Principle points for the application of control signals will be:

(a) control rods (or similar) to vary reactivity and hence power;
(b) primary coolant circulation, i.e. pump speed, to vary coolant temperature and perhaps indirectly reactivity;

(c) secondary steam flow rate via turbine throttle opening to either high or intermediate pressure stages and bypass valve dumping steam flow; boiler feed water flow (secondary pump) F_W.

Controllers are provided to take suitable measured control signals and compare them with set signals, to shape and amplify these difference signals and then to apply them to the selected points of application. These controllers in many cases are conventional three-term analogue process controllers which in the time scale we are considering readily provide proportional, derivative and integral term control, with parameters adjusted to obtain the desired dynamic response in respect of some criterion of efficiency but having to meet the various restrictive limits to temperatures and pressures in the system. Alternatively, direct digital control (DDC) may be instituted with a computer modelling the behaviour of the system and calculating desired control signals again against criteria of efficiency and state restraints. It should not be forgotten, however, that there is usually an appreciable delay in measuring a control signal (represented perhaps by a suitable simple lag in the controller response function) which may affect the dynamic stability of a proposed control system. We now describe some of the requirements the controllers must provide for.

PRESSURE/TEMPERATURE–LOAD VARIATIONS

Since the reactor must generally be capable of operating for extended periods at partial loads, there is an opportunity to prescribe the dependence of such states as average reactor temperature on the different partial loads. Such a relationship can be called the pressure/temperature–load (P/TL) relation and visualised through a P/TL diagram (sometimes called a "control program") as shown, for example, in Fig. 5.3.

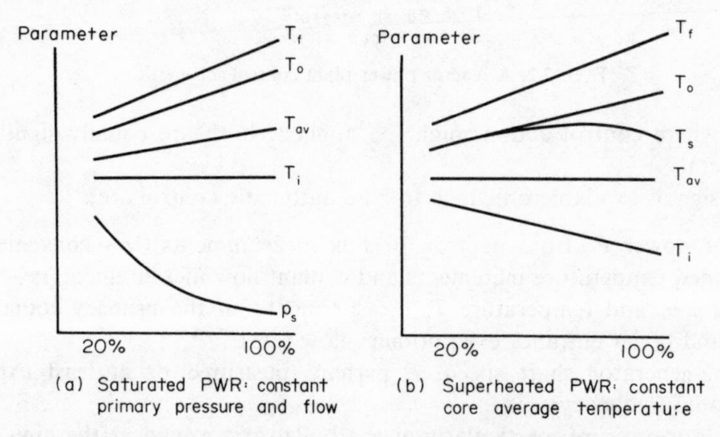

FIG. 5.3. Sample reactor pressure/temperature–load (P/TL) diagrams. T_f, fuel; T_o, T_i, coolant out and in; T_{av}, core; p_s, turbine inlet pressure; T_s, turbine inlet temperature.

It must be emphasised that a P/TL diagram illustrates only a steady state relation and is only a preliminary part, therefore, of the dynamic control design.

Since coolant flow rates and reactivity, etc., can be adjusted, there is some degree of choice available, and the P/TL diagram illustrates the choice made, in this case for a pressurised water reactor (PWR). The first example is based on a choice of constant primary pressure; the pressure vessel is expensive and it seems uneconomic to run the reactor at part load at lower pressure than designed for, and it is certainly convenient not to have to vary the primary coolant circulation rate. In the second example, the core average temperature is to be kept constant as well as the coolant flow rate.

The diagrams are not continued to zero load since these would really be meaningful only for run-up and run-down, and do not therefore represent a steady state situation.

LOAD COUPLING

In the context of a utility supply a reactor may be set to provide some power within its P/TL range and synchronised to the grid. If the control system provides for this set power to be maintained irrespective of the demands of the grid, then the reactor is designated as uncoupled. We have seen the necessity, however, of coupling the reactor to the grid so that its power may vary in accordance with the system demands. This may be achieved automatically by making use of the grid frequency variation as a signal to the controller, while noting the possibility of such feedback action promoting system instability. The grid frequency variation could be applied to the turbine throttle as the most immediate and therefore fastest-acting control point but equally (or in combination) be applied to varying the reactor power, hence the steam pressure and using this to vary the power delivered to the turbine. Coupling and uncoupling,† therefore, are not synonymous with a simple description of which "end" the plant is driven from.

Even in a decoupled system there will be safety reasons for monitoring the shaft speed and taking corrective action on the throttle to prevent a runaway, leaving the shear bolts as a final line of defence: so coupled/uncoupled is perhaps a relative term.

LOAD FOLLOWING

In general terms the provision of a negative coefficient of reactivity in a reactor design suggests that the reactor may be run at different power levels. Reactivity is added by the withdrawal of a control rod, the fission rate and power rise and in some general way the mean temperature increases until the reactivity is offset back to zero and a new steady state is feasible. However, this is only a steady state description and begs the question of whether a reactor will of its own nature respond to an increased demand as evidenced by an opening of the turbine throttle and move to a new, stable power. If it does, the reactor is said to be inherently load-following. If not, further control action is required.

Of course, even when a reactor is load-following, the transient behaviour may not be entirely satisfactory and so some supplement of reactivity control may be necessary. But a load-following plant has the attraction of decreasing the onus put upon the control system and eliminates unnecessary sophistication. It is therefore an attractive feature of the design and is seen to be complementary to the ability to meet a step or ramp demand transient. If the reactor is designated decoupled there will be a lessened emphasis on whether it is naturally load-following.

†Some writers use the terms "load-following/driving" for "coupled/uncoupled", but load following is generally understood in the sense of the next section.

SYSTEM STABILITY

Suppose that the reactor and its control system appear to meet all the design specifications for varying load, etc., within the materials restraints on temperatures and pressures. The isolated unit is stable as perhaps evidence by Nyquist diagrams evaluated at various power levels through a "linearisation" process. But when the plant is connected to the system, is the system stable?

An unstable system would be a catastrophe, but there are many combinations of supply units and possible demand to consider. It is helpful at least to have an empirical criterion on which one might expect a unit to contribute rather than detract from system stability. A useful statement therefore is to require each unit to be unconditionally and adequately stable when disconnected from the grid but running at synchronous speed under its own inertial load. (That is, decreasing loop gain will not decrease the already adequate margin of stability.)

CONTROL PHILOSOPHY

Amongst all these restraints to be reconciled one needs some simple guideline to plan the control system. It is perhaps best to approach the over-all design (or philosophy) on the basis of identifying three major automatic control loops:

(a) the reactor power;
(b) the boiler steam pressure and water levels;
(c) the turbine-generator speed.

By the "philosophy" we mean the conceptual choice made in the interconnection of these three loops. Thus we might approach the design of the control system on the basis of "reactor-follows-turbine" with additional reactivity changes to supplement any inherent load-following capability. Another approach might be that of "turbine-follows-reactor" where changes are induced in reactor power that vary the steam pressure in the boiler that in turn vary the admission of steam to the turbine. Depending on the importance to be given to steam pressure in the boiler, the philosophy may put more stress on this loop in the over-all control pattern.

Descriptions of actual plant that follow are meant to illustrate some of the various choices that can be made, but first some simple analysis is provided of the qualitative material presented so far in the context of a PWR.

Sample Control Calculations

In this section some illustrative calculations to support the choice of controls and the behaviour of the plant are presented, condensed and simplified of necessity but enough, it is hoped, to give the flavour and motivation of the control system design. For this purpose a much-simplified model of a PWR (no superheat) is adopted, having the full-load design parameters of Table 5.1.

The drum boiler in this design has a modest energy storage, $1\frac{1}{2}$ full-power seconds, but the mass of water slows down changes in the return core inlet temperature and thus

TABLE 5.1.

Sample PWR parameters

Primary			
Coolant pressure	p	16	MPa
Coolant flow rate	F	4300	kg/s
Inlet temperature	T_i	555	K
Outlet temperature	T_o	595	K
Secondary			
Steam pressure (HP)	p_s	5.5	MPa
Steam flow to turbine	F_s	490	kg/s
Steam temperature	T_s	543	K
Reactivity coefficient	α	−0.0025	K^{-1}

slows down load-following. We expect, therefore, to have to supplement throttle control with direct reactivity control, and the first consideration perhaps is to examine the steady state part-load relationships when the load is varied at the throttle and when the load is varied at the reactor.

Simple representations developed in Chapter 4 are used, particularly the mean temperature $T_{av}=\frac{1}{2}[T_o+T_i]$ and a single reactivity coefficient working to T_{av} (appropriate enough for a metallic fuelled PWR). In this preliminary analysis we suppose that operation at constant primary pressure (subject to surge variations) and constant primary coolant flow rate is desirable unless events prove that the complexity of varied flow is necessary.

PRESSURE/TEMPERATURE–LOAD RELATIONS

If the reactor power is varied by inserting control rods, the immediate result of an impressed reactivity drop is a fall in power and consequently in mean temperature. If the primary coolant inlet temperature remained constant (and thus for times short compared with the primary circulation time) then both the mean and core outlet temperatures fall. At least approximately the fall in power continues until the reactivity loss is offset by the temperature fall and the negative reactivity coefficient and would be stabilised when $\delta T_{av}=-1/\alpha\ \delta\rho$. Thus in the short term we might expect to find a power level established in accordance with

$$\frac{1}{P}\left(\frac{\partial P}{\partial \rho}\right)_{\text{th}} = \frac{1}{T_o - T_i}\left(\frac{\partial [T_o - T_i]}{\partial \rho}\right)_{\text{th}} = \frac{2}{T_o - T_i}\,\frac{\partial T_{av}}{\partial \rho} \qquad \text{transient}$$

$$= \frac{2}{T_o - T_i}\left[-\frac{1}{\alpha}\right] = 20\%\ \text{power}/\%\ \text{reactivity} \tag{5.1}$$

with $(\ )_{\text{th}}$ indicating *at constant throttle.*

However, the core inlet temperature does not remain constant and the above expression is at best an indication of a point reached during the transient. We can obtain from it, however, corresponding temperature changes and can see in this model that $\delta T_{av}=20\delta P/P$ and $\delta T_o=40\delta P/P$.

In the longer term, but short compared to changes in xenon levels, etc., which would otherwise complicate the reactivity balance, these temperature changes are reflected in a change in the core inlet temperature from the coupling through the boiler to the turbine and hence with the stream pessure admitted to the turbine. At fixed throttle, a reasonable approximation is for the fractional change in steam pressure to be equal to the fractional change of reactor power. Figure 5.4 sketches the thermodynamic relations for water along the saturation line.

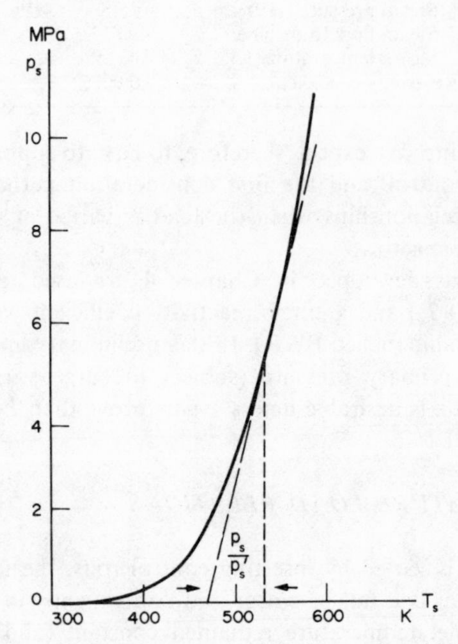

FIG. 5.4. Saturated water pressure and the inverse logarithmic derivative.

If we may assume an unchanged heat transfer coefficient in the exchanger, then the drop from primary side to secondary side $T_p - T_s$ is proportional to power and will change fractionally in the same way as power. We may therefore write

$$\frac{\delta P}{P} = \frac{\delta[T_p - T_s]}{T_p - T_s} = \frac{\delta p_s}{p_s} = \frac{p_s'}{p_s}\,\delta T_s \tag{5.2}$$

and hence establish the desired relation

$$\begin{aligned}\frac{1}{P}\left(\frac{\partial P}{\partial \rho}\right)_{\text{th}} &= -\frac{1}{\alpha}\frac{1}{T_p - T_s + p_s/p_s'} \qquad \text{equilibrium} \\ &= 6.1\,\%\ \text{power}/\%\ \text{reactivity}\end{aligned} \tag{5.3}$$

in this case, with $p_s' = dp_s/dT_s$.

It is notable, therefore, that the interconnection of the reactor to the turbine substantially decreases the feed forward relation of power to reactivity. More to the point, perhaps, we have $\delta T_{\text{av}} = \delta T_p = 20\delta P/P$ as before, but with the smaller power change there is, of course, a smaller change in the temperature rise across core or drop across heat exchanger. We may easily derive the formal relations, Table 5.2, from which at least the top end of the P/TL diagram can be sketched.

TABLE 5.2.

Pressure/temperature–load relations[a]

Coefficient	Constant throttle $(\ )_{th}$	Constant reactivity $(\ )_{\rho}$
$P\left(\dfrac{\partial T_o}{\partial P}\right)$	$T_p - T_s + p_s/p + \frac{1}{2}[T_o - T_i]$	$\frac{1}{2}[T_o - T_i]$
$P\left(\dfrac{\partial T_{av}}{\partial P}\right)$	$T_p - T_s + p_s/p_s'$	0
$P\left(\dfrac{\partial T_i}{\partial P}\right)$	$T_p - T_s + p_s/p_s' - \frac{1}{2}[T_o - T_i]$	$-\frac{1}{2}[T_o - T_i]$
$P\left(\dfrac{\partial p_s}{\partial P}\right)$	p_s	$-[T_p - T_s]p_s$

[a]Constant primary pressure and flow rate assumed.

The alternative is to leave the imposed reactivity unchanged but to vary the throttle. In the short term the power delivered to the turbine changes according to the throttle opening; suppose the immediate change is represented by $\delta A/A$, where A is the (linear) opening at constant pressure. At this time, conditions in the primary circuit and the core are, of course, unchanged. When, after a change $\delta A/A$ the system settles to a new equilibrium, the mean core temperature must (in our model) be unchanged. From the expression $P = Ap_s$ for the power delivered to the turbine we have at this later equilibrium $\delta P/P = \delta A/A + \delta p_s/p_s$ or

$$\frac{A}{P}\left(\frac{\partial P}{\partial A}\right)_{\rho} = \frac{1}{1 + \dfrac{T_p - T_s}{p_s} p_s'} \quad \text{equilibrium} \tag{5.4}$$

$$= 0.65 \text{ in this example}$$

With throttle variation, therefore, the longer term power change in the example is only 65% of the initial power variation, a result that does not depend on the reactivity coefficient.

For the P/TL diagram we have already shown that T_{av} is constant for throttle variation at constant reactivity while, of course, $T_o - T_i$ changes in proportion to power, results that are tabulated also in Table 5.2. It is also seen that the change of steam pressure in this model with load is given via

$$P\left(\frac{\partial p_s}{\partial P}\right)_{\rho} = -[T_p - T_s]p_s' = -2.9 \text{ MPa} \tag{5.5}$$

compared to +5.5 MPa for the corresponding slope in the constant throttle case. That is, with throttle variation the steam pressure *drops* as load increases whilst at constant throttle it *increases*, of course, in simple proportion to the load.

Clearly, therefore, there is some interesting scope to combine both forms of control to achieve certain goals in the P/TL relationship. Indeed, the concept of constant steam pressure may be so attractive as to make pressure control at the steam drum a principle control goal. However, this advantage would be purchased at the cost of the steeper rise in the temperatures which are likely, of course, to be constraining factors. Although some compromise may be possible, we are led to consider whether an over-all improvement can be achieved if we can enter into the additional complexity of varying primary coolant flow in the P/TL relation.

To a first approximation we might suppose that heat transfer coefficients remain unchanged with coolant speed and ignore additional pumping power. Both assumptions may have to be refined; primary and secondary coolant speed increases will generally have the advantage of increasing heat transfer and thus diminish the drop in temperature from the constrained maxima and thus lead to improved steam conditions: it is, of course, the drop across the heat exchanger that depresses the steam pressure in eqn. (5.5), while in gas cooled reactors primary coolant circulation absorbs substantial power. However, as a first look we make this assumption and it follows that if we arrange for an increasing primary coolant flow rate with load, then the temperature drop across the core and across the heat exchanger can be reduced or even made independent of load. Figure 5.5 illustrates such an ideal where increasing primary flow offsets the core

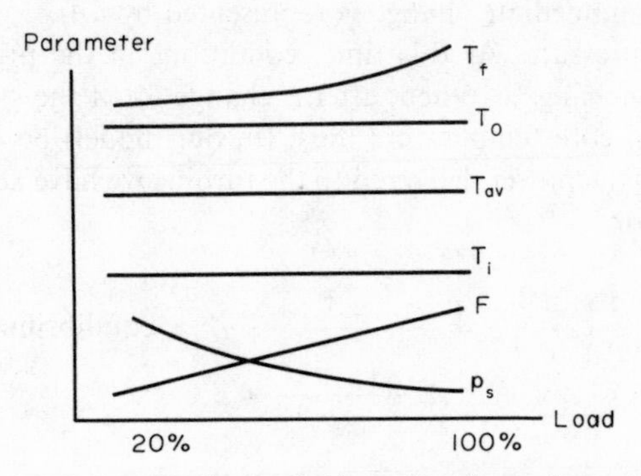

FIG. 5.5. Idealised variable coolant flow relation.

temperature changes; steam pressure will still drop, of course, somewhat because of the drop across the heat exchanger.

In our calculations we have not distinguished between the thermal power of the reactor plant and the net electrical or mechanical output at different loads; that is we have implicitly assumed a constant net thermodynamic efficiency. Of course, this assumption should be refined in any computation involving final output.

TRANSIENT STUDIES

Considerations such as the foregoing suggest a philosophy for the control system of how and where control shall be applied. The next step might be to establish a dynamic model of the plant and typical three-term controllers to simulate the time dependent behaviour and to study the response to the various specifications of ramp and step changes at, of course, a host of different normal part loads and, indeed, abnormal conditions.

Figure 5.6 illustrates a typical result for a step demand in our PWR acting principally

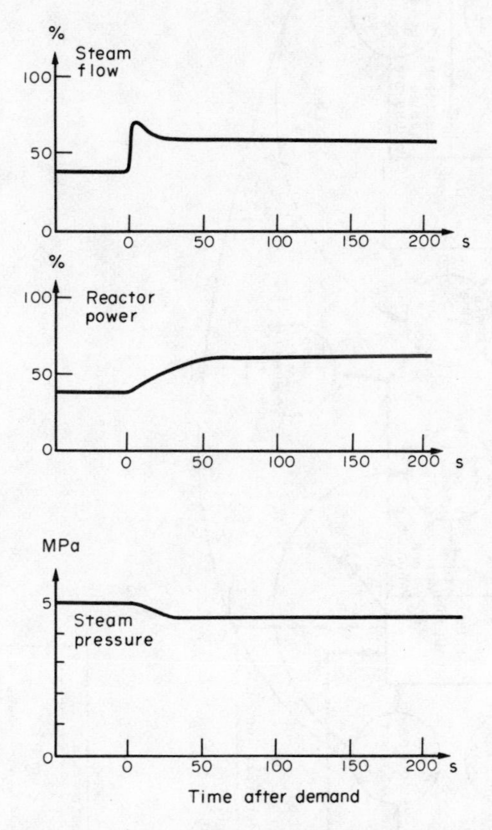

FIG. 5.6. PWR simulation: step demand from 40% to 60% full power.

through the throttle. It is seen that (1) the steam flow increases rapidly at first on the initial throttle opening but then falls away in accordance with eqn. (5.4) (2) that the reactor power increases more slowly due to delays between the boiler secondary side and changes in the core temperature; it is not wise to accelerate this change via reactivity control too quickly because of temperature rate restraints, whilst (3) steam pressure drifts downwards at higher load.

These and other studies are checked to see that all specifications and restraints can be met by some setting of the controller parameters; if so, then there is scope to optimise the behaviour against some further objective criterion or cost function. If not, the control design is unsatisfactory and a new philosophy must be adopted.

Further studies must be carried out to cover safe startup, both from the hot and cold situations. Figure 5.7 shows the sort of considerations (for the Heysham AGR) that

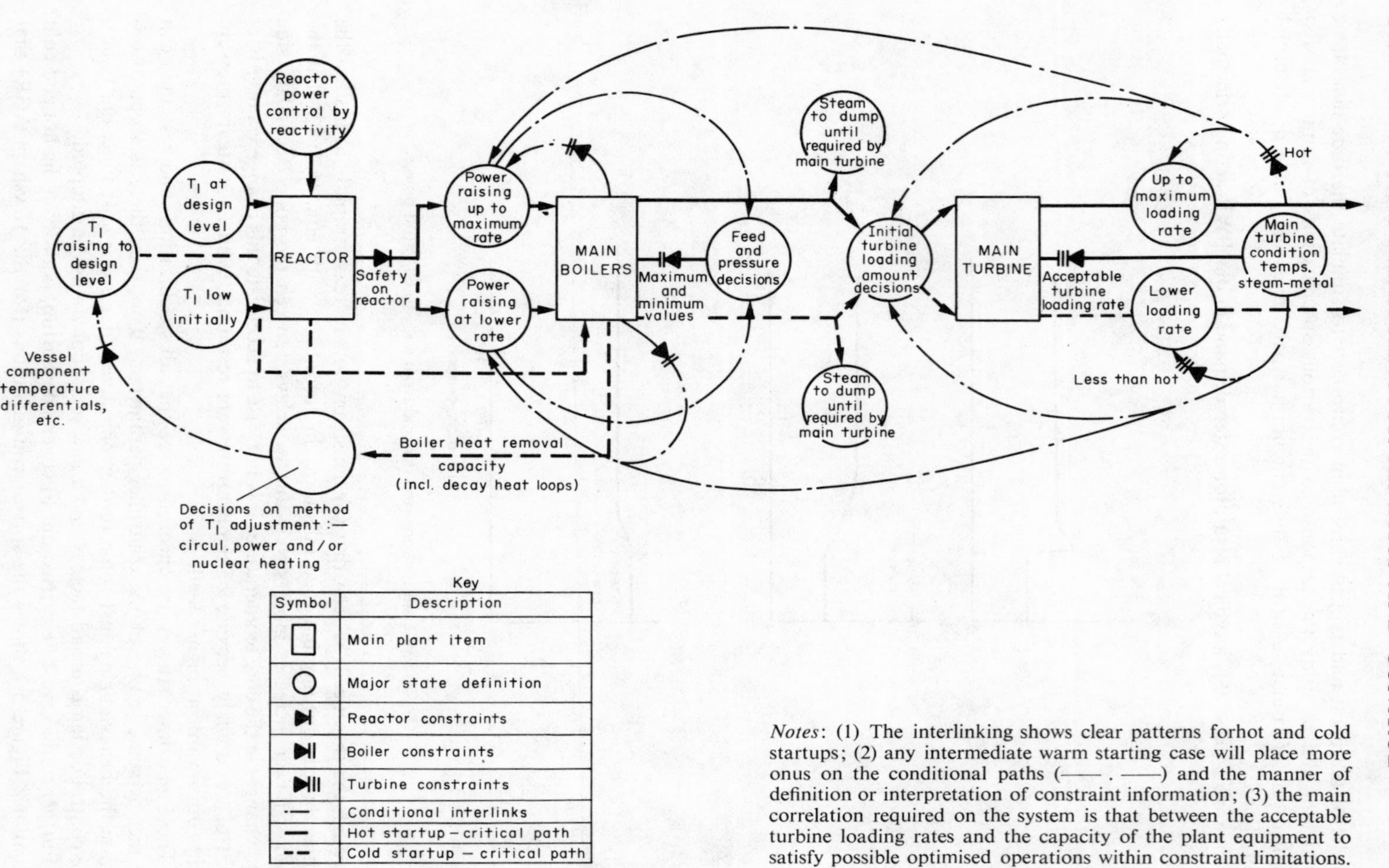

Notes: (1) The interlinking shows clear patterns forhot and cold startups; (2) any intermediate warm starting case will place more onus on the conditional paths (—— . ——) and the manner of definition or interpretation of constraint information; (3) the main correlation required on the system is that between the acceptable turbine loading rates and the capacity of the plant equipment to satisfy possible optimised operations within constraint limitations.

FIG. 5.7. Reactor startup considerations (Heysham AGR).

must be given to such a problem. The figure represents the starting point for the design of a (digital) control program.

BOILER DEMAND TRANSIENT

When the throttle is opened the immediate effect is an increase in steam flow but unless the power received by the boiler changes this cannot ultimately lead to an increased power flow. Instead, the pressure and temperature drop off until, if a steady state is recovered, the net effect is a slightly increased steam flow at a lower pressure, carrying the same enthalpy (although the decreased thermodynamic availability will lead to a smaller work output from the generator).

Figure 5.8(a) illustrates this sequence. Clearly the drop-off is undesirable and either

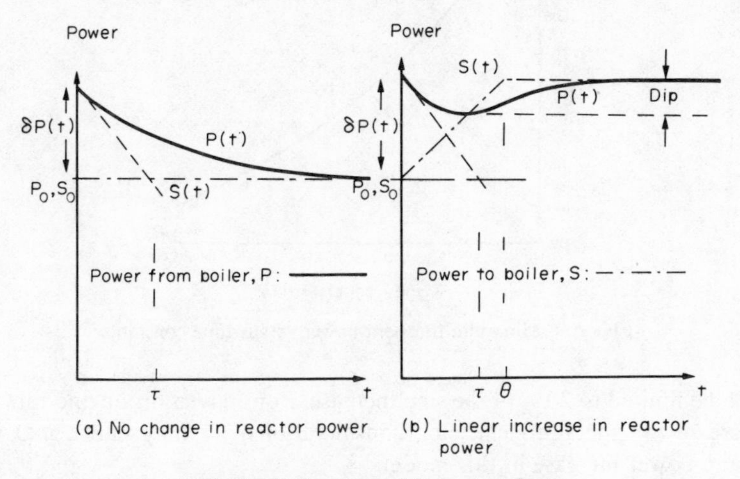

FIG. 5.8. Boiler power delivery transients.

naturally or by applying impressed reactivity changes we would wish to raise the reactor power, presumably to a new level implied by the throttle opening. This power input will take time and the power transient may still pass through a temporary dip before settling to the new level, as shown in Fig. 5.8(b).

There is some interest, therefore, in determining the size of this droop in terms of the natural time constants of the boiler and the rate at which reactor power is increased, assumed *linear* here. This rate will be limited and in turn therefore the calculations will indicate the maximum step increase of power that can be accommodated without the transient power dip exceeding some specified value, such as not more than 20% of the step demand $\delta P(0)$.

A simple model can be constructed along the lines of the derivation of the boiler transfer function, Chapter 4, and for the moment we can identify the time constant τ^+ as that of the boiler whose power excess will run down with an exponential fall-off in the absence of increased energy input. Suppose the ramp input, however, is concluded in θ s. It is then readily shown that the minimum boiler output power compared to the demand power is given by the solution to

$$\frac{d\delta P}{dt} = -\frac{\delta P}{\tau^+} + \delta S; \quad \delta P(0) = \delta P_0 \tag{5.6}$$

with δS the reactor power. The solution is shown in the form of a non-dimensional curve in Fig. 5.9. For example, the ramp input must be completed in time $\theta < \frac{1}{2}\tau^+$ if the

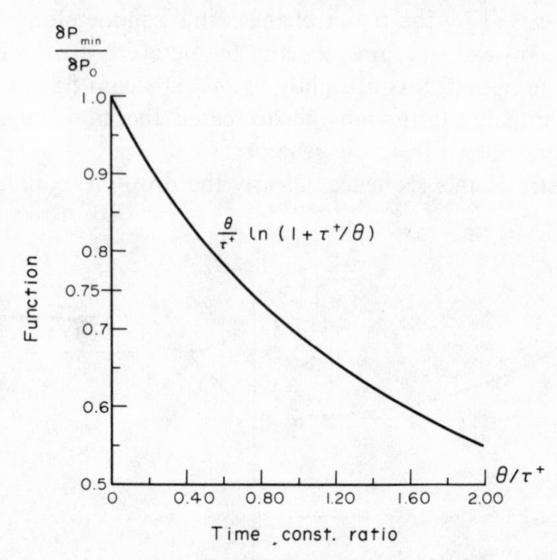

FIG. 5.9. Minimum transient power versus time constants.

droop is to be limited to 20% of the step increase. For a given droop and rate of reactor power increase (on metallurgical, etc., considerations) we may find a maximum permissible step power increase in this model.

In addition, the function $1-\delta P_{\min}/\delta P_0$ gives the fraction of the available energy "borrowed" by the transient before the increasing reactor power starts to pay energy back. In this context, however, stored energy means the amount of energy available for release until the steam pressure and temperature have dropped down to the new conditions that would be taken up at the large throttle opening in the absence of a change in reactor power. This has been allowed for in the definition of the boiler time constant τ^+ (Table 4.3).

STABILITY STUDIES

If the dynamic simulation included the inertial load of the turbine-generator and the frequency feedback, then some information on stability has been forthcoming. Alternatively, the stability of the unit, disconnected from the grid, will be studied via a Nyquist diagram for the plant $G_f(j\omega)$, defined as the transfer function for mechanical power with respect to the frequency feedback signal, wherever that may be applied. Figure 5.10 shows a typical relationship for power variations with frequency variations as delivered to the turbine and with the inertial transfer function $1/j\omega I$ for the turbine-generator itself.

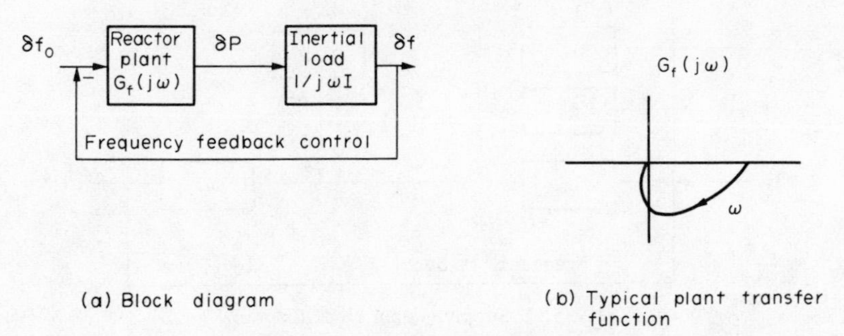

FIG. 5.10. Isolated unit stability study.

The stability of the unit and gain margins can be determined by combining these transfer functions in the usual form of a scaled Nyquist diagram (Fig. 5.11). It appears in this example that if the inertial load is large enough, the unit will be stable (uncon-

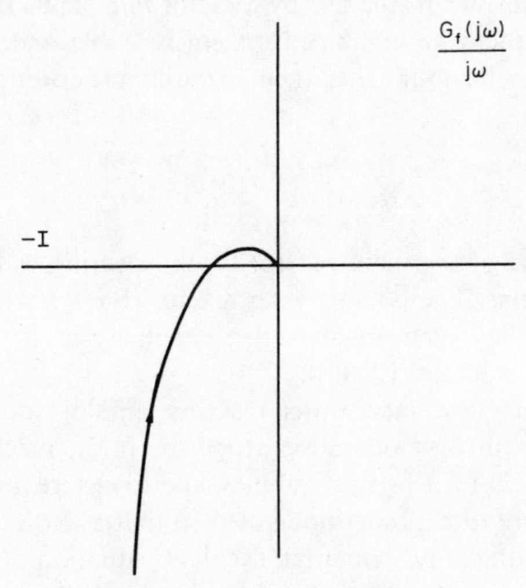

FIG. 5.11. Unit stability diagram.

ditional) and if larger still it will be adequately stable. If the shape of $G_f(j\omega)$ is right but I is too small, further design changes are indicated. This simple picture conceals the detailed studies of $G_f(j\omega)$ necessary at different powers, to allow for nonlinearities, different fuel loadings, etc., and consequent changes in G_f; stability of the independent unit must be ensured at all powers. However, there is the further question of the system stability when this and other units are synchronised and carry the system electrical load. An assumption that all the supply units are identical allows a naive argument to justify the specification for unit stability satisfying system stability.

Consider the combined supply system, Fig. 5.12, where the combined inertial load I_{tot} includes not only the inertia of the supply units but also of the various synchronous motors on load in the grid.

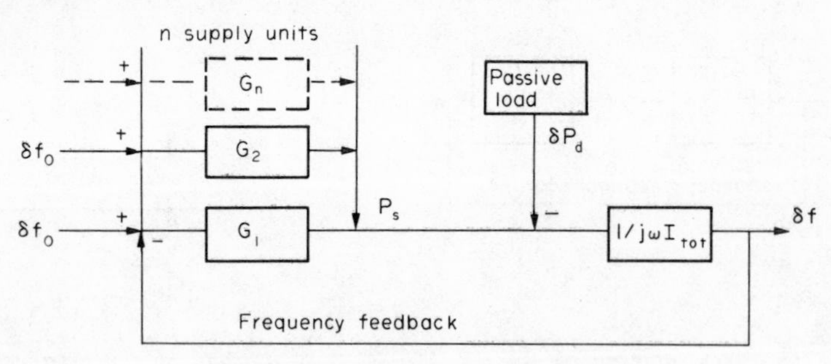

FIG. 5.12. Supply system block diagram.

If there are n identical supply units then the combined transfer function for total power with respect to frequency is $nG_f(j\omega)$, n times greater than the transfer function of one unit. The inertial load, however, is increased by a number m, say, greater than n due to the additional motors on load. Thus the combined system calls for a study of the return difference $1+nG_f/mj\omega I$, and clearly we may make use of Fig. 5.11 with reinterpretation that I moves further to the left by a factor m/n. Thus if the isolated unit was unconditionally stable, then the combined system is stable and with a bigger margin of stability for transient behaviour. This, then, is the simple criterion for system stability.

Light Water Reactors

We now describe some of the main reactor types, covering water cooled, gas cooled and sodium cooled systems. The boiling water reactor (BWR), where the steam is taken directly from the core, has such an attractive simplicity at first sight that we start, indeed, with light water reactors (LWRs).

Light, or more properly ordinary, water reactors, employ water with its good heat transfer properties and neutron-moderating properties in the reactor core. As the water heats up it expands or even forms steam bubbles, and to ensure a safe design it is fundamental to require the core to be undermoderated so that a loss of moderator leads to a fall in reactivity offsetting any potential accident situation. (Correspondingly, the amount of any soluble poison for control will need to be limited to keep a negative void coefficient such that the ejection of moderator on the formation of a bubble does not add reactivity; hence soluble poisons are limited to shim/safety applications—they would anyway be too slow for the present purposes.) This overriding requirement will affect the dynamic behaviour of LWRs. In both PWRs and BWRs it is customary to use Zircaloy clad fuel elements. To prevent a self-perpetuating exothermic reaction with water, it is essential to limit the temperature of this cladding to about 1800 K (1500°C) and this is a major control system restraint.

BOILING WATER REACTOR (BWR)

In the simple water boiler as first developed (EBWR) the storage of energy in the form of hot water within the core certainly promises good response to changes in load and throttle setting. However, such a design is not load-following, as we may see. An opening of the throttle in response to an increase in demand allows freer passage of the steam and thus drops the steam pressure. This in turn allows some of the hot water in the core to flash to steam but in so doing, drops the temperature and—more important still—increases the voidage in the core. Since the coefficient of reactivity is negative, the reactivity—and consequently the reactor power—fall at the time the demand has increased. Ultimately a steady state may be reached but only when the delivered power is lower than before.

Some early modifications of the BWR therefore combined this direct cycle with what can be called an indirect cycle, supplying steam through a heat exchanger to an intermediate pressure turbine stage, thus introducing some of the characteristics of the PWR (see next section).

In the current designs now in use, however, the attractiveness of the simple direct cycle is maintained by varying the recirculation rate of water in the reactor core in response to load demand changes (Fig. 5.13). The rate at which coolant is recirculated may be three or more times that at which steam is drawn off to the turbine.

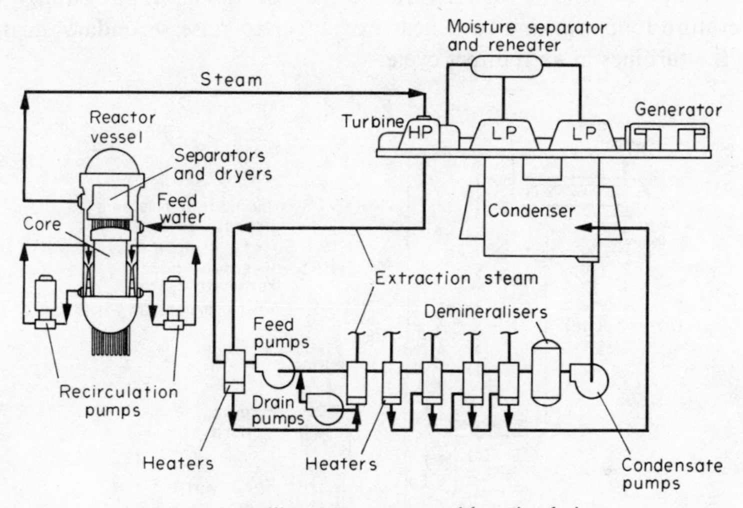

FIG. 5.13. Boiling water reactor with recirculation.

Load-following is engineered into the system by arranging through automatic controls that the recirculation flow increases when the demand is felt. The faster flow through the core sweeps out bubbles (as well as improving heat transfer) and thus lessens net voidage. The reactivity is thus *increased* in accordance with the demand and the system is now load-following. Considerations of the P/TL relationship now affect the choice of recirculation rate with load.

Load-following may also be promoted by restricting throttle movement to act to maintain steam pressure rather than directly with demand. In this philosophy the frequency drift signal is applied to the core reactivity (and recirculation pumps of course) and not until the power in the reactor has changed in accordance with demand to produce the appropriate change in steam pressure will the throttle be opened. Evidently this slows down the rate of response of the system but tight pressure control opposes the natural tendency of the BWR not to follow load.

Questions of BWR stability are complex and not suited to simple analytical forms, involving as they do hydrodynamic aspects as much as heat transfer and reactor physics. As some compensation, questions of xenon stability are less significant in BWRs since the ready streaming of neutrons through the steam voids couples the parts of the reactor together and demotes spatial oscillations. Thus soluble poisons can be used simply for shutdown/safety purposes.

To achieve good vertical flux distribution, it is common to use bottom-acting control rods in BWRs which thus compensate for the loss of moderator, due to steam voids, further up the rising coolant channels. Temperature distributions are sensitive to these flux shapes and as the rods move to adjust for gross xenon changes and fuel changes after a power disturbance, the flux *shaping* may, indeed, be a limiting factor in the speed of response to a power demand.

PRESSURISED WATER REACTORS (*PWRs*)

Figure 5.14 shows a scheme for a PWR and the essential concept of a high pressure–high temperature loop working to a heat exchanger to raise secondary steam for admission to the turbines in an indirect cycle.

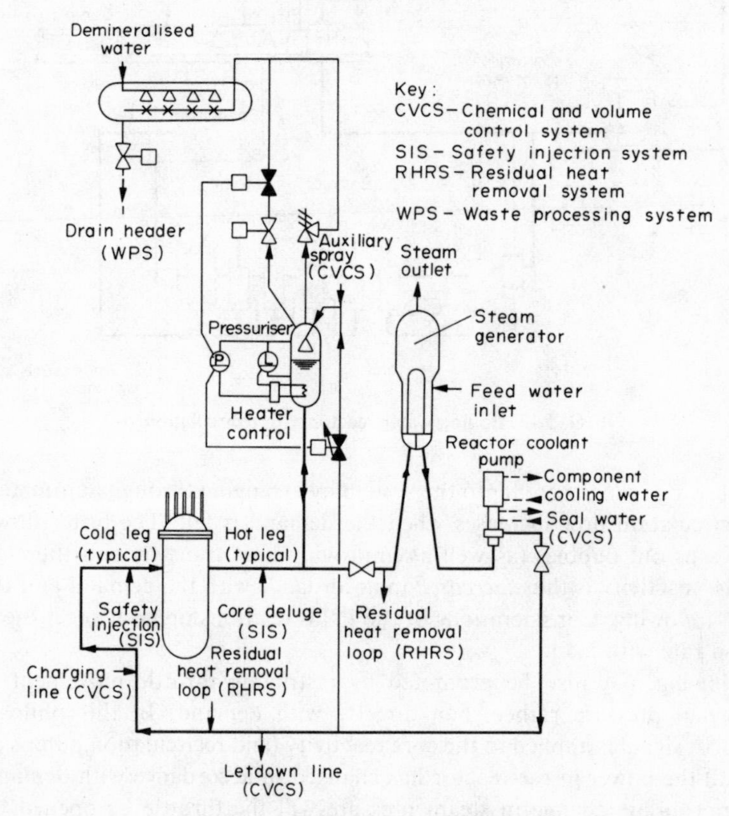

FIG. 5.14. Pressurised water reactor.

Note the pressuriser provided with a cold spray and an electric heater and thus able to maintain a constant primary pressure.

In a majority of PWR designs the secondary side provides dry saturated steam via a recirculating or La Mont boiler with appropriate moisture baffles. Not shown is an alternative PWR design in which superheated steam is obtained from a "once-through" heat exchanger.

The recirculating steam-generator has a substantial energy storage, say 10 full-power seconds, in the form of a reservoir of hot water at boiling point. When the throttle is opened there is an additional steam flow leading to a lowered secondary steam pressure and the flashing of the required additional steam. The system is thus well positioned to meet a step increase in power. The immediate result of such a change is a cooling of the secondary water in the boiler and hence a greater flow of heat from the primary side (an attractive extra reservoir of energy) and the lowering of the primary side temperature. Thus the temperature of the return coolant to the core is lowered and the negative temperature coefficient ensures that the reactor power moves in the direction of load demand. Thus, in contrast to BWRs, the PWR is naturally load-following. However, the initial increase in steam flow and power is not entirely maintained and further reactivity control is appropriate to improve on an inherently suitable performance to meet typical step and ramp specifications.

The PWR with superheat has a higher thermodynamic efficiency (and less problems in disposing of cooling water waste heat) but also a smaller energy reservoir in the boiler. Thus the response to a load change is perhaps slower although still naturally load-following.

Appropriate P/TL diagrams have already been given (Fig. 5.3). The calculations of the last section illustrated the need for further reactivity control to make up the initial response droop. Constant primary flow is usually chosen for the part load specification. Boron absorbers dissolved in the primary circuit are used for shim control to offset gross xenon poisoning and fuel burn-up, as well as for safety purposes.

Figure 5.15 gives some idea of the interconnections in the control system to secure the desired behaviour. Since the flow of steam through the throttle is a function of steam pressure and throttle opening, a linearising circuit is helpful to cover the full range. The influence of the pressuriser on stability is significant and care is needed in design to see that surges of water into the pressuriser do not lead to oscillations at frequencies promoting dynamic instability or, indeed water hammer damage.

The use of PWRs for naval and marine applications should not be forgotten. Submarines certainly have their own operational problems and needs for reliability but fortunately avoid most of the problems of an interconnected grid.

The foldout at the end of the book reproduces the control schematic for the Kraftwerke Union Biblis power reactor which at 1000 MW(e) is one of the largest (single unit) operating nuclear power plants in the world.

Heavy Water Reactors (HWRs)

Heavy water, with equally good heat transfer properties and nearly as good neutron-moderating properties as ordinary water, has substantially better neutron absorption

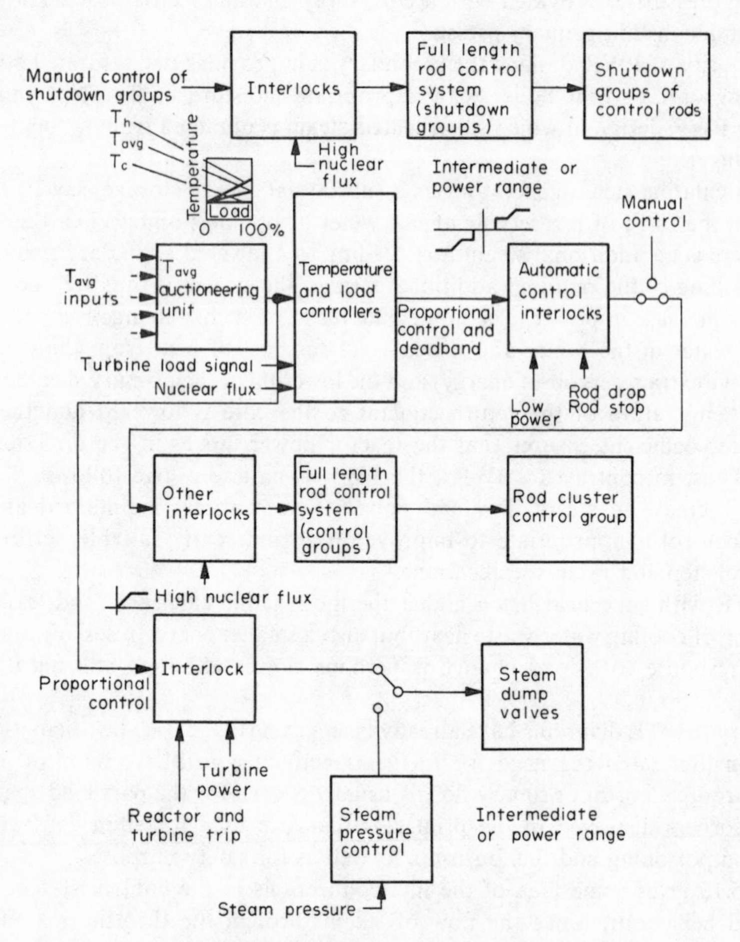

FIG. 5.15. PWR control schematic.

properties. Thus HWRs can be designed with larger spacing between fuel elements which in turn enables pressure tubes to be provided around individual fuel elements rather than the pressure vessel of LWRs. Pressure tubes can be thinner and therefore cheaper than the large diameter pressure vessel for the same duty. Coolant within the tubes can be either heavy or ordinary water; the Canadian practice (CANDU) has been mostly to employ heavy water while the UK design (steam generating or SGHWR) has been directed to ordinary water thus providing some of the features of a BWR in the design. The Canadian Gentilly reactors are of the boiling light water type however.

STEAM GENERATING HEAVY WATER REACTOR (*SGHWR*)

In the prototype SGHWR, steam is available in a steam drum of substantial capacity, promising good immediate response to demand changes. The design of a separate drum, however, means there will be limited response in the core to throttle changes, i.e. that

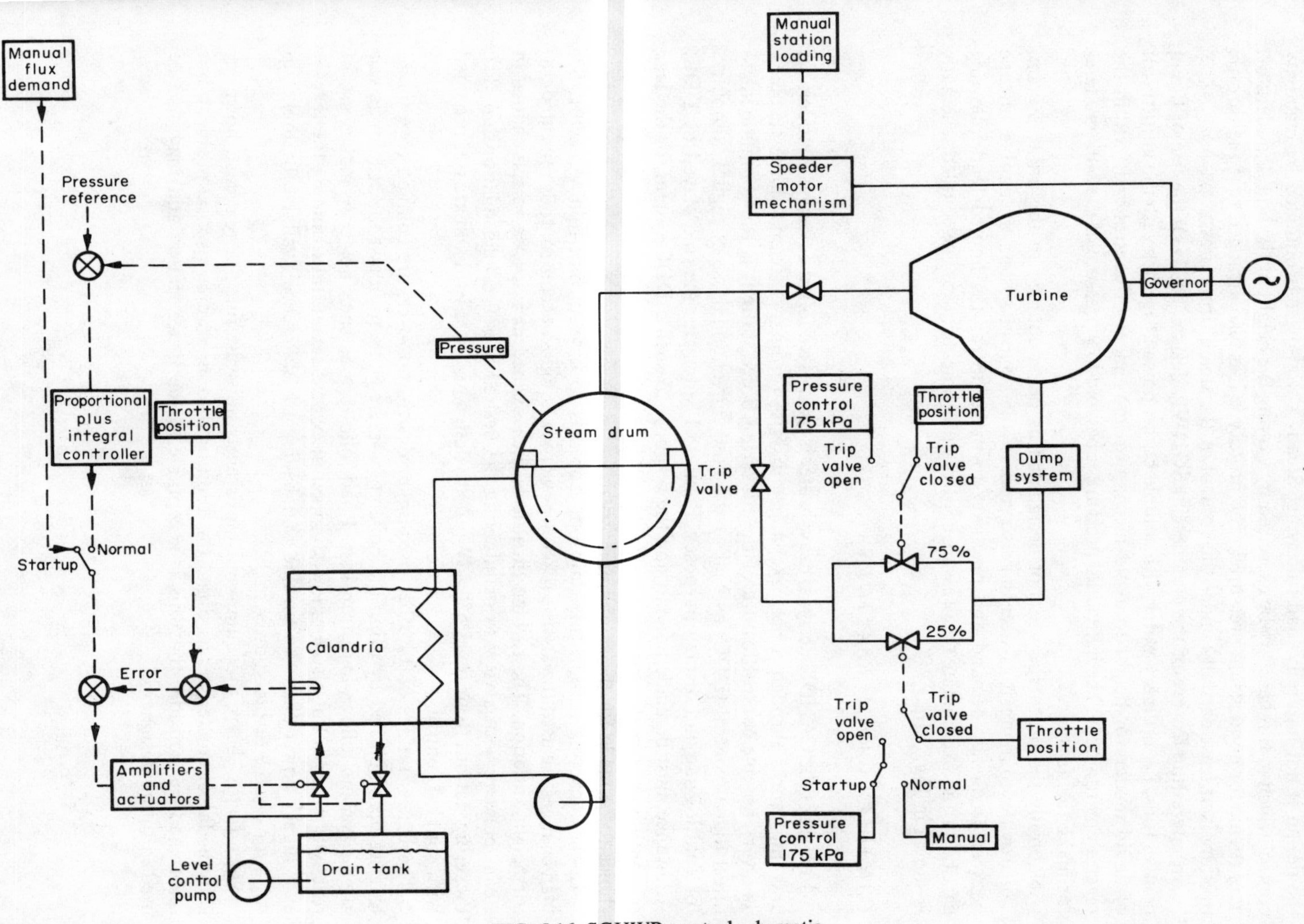

FIG. 5.16. SGHWR control schematic.

the design is not naturally load-following. Steady core power is obtained by reactivity control which in this case can be exercised by varying the height of the liquid moderator. To provide for good steam pressure in the prototype it is also arranged that the throttle opening is at least partially under the control of the steam drum pressure.

An interesting feature of the commercial SGHWR designs has been the use of liquid control rods, i.e. ordinary water with soluble boron poison together with the adjustment of liquid moderator height to avoid mechanical movement of control rods within the reactor environment. The individual pressure tubes and low pressure of the bulk moderator make this feasible.

Naturally the prototype was not designated as being coupled to the utility system. To provide coupling it is necessary to add to the control system a suitable linkage between frequency demand and reactivity to make up for the limited load-following capability and to eliminate the pressure control of turbine throttle. The coupled schematic is shown in Fig. 5.16.

CANADIAN CANDU REACTORS

In most of the CANDU designs, heavy water is used as a coolant as well as a moderator and thus works through a heat exchanger to raise (ordinary) steam for the turbines. The very notable control feature of the Pickering/Bruce stations is the commitment to direct digital control (DDC) as well as the now conventional use of digital computers for data logging and display purposes. Figure 5.17 indicates the role played by DDC. Allowance must be made in analysing the response of such a DDC system for the lack of continuous signals and that the state of the system is analysed discretely, with a choice to be made for the sampling interval.

In the Pickering station the philosophy used is to have turbine-follow-reactor. Set load changes are used to vary the reactor power via liquid zone control rods plus moderator height variation. The turbine throttle is then opened in response to the change in steam pressure with a view to maintaining the desired pressure–load relationship. The Pickering stations (four times 550 MW(e)) are designated base load and are not provided with coupled control.

The Bruce stations of an otherwise similar design, however, are coupled to respond to grid, accomplished by using the boiler pressure signals to adjust the reactor power after a turbine-follows-load transient. Load-following as such not only makes use of reactivity changes but also employs dumped or bypass steam, thus running the reactor at a higher power than the turbine and having the difference available to make up variations in the turbine power.

Since fuel is cheap, this dumping is apparently not wasteful. The real cost, however, is in the larger condensers used and the additional cooling water disposal problem, but the Canadians consider that this is acceptable at the Bruce station and may also be used in future designs.

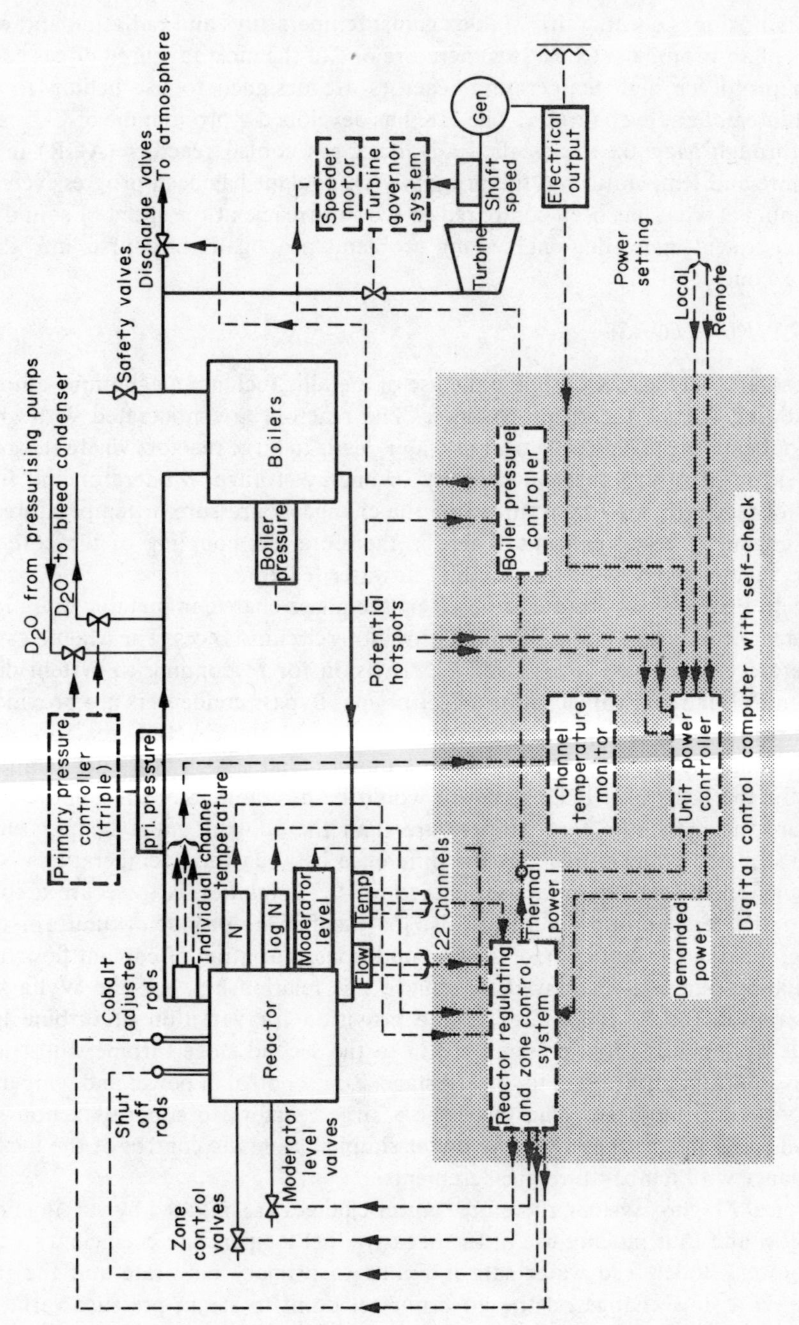

FIG. 5.17. CANDU HWR with direct digital control schematic.

Gas Cooled Reactors

Carbon dioxide is cheap, fairly stable against temperatures and radiation and without problems of absorbing neutrons. It is therefore one of the most favoured of gas coolants, although prototype high temperature reactors are designed to use helium to reduce chemical interactions even further. The UK has developed a programme of CO_2 reactors passing through Magnox to so-called advanced gas cooled reactors (AGR) in which the pressure and temperature of the primary CO_2 coolant has been progressively raised until running at what has been compared to red hot treacle at the speed of sound. Some of the consequent interesting engineering problems propagate, of course, into the field of reactor control.

MAGNOX REACTORS

The first commercial UK design make use of metallic fuel in a magnesium–aluminium alloy cladding—hence the name Magnox. The reactors are moderated with graphite which, not being as good a moderator as water, leads to large reactors where the graphite also provides for a high thermal capacity or energy storage. Moderator and fuel coefficients of reactivity are important while the change of pressure or temperature of the primary coolant is less signficant. There is therefore less coupling of the core to the secondary side of the heat exchanger than in water reactors.

Limits on the fuel temperature imposed by the phase change in metallic uranium lead to poor steam conditions under the most propitious circumstances. The Magnox stations are therefore run on base loads with no provision for responding to system demand other than through manual variation of set points. Bypass condensers are provided and the same considerations apply to their provision as with the CANDU reactors.

The limiting fuel temperature constraint implies that a constant fuel temperature relation for the basis of a P/TL diagram would be desirable. Since it is more difficult to measure the maximum fuel temperature than the coolant outlet temperature and since, for a metallic-fuelled reactor, the difference between these temperatures is not a strong function of load, the practical choice for P/TL relation is to secure a constant coolant outlet temperature. This can also be justified as producing uniform quality steam, i.e. making the best of poor steam conditions. Variation of coolant flow rate via the circulating pumps is employed to achieve this relationship, e.g. the Wylfa station allows variation for 50–100% flow. Some provision for variation of turbine load is also made by throttling low pressure steam to the second stage turbine whilst leaving high pressure steam unthrottled to the HP stage. Zone control of power and temperatures in the core is an important feature of these large reactors to suppress xenon spatial oscillations, and this action is usually under simple automatic control of the local rods in accordance with temperature measurements.

In a typical Magnox system, therefore, power changes are initiated by varying primary coolant flow and thus making use of the negative fuel temperature coefficient to change reactor power. Boiler feed water rate is linked to primary flow rate and the turbine responds via a slow change on the governor set point to steam pressure variation in the boiler. This leaves the governor available for fast action in response to grid frequency changes. The once-through boilers of the later Magnox stations means that the primary coolant inlet temperature responds rapidly to variations of the boiler feed water.

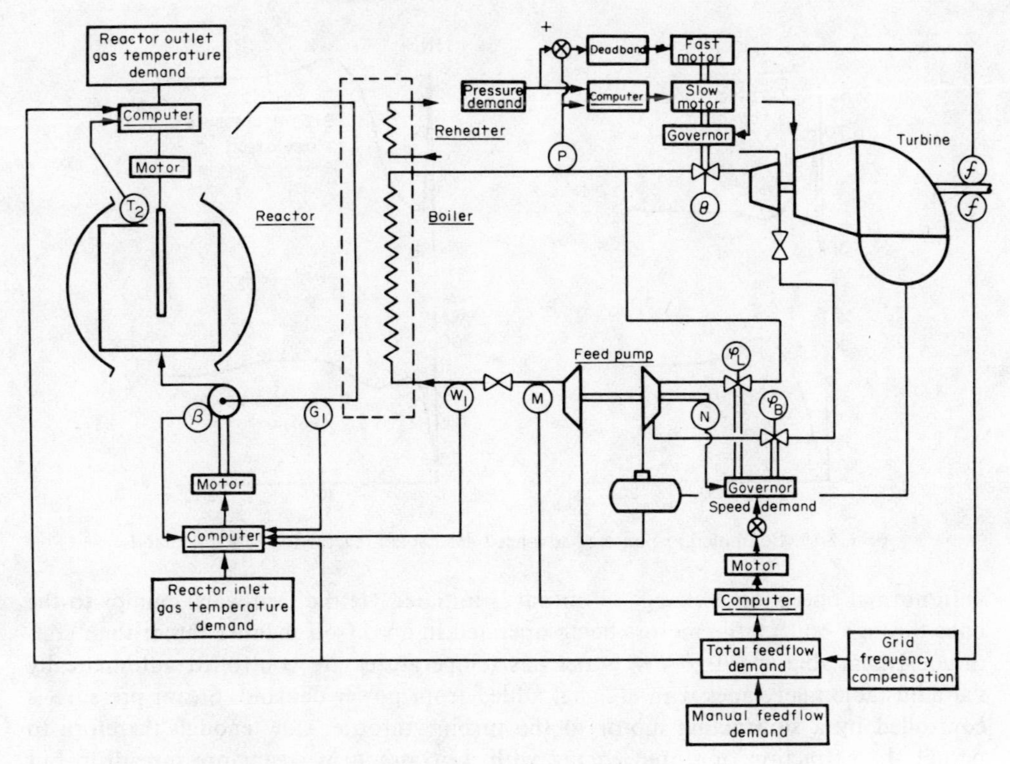

FIG. 5.18. Advanced gas cooled reactor control schematic (Hartlepool).

ADVANCED GAS COOLED REACTOR (*AGR*)

The AGR uses oxide fuel and the stations work to substantially higher temperatures with conventional turbine-generator sets of good thermal efficiency. Figure 5.18 shows an AGR schematic with its control arrangements.

The heat exchanger provides a reservoir of energy that offers a capability of meeting load demand changes, some 10 full-power seconds being available. There is some benefit from the energy of the hot graphite. The rapid transit of the feed water into steam through the secondary tubes has the feature of making the primary coolant inlet temperature respond more quickly to demand changes experienced at the turbine end and hence in a change of fuel temperature and consequent reactivity effect. Thus although the present AGRs are not considered to be other than base load and hence decoupled, there is an inherent capability that can be exploited.

Figure 5.19 shows computed transients for the Hartlepool AGR taking up a 10% load change. The AGR, like the Magnox, has variable primary coolant (gas) flow and it can be noted in passing that gas cooling requires substantial pumping power, in some cases as much as 10% of the gross mechanical output.

The Hartlepool AGR has been provided with DDC, and although it is taken to provide base load and hence decoupled, the flexibility of DDC permits a switching to a coupled configuration with frequency correction action.

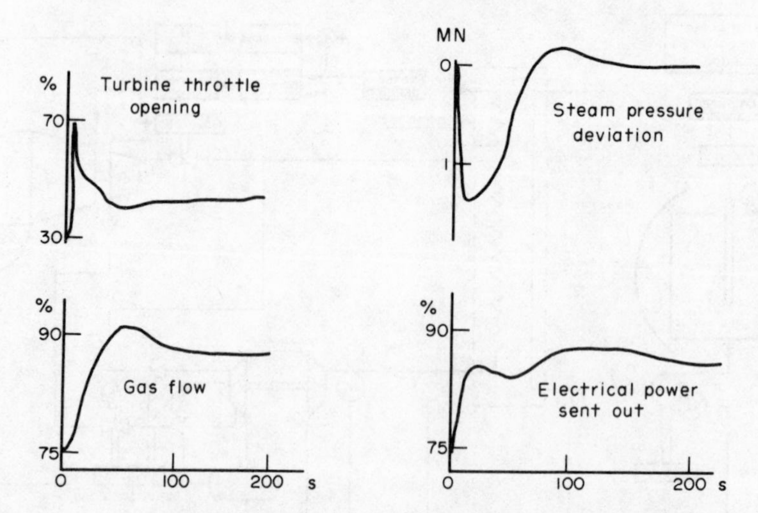

FIG. 5.19. Computed response of advanced gas cooled reactor to +10% demand.

In normal operation a change in output is initiated via the feed-water pumps to the once-through boilers, the motors being operated in an on–off manner rather than continuously variable speed. *Boiler* outlet gas temperatures are controlled automatically via adjustable inlet vanes with a signal added from power demand. Steam pressure is controlled by a slow-acting motor to the turbine throttle, slow enough therefore to permit the extraction of stored energy with its consequent steam pressure drop but with an overriding fast motor to limit such pressure transients. Reactor gas outlet temperature is controlled via a flux reactivity loop, suitably modified by the station power demand signal. Thus although there is some immediate response to grid frequency variations, the throttle movement is ultimately restored to enable the reactor to remain at constant set power and vary both more slowly and with smaller amplitude than the turbine power.

HIGH TEMPERATURE GAS REACTORS (HTRs)

HTRs have been constructed as prototypes. Commercial designs offer the prospect of using thorium fuel and direct cycles in which the helium is used in a gas turbine combined perhaps with a secondary, indirect steam cycle.

Figure 5.20 illustrates the choice of controllers (proportional, integral and differential) for a HTR with reheat between high and intermediate turbine stages. Note also the turbine-generator-motor system for primary coolant flow and its control based on set point demand and actual HP steam, making use of a reactor-follows-turbine concept.

Both AGRs and HTRs are built using reinforced concrete pressure vessels and it is important to control the maximum pressure exerted by the primary coolant, CO_2 or helium. This and the control of secondary steam conditions will be to the fore in selecting control philosophies. Only when the HTR is commercially established, however, is it likely that it will be used for other than base load and therefore provided with coupled controls.

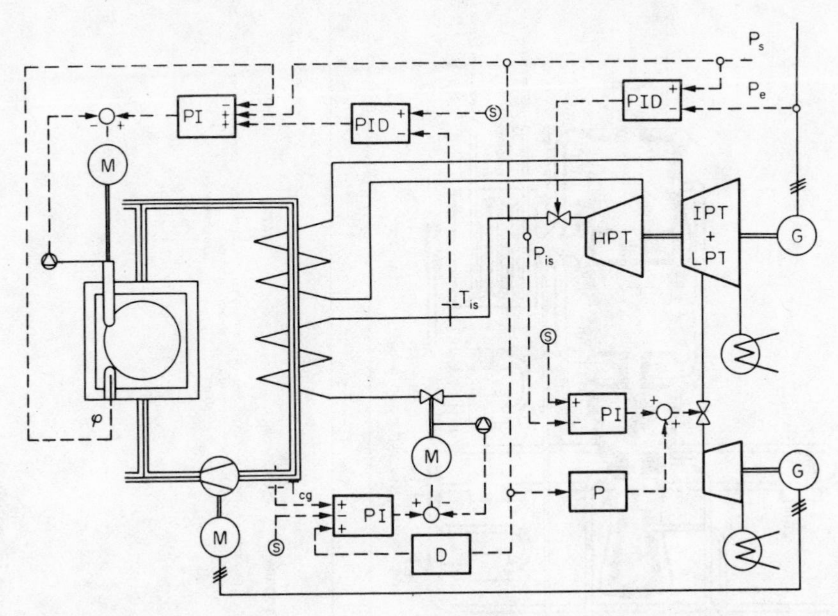

FIG. 5.20. Thorium high temperature reactor control schematic. *P*, proportional; *D*, derivative; *I*, integral control; *M*, motor; *G*, generator: high, intermediate and low pressure turbines.

Sodium Cooled Reactors

Coolants discussed so far have appreciable moderating properties. It follows that they are not suitable for a fast reactor in which neutrons are to remain unmoderated, yet it is these fast reactors that offer better transmutation ratios to breed fissile material from fertile. Thus it has been necessary to develop the use of good metallic coolants, particularly sodium although lithium and the lower melting eutectic alloy NaK has been used. These coolants will have a heavier atomic weight and hence low moderating properties; in the fast neutron spectrum envisaged, their neutron absorbing properties are less important but this still leaves few other competitors that will be liquid at working temperatures and compatible with the remaining materials. Such reactors may be called fast breeders (FBR) or liquid metal fast breeders (LMFBR).

Sodium is activated by neutron capture, and while the loss of neutrons can be overcome the activation radiation is potentially a drawback. It is currently thought desirable to separate the radioactive sodium from the steam by an intermediate loop of inactive sodium. Thus if there should be leaks in the heat exchangers (and security of tubes for sodium is now notoriously difficult to achieve) at least one does not have in the same place the troubles of a sodium–water reaction *and* radioactive contamination.

Prototype fast reactors in the range of 300 MW(e) have been commissioned in several countries, so that there is a degree of experience of breeder reactors approaching commercial practice. In the absence of moderator to bulk out the station and considering the capital value of the fuel, all these designs are of high power density, yet another demand made on the coolant. With this in mind, the integrity of the primary system is of great importance and there are two main designs to achieve this: pool or loop.

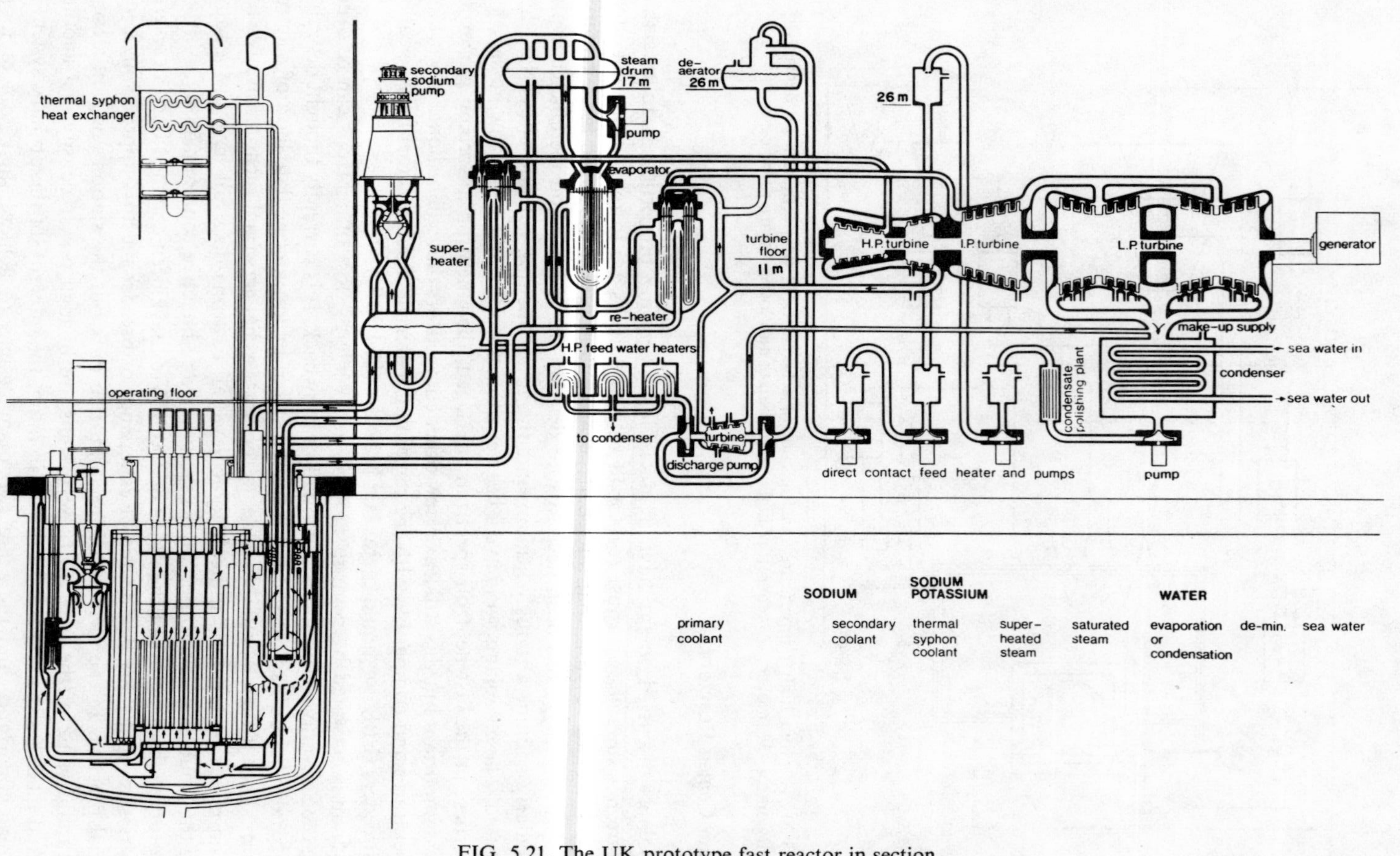

FIG. 5.21. The UK prototype fast reactor in section.

Fortunately the initial fears that special electromagnetic pumps for liquid metals would be essential has been reduced with the development of satisfactory glanded pumps, making the controlled variation of coolant flow possible. Figure 5.21 shows a cross-section of the UK prototype (PFR).

In the prototype certainly, integrity of fuel elements has been important and there is a general need to restrict temperature changes both in the load diagram and during transients. A scheme chosen for the Federal German SNR 300 reactor is to have a constant steam admission temperature and to minimise other temperature variations by using variable flow coolant in primary and in intermediate loops.

Figure 5.22 shows the temperature–load diagram. The commercial versions will probably also use an intermediate exchanger which brings the additional drawback of two time lags between reactor core and turbine throttle, making rapid contributions to load variations difficult if temperature excursions are to be avoided.

Due to the fast spectrum of neutrons and lack of moderator effects, many of the natural control processes of the thermal reactors are absent in fast reactors and their performance is more critically dependent on the reliability of conventional control rods working in the hostile sodium environment. Figure 5.23 shows the UK prototype fast reactor's control system.

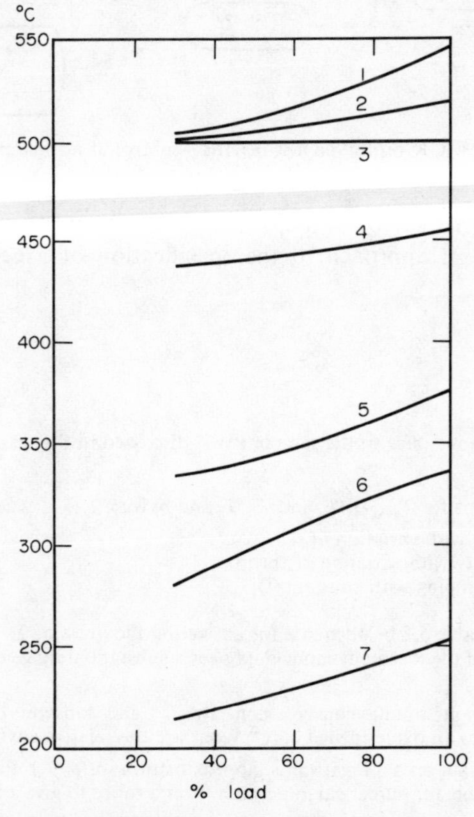

FIG. 5.22. SNR 300 temperature–load diagram. 1, reactor outlet T_o; 2, superheater sodium inlet; 3, steam T_s; 4, evaporator sodium inlet; 5, reactor inlet T_i; 6, evaporator outlet; 7, feed water.

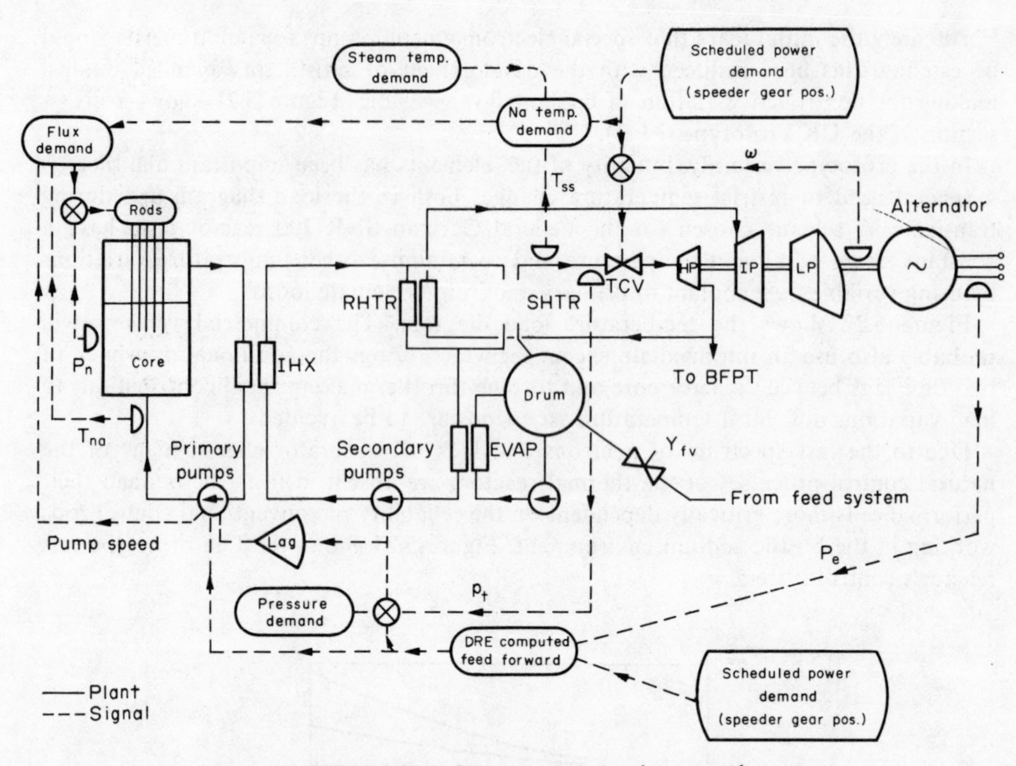

FIG. 5.23. UK prototype fast reactor—control at normal power.

This completes our brief survey of major reactor types and the special characteristics that illustrate the general approach to the specification of a control system and its implementation.

Problems for Chapter 5

The following four questions refer to the sample PWR discussed in the text and should be answered accordingly.

5.1. Sketch P/TL diagrams for T_{av}, T_i, T_o and T_f, T_s and p_s for:

(a) constant throttle with variation of reactivity;
(b) constant reactivity with variation of throttle;
(c) bypass steam dumping with constant T_i;

over a range 20–100% load.

Will the expressions of Table 5.2 be adequate for answering the question?

How will you make use of the thermodynamic data over a substantial range of loads? (*Hint*: Consider enthalpy changes.)

5.2. Is it possible to obtain simultaneously a constant T_{av} and constant T_f P/TL relation in (a) a metallic-fuelled core, and (b) an oxide-fuelled core? What are the relative advantages of the two cases?

5.3. The form of Fig. 5.4 suggests a logarithmic approximation for $p_s(T_s)$. Fit such a curve and hence produce an expression suitable for numerical integration over a range to give a P/TL diagram, 20–100%.

5.4. What are the relative advantages in P/TL relations of constant primary flow and variable primary flow? Suggest a compromise and sketch the general pattern of the P/TL diagram.

5.5. Modify the simple model of a boiler transient response to a step demand on power by representing the reactor power input as an exponential type build up (simple lag) rather than a ramp build-up.

What value of τ/θ (where θ is now the characteristic power input time) secures a minimum power increase of 80% throughout the transient?

What is the steepest rate of reactor power increase in this model?

Typical values for the SGHWR might be $\tau = 35$ s and 150 s power doubling time on a ramp increase. What is the maximum step power that can be met at the 80% level?

5.6. The demonstration of system stability discussed around Fig. 5.12 depended on identical units. If they are not identical, we are involved in carrying out the (vector) functional sum for the combined system

$$(G_f)_{\text{tot}} = \sum_i G_{fi}(j\omega)$$

Can you construct G_f functions for a system of two dissimilar units which satisfy the empirical prescription for system stability but whose combination has a *smaller* margin of stability if there is no synchronous inertia other than the supply units? (If so, perhaps you can suggest a better empirical specification.)

5.7. For a term project or a group design study, select an existing commercial reactor design and prepare a model for analogue or digital computer simulation. Using an appropriate part load relationship and within such temperature restraints as you can determine from the literature and as lie within the scope of a "lumped" model, develop the parameter settings for three-term controllers. Ensure that the unit meets the empirical stability requirements and produce result of studies of the transient response to a step change of ±10% in power at (a) 50%, and (b) 90% load.

References for Chapter 5

1. M. A. Schultz, *Control of Nuclear Reactors and Power Plants* (2nd edn.), McGraw-Hill, New York, 1961.
2. H. Etherington (ed.), *Nuclear Engineering Handbook*, McGraw-Hill, New York, 1956.
3. K. C. Lisk, *Nuclear Power Systems and Equipment*, Industrial Press, New York, 1972.
4. Boiler dynamics and control in nuclear power stations, *BNES Conference Proceedings, London*, 1973.
5. Fast reactor power stations, *BNES Conference Proceedings, London*, 1974.
6. *Shippingport Pressurised Water Reactor*, Addison-Wesley, New York, 1958.
7. I. Gaber, Application of control systems for pressurised water reactors, *Proc. 11th Int. Symp*, 11 (*Instrumentation in Power Reactors*), ISA.
8. Nuclear power plant control and instrumentation, *Symposium Proceedings IAEA, Vienna*, 1973.
9. Computer applications in power plant control, *Nucl. Engr. Int., IPC, London*, 1973.
10. *Engineering for Nuclear Power* (Bulleid Memorial V), Univ. of Nottingham, 1971.
11. Nuclear engineering maturity (Plenary Session), *ENS Conference, Paris*, 1975.
12. L. S. Tong and J. Weisman, *Thermal Analysis of Pressurised Water Reactors, Am. Nucl. Soc.*, Illinois, 1972.
13. M. W. Jervis, Computers in power stations, *IEE Rev.* 119 8R (1972); and On-line computers in nuclear power plants, *Adv. in Nucl. Sci. and Engr.* **11**, Plenum, 1978.

CHAPTER 6

Fluctuations and Reactor Noise†

Introduction

In many cases we may express the kinetic behaviour of a reactor or its associated plant *deterministically* in terms of the mean behaviour of the neutron population, etc. That is, the equations used are based on the expected or mean behaviour averaged either over time or in the sense of an ensemble average, over many repetitions of the circumstances. However, observed variations about this mean can be expected to provide additional information on the nature of the system. A further point is the possibility that on a particular occasion a system may develop behaviour noticeably different to the mean or expected behaviour so that questions of performance and therefore safety arise.

A naive justification for ignoring variations about the mean is to suppose that if there are "enough" neutrons present, the observed behaviour is unlikely to be significantly different from the mean behaviour. That would suggest that a power reactor may be analysed in the mean without regard to fluctuations. Such a viewpoint is possibly true (although not to be justified so casually) but conceals the possibility of obtaining information from power reactors as well as low power reactors from fluctuation studies. If *detectors* sample a small number of neutrons (they would hardly remove them all) then the fluctuations in these detection events may be large about their mean values and convey information even with a large population of neutrons being sampled.

In this chapter we briefly develop a *stochastic* model for the "lumped reactor", developing equations that measure the departure from the mean. This is followed by a short account of three representative experiments or applications of the theory to neutron detection. We can only point out the scope for measuring the fluctuations in other state variables such as coolant pressures, bubble fluctuations, etc., and the reader may pursue these aspects further in Williams.[1] We also omit the interesting topic of *inducing* fluctuations in a known manner and thus making the analysis of observations both easier and more rewarding, a method known as pseudo-random fluctuation measurements. All these may yet develop into valuable techniques.

Measurements of the behaviour of a reactor, such as measurements of flux in an ion chamber or fission chamber, will show fluctuations. Some of these variations are due to the inherent stochastic nature of the process itself; others are due to fluctuations in the detector equipment. At one level we may regard these fluctuations as a nuisance like background, to be eliminated. At a deeper level the fluctuations themselves contain information which may be of interest.

†This chapter contains advanced material which may be omitted on a first reading.

"Noise" itself is taken from acoustical terminology, and implies a gross mixture of frequencies and phases in the observed signals. Nevertheless, hidden within these fluctuations may be a dependence on frequencies related to the system equations.

A particularly happy characteristic of the analysis of systems through noise is the ability to take the measurements on the system "as it is". The ion chamber, etc., is a small perturbation in the system and, indeed, may be of necessity present for other reasons. The reactor does not have to be shut down to have measurements made upon it; the power does not have to be varied grossly as in measuring transfer functions by oscillating a control rod, etc. The information is potentially available in a continuous fashion during the operation of the reactor. A further corollary is that continuous monitoring of this information may show up changes in the reactor behaviour that herald some gross characteristic. For example, a fuel element may have a freedom to vibrate at a characteristic frequency. If the element deteriorates in use, the amplitude may grow, leading to a detectable change in the noise spectrum before any other warning of potential failure.

It must be said that noise studies usually involve substantial time and computing effort to record the fluctuating data, to perform correlation calculations and then perhaps to take Fourier transforms. The availability of dedicated on-line minicomputers, however, particularly with fast Fourier transform capability, has made it feasible to propose noise studies in operational as well as research situations.

Probability Equations

In this section we derive stochastic equations for a reactor-detector model and from these equations obtain further equations describing the behaviour of the first and second moments of the detection events, equivalent to mean and variance equations for the observables. We neglect delayed neutrons, in a "lumped", constant property model with the following elements:

τ_f is the mean time between fissions, called by Hurwitz the generation time. In this model we have $\tau_f = 1/v\Sigma^f = \bar{\nu}\Lambda$.

τ_c is the mean time between *other* capture or loss events, so τ_c is not the conventional neutron lifetime. In this model, $\tau_c = 1/v[\Sigma^a + DB^2 - \Sigma^f]$.

ε/τ_f is a conventional form for the mean rate with which one neutron causes a count in a detector associated with the reactor. This is a neutron removal term, of course, and has been included also in the capture events above.

p_ν as before is the distribution of neutrons is fission probability.

Let $P_{n,m}(t)$ be the probability that at time t there are exactly n neutrons in the reactor *and* exactly m counts have been accumulated in the detector, starting from zero counts at time zero. Figure 6.1 illustrates the processes that may lead to a change in $P_{n,m}$ in a small interval δt.

Such matrices govern the conditional probability, from which we may obtain $P_{n,m}$ $(t+\delta t)$ *given* $P_{n,m}(t)$:

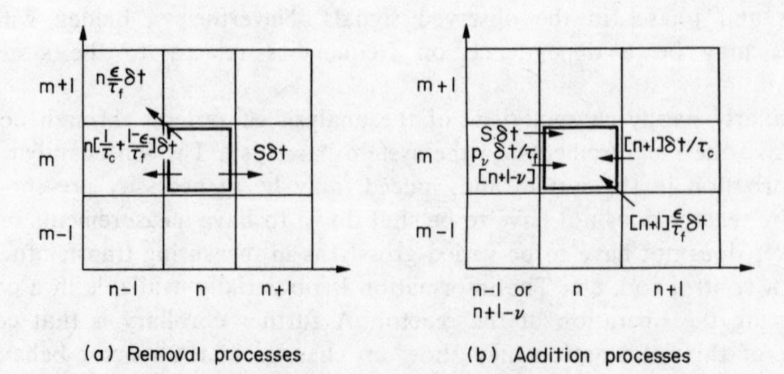

FIG. 6.1. The neutron-detector transfer event matrices.

$$P_{n,m}(t + \delta t) = \left[1 - n \left[\frac{1}{\tau_c} + \frac{1}{\tau_f} \right] \delta t - S\delta t \right] P_{n,m}(t) + [n + 1] \frac{1}{\tau_c} \delta t P_{n+1,m}$$

$$+ [n + 1] \frac{\varepsilon}{\tau_f} \delta t P_{n+1,m-1} + \frac{1}{\tau_f} \sum_{\nu=0}^{n} [n + 1 - \nu] p_\nu \delta t P_{n+1-\nu,m} + S\delta t P_{n-1,m} \tag{6.1}$$

Note the assumption that δt is sufficiently small that at most one neutron is emitted by the independent source, strength S, but the fission produces a probability spread of additional neutrons. By the usual limiting process

$$\frac{dP_{n,m}}{dt} = - n \left[\frac{1}{\tau_f} + \frac{1}{\tau_c} \right] P_{n,m} - SP_{n,m} + [n + 1] \frac{1}{\tau_c} P_{n+1,m} + [n + 1] \frac{\varepsilon}{\tau_f} P_{n+1,m-1}$$

$$+ \frac{1}{\tau_f} \sum_{\nu=0}^{n} [n + 1 - \nu] p_\nu P_{n+1-\nu,m} + SP_{n-1,m} \tag{6.2}$$

Suitable boundary conditions for exactly N initial neutrons and zero counts in the detector are $P_{n,m}(0) = \delta_{nN}\delta_{m0}$.

We have a (doubly) infinite set of ordinary differential equations. There are elegant ways of condensing these to a single partial differential equation from which we may in principle obtain all the information we need about the system. In practice, exact solutions of the resulting equation are not readily obtained and so we chose to go directly to equations to describe mean and variance in terms of the first and second moments.

Recollect from Chapter 1 the concept of expected values so that we might write

$$\langle n \rangle = \sum_{n=0}^{\infty} \sum_{m=0}^{\infty} nP_{n,m}; \quad \langle m \rangle = \sum_{n=0}^{\infty} \sum_{m=0}^{\infty} mP_{n,m} \tag{6.3}$$

and similar expressions for higher moments in $\langle m(m-1) \rangle$, $\langle mn \rangle$ and $\langle n(n-1) \rangle$. Multiply eqn. (6.2) first by n and sum over-all n and m to obtain

$$\frac{d}{dt}\langle n\rangle = \left[\frac{1}{\tau_f}[\langle\nu\rangle - 1] - \frac{1}{\tau_c}\right]\langle n\rangle + S; \quad \langle n\rangle|_{t=0} = N \tag{6.4}$$

Most of these sums are straightforward. The hardest, the triple sum over n, m and ν, can be verified by writing out the first few terms (see problem 6.12). Recollect that $\langle\nu\rangle$ is the measured value $\bar{\nu}$ so that we can identify from the model

$$\frac{1}{\tau_t}[\bar{\nu} - 1] - \frac{1}{\tau_c} = \frac{\rho}{\Lambda} \tag{6.5}$$

and write the equation for the mean neutron behaviour as

$$\frac{d}{dt}\langle n\rangle = \frac{\rho}{\Lambda}\langle n\rangle + S \tag{6.6}$$

Happily, this is exactly the same result as we obtain in a discussion from the deterministic point of view in Chapter 3.

Similarly, multiply the probabilistic equations by m before summing over all n and m to obtain the equation for the mean detector behaviour

$$\frac{d}{dt}\langle m\rangle = \frac{\varepsilon}{\tau_f}\langle n\rangle; \quad \langle m\rangle|t_{=0} = 0 \tag{6.7}$$

a result we might again have anticipated.

Second moment equations are obtained in a similar fashion, multiplying in turn by $n^2 - n$, nm and $m^2 - m$ before summing over all n and m to yield the following equations:

$$\left.\begin{aligned}
&\frac{d}{dt}\langle n(n-1)\rangle = 2\frac{\rho}{\Lambda}\langle n(n-1)\rangle + \left[\frac{\langle\nu(\nu-1)\rangle}{\tau_f} + 2S\right]\langle n\rangle; \ \langle n(n-1)\rangle|_{t=0} = N^2 - N \\
&\frac{d}{dt}\langle mn\rangle = \frac{\rho}{\Lambda}\langle mn\rangle + \frac{\varepsilon}{\tau_f}\langle n(n-1)\rangle + S\langle m\rangle; \quad \langle mn\rangle|_{t=0} = 0 \\
&\frac{d}{dt}\langle m(m-1)\rangle = 2\frac{\varepsilon}{\tau_f}\langle mn\rangle; \quad \langle m(m-1)\rangle|_{t=0} = 0
\end{aligned}\right\} \tag{6.8}$$

Again the triple sum is tedious but a few terms written out may convince the reader. We may write the second moment of the neutron distribution if fission in terms of the Diven parameter, $D = \langle\nu(\nu-1)\rangle/\langle\nu\rangle^2$.

DEVELOPED SOLUTIONS

We have allowed for an initial condition with an arbitrary neutron number N. In practice what happens is that the reactor develops a steady population distribution which may fluctuate about a mean value determined by the source, reactivity, etc., but which has no dependence upon N, the initial value. Although the rate of *detection* of neutrons is not zero, of course, we may assume the developed solution for the *neutron* behaviour and put the left hand side of the pure neutron moment equations to zero to obtain

$$\langle n\rangle \to \frac{S\Lambda}{-\rho}; \quad \langle n(n-1)\rangle \to \langle n\rangle^2\left[1+\tfrac{1}{2}\frac{D\bar{\nu}}{S\Lambda}\right] \tag{6.9}$$

while from the first of these we obtain the elementary result that $\langle m\rangle \to \langle n\rangle \varepsilon t/\tau_f$, i.e. the expected counts rise linearly with time in the developed solution.

Solutions of the remaining equations are readily found to be

$$\left.\begin{aligned}\langle mn\rangle &= \frac{\varepsilon}{\tau_f}\langle n\rangle^2\left[t+\tfrac{1}{2}\frac{D\tau_f}{\rho^2\langle n\rangle}[1-\exp(\rho t/\Lambda)]\right]\\ \langle m(m-1)\rangle &= \left[\frac{\varepsilon}{\tau_f}\langle n\rangle\right]^2\left\{t^2+\frac{D\tau_f}{\rho^2\langle n\rangle}\left[t-\frac{\langle n\rangle}{S}[1-\exp(\rho t/\Lambda)]\right]\right\}\end{aligned}\right\} \tag{6.10}$$

and from these we may find, for future reference, the variance-to-mean ratio of detected counts:

$$\frac{\langle m^2\rangle-\langle m\rangle^2}{\langle m\rangle} = 1+\frac{\varepsilon D}{\rho^2}\left[1-\frac{1-\exp(\delta t/\Lambda)}{\delta t/\Lambda}\right] \tag{6.11}$$

Correlation Functions

How may we observe experimentally whether events in a sequence are related or whether they are drawn from a random set of observations? The technique used is that of correlation functions, with two main variants: auto-correlation and cross-correlation.

AUTO-CORRELATION

Suppose the events are numbered in their sequence $\{x: x_1, x_2, x_3, \ldots x_n\}$. It is simplest to think of these results being uniformly spaced in time with the same interval between successive terms in the sequence. We consider in turn the connection between pairs of adjacent observations, between pairs one interval apart, then two additional intervals, etc. It would be natural to consider the sum of pairs i apart in the form $\sum_{i'}[x_{i'}x_{i'+i}]$. Since the data is finite, however, the evaluation of such a sum in practice is complicated by the lack of pairings for the last $i-1$ items of data. To allow for this it is convenient to define the *auto-covariance* function with a finite data correction term, forming a suitable normalisation or average:

$$S_{xx}(i) = \frac{1}{n-i}\sum_{i'=1}^{i'=n-i} x_{i'}x_{i'+i} \tag{6.12}$$

with i taking the successive values 1, 2, 3, . . ., n for the intervals. It is often convenient to normalise this auto-covariance function $S_{xx}(i)$ by dividing through by the value $S_{xx}(0)$ (which is necessarily positive) to form the *auto-correlation* function $\phi_{xx}(i)$:

$$\phi_{xx}(i) = \frac{S_{xx}(i)}{S_{xx}(0)} = \frac{n}{n-i} \frac{\sum_{i'} x_{i'} x_{i'+i}}{\sum_{i'} x_{i'} x_{i'}} \tag{6.13}$$

Normalised or not, the results will tend to show whether the observations do depend on each other's place in the sequence. To argue this naively it is convenient to suppose for the moment that the x_i refer to observations around a mean, i.e. that the mean of x_i is zero. For a random set of observations we could expect that for every $x_{i'}$, there would be just as many $x_{i'+i}$ above the zero mean as below, so that the sum of the terms over $_{i'}$ would vanish (except in the positive definite case that $i=0$). We might therefore suppose† that for a set of observations of random events about a zero mean, both $S_{xx}(i)$ and $\phi_{xx}(i)$, will vanish for $i \neq 0$. The normalised auto-correlation function will have $\phi_{xx}(0)=1$, however.

When the data is in continuous form, the corresponding argument $x(t)$ is a *density* function for the data. Integral forms can be constructed by analogy over the data range T:

$$S_{xx}(\tau) = \frac{1}{T-\tau} \int_{-\frac{1}{2}(T-\tau)}^{\frac{1}{2}(T-\tau)} x(t)x(t+\tau)dt \tag{6.14}$$

and correspondingly $\phi_{xx}(\tau) = S_{xx}(\tau)/S_{xx}(0)$, the auto-correlation.

In discussing *theoretical* meanings of covariance or correlation in any application, it is convenient to suppose the data collection period becomes unbounded, $T \to \infty$. Then for any finite spacing τ the correction for finite data becomes negligible and we have

$$S_{xx}(\tau) = \mathop{\mathscr{L}}_{T\to\infty} \frac{1}{T} \int x(t)x(t+\tau)dt; \quad \phi_{xx}(\tau) = \frac{\int_{-\infty}^{\infty} x(t)x(t+\tau)dt}{\int_{-\infty}^{\infty} x(t)x(t)dt} \tag{6.15}$$

$S_{xx}(i)$ is thus an average value of $\overline{x_i x_{i'+1}}$ and $S_{xx}(\tau)$ an average value of the corresponding densities and could be written $\overline{x(t)x(t+\tau)}$ (observed).

Such a continuous signal might be the current from an ion chamber. The discrete signals might be the discretely sampled amplitude of such a current or as basic as binary information in an "on–off" or polarity (+ or −) form. For discrete or continuous form it is expected that where the information is the variation about a zero mean then the auto-correlation function vanishes for random sequences (except for the normalisation point i or $\tau=0$) as the amount of information available for processing tends to infinity.

In the case of a Markov system, however, the adjacent results do by assumption affect each other and will show a non-zero correlation for $i=1$, etc. Then the results two apart must influence each other, more weakly of course, through the intermediate

†We have not proved the assertion which for special cases is not actually true (see Feller (ref. 10) **2**, 133). It is true if $\langle u(x_i)v(x_i)\rangle = \langle u(x_i)\rangle \langle v(x_i)\rangle$ for *all* continuous functions u and v vanishing outside a finite range.

result. The consequence is that the *expected* fall off of ϕ_{xx} in a Markovian system has one or more exponential terms, as will be demonstrated. This exponential period is a rough measure of the "memory" of the system and after a few e–folding periods, the correlation function may again be expected to fall to zero (subject again to the vagaries of finite amounts of data) for observations around a mean.

If we assume an ergodic property or a *stationary* system, then the time averages of the data are the same as the expected values. (If not, the expected values are to be understood as an average over an ensemble of many "identical" reactors, etc.) In particular, the auto-covariance function will be the same as the *second moment* of the probability distribution (see Chapter 1).

CROSS-CORRELATION

Here we suppose that we have two sets of observational data in sequences $\{x: x_1, x_2, x_3, \ldots, x_n\}$ and $\{y: y_1, y_2, y_3, \ldots, y_n\}$. We form the cross-variance of x upon y

$$\left.\begin{aligned} S_{xy}(i) &= \frac{1}{n-i} \sum_{i'}^{n-i} x_i' y_i'{}_{+1} && \text{discrete} \\ S_{xy}(\tau) &= \frac{1}{T-\tau} \int_{-\frac{1}{2}(T-\tau)}^{\frac{1}{2}(T-\tau)} x(t)y(t+\tau)dt && \text{continuous} \end{aligned}\right\} \tag{6.16}$$

which again may be normalised by dividing by the value at $i=0$ or $\tau=0$ and called the cross-correlation. The continuous theoretical form for finite τ is then $\phi_{xy}(\tau)=\int_{-\infty}^{\infty} x(t)y(t+\tau)dt/\int_{-\infty}^{\infty} x(t)y(t)dt$ (as a suitable limit).

For random systems in the sense of no dependence of the $\{x\}$ upon $\{y\}$ (and of course vice versa) and where the mean values of x and y are zero, the expectation of the cross-correlation function is again zero. If ϕ_{yx} does not vanish we must ask ourselves "is this because of the finiteness of our data or is it because there is indeed a connection"? The nature of this connection may, of course, be positive or negative (correlated or anti-correlated). Note also that it makes sense to consider negative i or negative τ to see if the connection is between x- and y-values at an earlier time.

However, we must be careful not to interpret the results of a non-vanishing cross-correlation function as proving that x *causes* y or y *causes* x. It may equally be that both x and y depend on some common cause. Thus both umbrellas and raincoats appear on the streets correlated but a rational hypothesis is not to say that raincoats cause umbrellas, but that both are caused by the prospect of bad weather. Of course, a hypothesis of cause and effect would be disproved if there were no correlation, and statistics must frequently work in such a negative way.

FOURIER TRANSFORMS OF CORRELATION FUNCTIONS

Having thought of correlation functions as dependent upon time, either uniformly spaced intervals or continuous measurement, it becomes possible to take Fourier transforms and examine the correlation in the frequency space. We briefly derive certain useful results, assuming unnormalised covariance functions.

CONVOLUTION

Recollect the Green function convolution solution to a linear response problem. If $g(t)$ is the response of the system to an impulse source at time zero, then for any source $s(t)$ the output $n(t)$ is

$$n(t) = \int_0^t g(t - t')s(t')dt'$$

We wish to extend the range of integration to infinity in order to make use of Fourier transforms. We may do this by first extending the definition of $g(t)$ such that it is zero before the impulse is introduced (which is physically obvious) and extend the source to account for all initial conditions so that we may write formally

$$n(t) = \int_{-\infty}^{\infty} g(t - t')s(t')dt' \tag{6.17}$$

Take the Fourier transform and change order of integration:

$$\bar{n}(j\omega) = \int_{-\infty}^{\infty}\int_{-\infty}^{\infty} \exp[-j\omega(t-t')]g(t-t')\exp(-j\omega t')s(t')dt'dt$$

$$= G(j\omega)\int_{-\infty}^{\infty} \exp(-j\omega t')s(t')dt' = G(i\omega)\bar{s}(i\omega) \tag{6.18}$$

This may be done with certain provisos in the integrability of n, s, $|n|^2$ and $|s|^2$ that are usually met. Whilst $g(t)$ is real in most of our applications, we shall require in a moment the formal relation for real $g(t)$ that $G^*(j\omega) \equiv G(-j\omega) = \int_{-\infty}^{\infty}\exp(j\omega t)g(t)dt$.

CROSS-COVARIANCE

Assume unnormalised covariance functions. Then as $T \to \infty$

$$S_{ns}(\tau) \to \frac{1}{T}\int_{-\infty}^{\infty} s(t)n(t + \tau)dt = \frac{1}{T}\int_{-\infty}^{\infty}\int_{-\infty}^{\infty} s(t)g(t + \tau - t')s(t')dt'dt$$

so if $P_{ns}(j\omega)$ is the cross-spectral density, Fourier transform of the cross-covariance function of n upon s, then

$$P_{ns}(j\omega) \to \frac{1}{T}\iiint_{-\infty}^{\infty} \exp(-j\omega[t + \tau - t'])g(t + \tau - t')\exp(-j\omega[t - t'])s(t)s(t')d\tau\, dt'dt$$

$$= G(j\omega)\frac{1}{T}\iint_{-\infty}^{\infty} \exp(-j\omega\tau)s(t)s(t + \tau)dt\, d\tau$$

$$= G(j\omega)P_{ss}(j\omega) \tag{6.19}$$

under a suitable limiting process for $T \to \infty$.

AUTO-COVARIANCE

We have again as $T \to \infty$.

$$S_{nn}(\tau) \rightarrow \frac{1}{T} \int_{-\infty}^{\infty} n(t)n(t+\tau)dt$$

$$= \frac{1}{T} \iiint_{-\infty}^{\infty} g(t-t')g(t+\tau-t'')s(t')s(t'')dt'dt''dt$$

Then taking Fourier transforms to yield the power spectral density,

$$P_{nn}(j\omega) \rightarrow \frac{1}{T} \iiiint_{-\infty}^{\infty} \exp(-j\omega[t+\tau-t''])g(t+\tau-t'') \exp(j\omega[t-t'])g(t'-t)$$

$$\exp(-j\omega[t''-t'])s(t')s(t)\, d\tau\, dt\, dt'dt''$$

$$= G(j\omega)G^*(i\omega) \frac{1}{T} \iint_{-\infty}^{\infty} \exp(-j\omega[t''-t'])s(t')s(t'')dt'dt''$$

$$= G^2(j\omega)P_{ss}(j\omega) \tag{6.20}$$

If the functions (i.e. $g(t)$) are not real, correlation definitions are modified to correlate a function with its complex conjugate, and results of similar form will be obtained.

Experimental Applications

Three experimental situations will be analysed. In the first the stochastic nature of the reactor is ignored and the noise arises from an assumed random source, modified deterministically by the reactor as a multiplying, linear system, and detected in the form of random output of detectable neutron or flux measurements. Such a measurement may be made to yield some information on the reactor transfer function. In the remaining two experiments the focus is on the reactor as a stochastic process and the fluctuations that arise from the branching process of neutron multiplication are analysed to yield the so-called Rossi-α parameter, $\alpha = [\rho - \beta]/\Lambda$ (or the prompt reactivity if delayed neutrons are ignored on the time scale used).

SPECTRAL DENSITIES AND TRANSFER FUNCTIONS

The block diagram (Fig. 6.2) indicates a random input variable passed through a determinate system and producing a stochastic but modified output variable. We can identify the input variable as being the variation of source, or perhaps the reactivity term if this fluctuates about its mean, and the output variable as being the variation of the measured neutron density or flux about its mean. If it were feasible to measure input as well as output, the correlation between them, $S_{ns}(\tau)$, would depend on the nature of the transfer function. It was established in the last section in fact that the output may

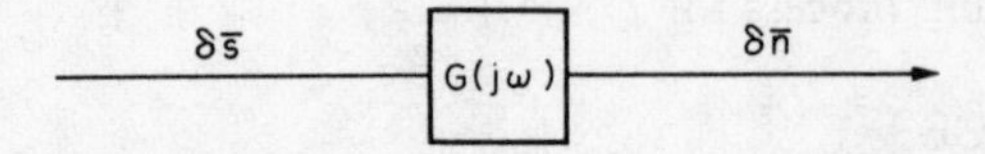

FIG. 6.2. Block diagram for reactor fluctuations.

be obtained as the convolution of the input and the system dynamics, even though the state variables are fluctuating. On taking Fourier transforms† and disregarding the particular initial conditions, we have the following relations discussed in the last section:

$$P_{ns}(j\omega) = G(j\omega)\, P_{ss}(j\omega) \tag{6.21}$$

where $P_{ss}(j\omega)$ is the power spectral density (PSD), the Fourier transform of $S_{ss}(\tau)$, and $P_{ns}(j\omega)$ is the cross-spectral density, the Fourier transform of $S_{ns}(\tau)$.

In fact we may not have a way of measuring the input, but if we may make some assumptions about its nature we can utilize a similar relation involving the PSD of input and output (again established in the last section):

$$P_{nn}(j\omega) = G^2(j\omega)\, P_{ss}(j\omega) = |G|^2 P_{ss} \tag{6.22}$$

A significant difference here is that the only information apparently available arises in the form of the *magnitude* of the transfer function, i.e. amplitude as a function of frequency, but not phase.

To interpret eqn. (6.22) we must make some assumption about the spectral density of the input $P_{ss}(j\omega)$. The simplest assumption that can be made is that the source variation arises from a purely random distribution or "white" noise. The auto-covariance function of this distribution would be expected to show no dependence on correlation time and would be proportional to a (Dirac) delta distribution $\delta(\tau)$. Correspondingly, the Fourier transform or PSD will have all frequencies equally represented in amplitude. Since the covariance function is symmetric in τ, the Fourier transform is real, with zero phase. The assumption of white noise is acceptable in the range of frequencies of interest.‡

With this assumption, the PSD of the output, the measurable signal, is proportional to the square of the magnitude of the transfer function itself and eqn. (6.22) can be interpreted accordingly. In the one-precursor model

$$G^2(j\omega) = \frac{\lambda^2 + \omega^2}{\omega^2[\lambda - \beta/\Lambda]^2 + \omega^2} \rightarrow \sim \frac{1}{[\beta/\Lambda]^2 + \omega^2} \tag{6.23}$$

where the final term covers the behaviour at *sufficiently* high frequencies and we have assumed the reactor is at delayed criticality,

$$\rho = 0, \quad \alpha_d = \beta/\Lambda$$

Figure 6.3 shows a plot of the PSD in the University of London Research Reactor (Silwood Park) against frequency. It may be seen that the characteristic plateau at intermediate frequencies is well established together with a "roll-off" at high frequencies with slope 20 db per decade corresponding to the term in ω^2. (That is, for *power*, Shandasini uses 1 db = 10 log.)

The Rossi-α may be deduced either from the break frequency method (shown as I

†We continue to interpret these formulae as variations about a stationary mean. Otherwise the Fourier integrals would give difficulties if investigated too carefully. The interpretation is quite natural, however, in a linear system.

‡"White" noise is usually taken to have random phases but this point is immaterial in the present context. The discrepancy is because we are taking the Fourier transform of the auto-covariance and not of the true behaviour of the noise itself.

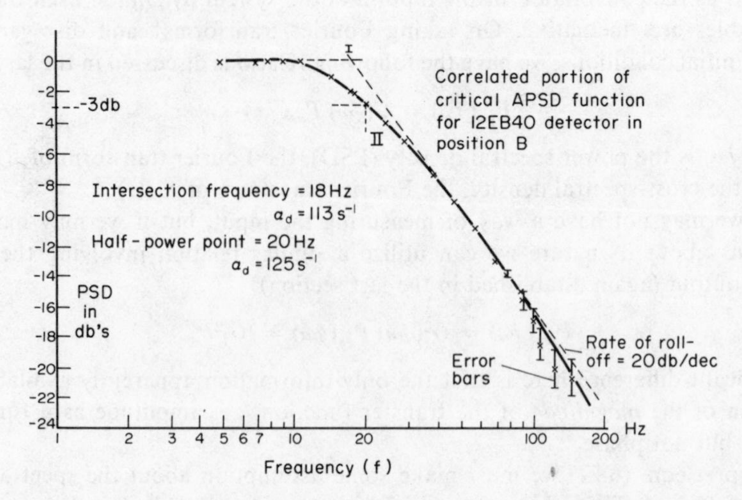

FIG. 6.3. Power spectral density in the University of London Research Reactor (Shandasini, R.).

in the figure) or by noting that it should also be the frequency at which the PSD is halved from the plateau value (see eqn. (6.23)) (shown at II in the figure).

Limitations of the experiment arise chiefly towards the lower end of the frequency scale with distortion in the detector equipment.

This experimental method ignores stochastic processes in the reactor itself and would be most suitable therefore when a strong source dominates the fission stochastic process, i.e. at lower power and low reactivities. We turn therefore to two experiments directed to the reactor itself as a stochastic process.

ROSSI-α EXPERIMENT

Bruno Rossi suggested the auto-correlation of a sequence of counts as a way to determine $\alpha = [\beta - \rho]/\Lambda$. We "gate" a detector into sufficiently small width channels or intervals ΔT such that at most one count is recorded. We then have a sequence of counts available which may be auto-correlated; note that in this usage the results will be normalised to give $\phi_{xx}(0)$ but that the mean value $\langle x \rangle$ is not itself zero. The experiment was first carried out by Orndoff.[4]

To demonstrate the significance of the auto-correlation function in this instance it may be helpful to take a numerical example. Suppose, therefore, we have a sample sequence of sixteen intervals ($n = 16$) in which there are six counts ($m = 6$) in the form of a sequence $\{x: 0,1,1,0,0,0,1,0,1,0,1,0,0,0,1,0\}$ (although in practice many more intervals are needed). We may tabulate $\sum_i x_{i'}x_{i'+i}$ against i to obtain Table 6.1.

Consider the corrected and normalised correlation function

$$\phi_{xx}(i) = \frac{1}{n-i}\sum_{i'} x_{i'}x_{i'+i} \Big/ \frac{1}{n}\sum_{i'} x_{i'}^{\prime 2} = S_{xx}(i)/S_{xx}(0) \tag{6.24}$$

Now the actual number of pairs of counts i apart is $\sum_{i'}$. The *potential* number of pairs i

TABLE 6.1.
Example auto-correlation sequence

i	0	1	2	3	4	5	6	7	...
Σ_i'	6	1	2	0	3	1	...		
$n-i$	16	15	14	13	12	11	...		
$S_{xx}(i)$	$\frac{6}{16}$	$\frac{1}{15}$	$\frac{2}{14}$	0	$\frac{3}{12}$	$\frac{1}{11}$	...		
$\phi_{xx}(i)$	1	0.178	0.381	0	0.667	0.242	...		

apart is $n-i$. Thus $S_{xx}(i)$ is the probability of a count in one channel *and* a count in another channel i forward; it is thus a joint probability. When $i=0$, however, we have a joint probability of identical events which therefore reduces simply to the single probability of a count in a channel, $S_{xx}(0)$. It follows that $\phi_{xx}(i)$ is the joint probability of two events divided by the probability of one of the events, i.e. it is the *conditional* probability of a second count given the first. The unnormalised auto-covariance function $S_{xx}(i)$ can be interpreted as the joint probability itself.

To relate the experimental measure to theory, note first that the total number of potential coincidence counts is given by $\frac{1}{2}n[n-1]$ and that the actual number is given by $\frac{1}{2}m[m-1]$ or by 150 and 15 respectively. It may be seen directly, of course, that the probability of a count is, indeed, 6/16 in this example, consistent with the interpretation of $S_{xx}(0)$.

That is, the expected number of coincidence counts will be $\frac{1}{2}\langle m(m-1)\rangle$ where $m(m-1)$ gives the total number of pairs possible, the half suppresses the ordering of pairs and the expectation value reduces the possible values to expected values. Equally well, the same result must arise from a suitable double integration of the joint probability distribution function (PDF) $\Gamma(t_1,t_2)$ (a double differential density) for seeing a count in dt_1 about t_1 *and* a count in dt_2 about t_2. Thus

$$\tfrac{1}{2}\langle m(m-1)\rangle = \int_0^{\tau}\int_0^{t_1} \Gamma(t_1,t_2)dt_1dt_2 \tag{6.25}$$

where τ is now the counting interval 0 to τ. It is part of the hypothesis for stationary systems (developed solutions to the neutron equations) that the PDF $\Gamma(t_1,t_2)$ is a function of the difference $t_1-t_2=\tau$ only, i.e. becomes $\Gamma(\tau)$. Using the result of eqn. (6.10), therefore, we can double differentiate to obtain

$$\Gamma(\tau) = \tfrac{1}{2}\frac{d^2}{d\tau^2}\langle m(m-1)\rangle = \left[\frac{\varepsilon\langle n\rangle}{\tau_f}\right]^2\left[1+\tfrac{1}{2}\frac{D\bar{\nu}}{S\Lambda}\exp(\rho\tau/\Lambda)\right] \tag{6.26}$$

where $\varepsilon\langle n\rangle/\tau_f$ is the mean counting rate of course, $d\langle m\rangle/dt$, so (6.26) shows an uncorrelated and a correlated component, the latter diminishing at higher source S.

From the interpretation of the auto-covariance function as a joint or coincidence probability, then $S_{xx}(i)/\Delta T^2$ is an estimator for $\Gamma(\tau)$. Also we have $\Gamma(\tau)\Delta T^2 \simeq (d\langle m\rangle/dt)\,\Delta T\phi_{xx}(i)$, if we prefer to work with the normalised auto-correlation function.

Figure 6.4 illustrates the experimental results carried out on the Godiva fast assembly[4]

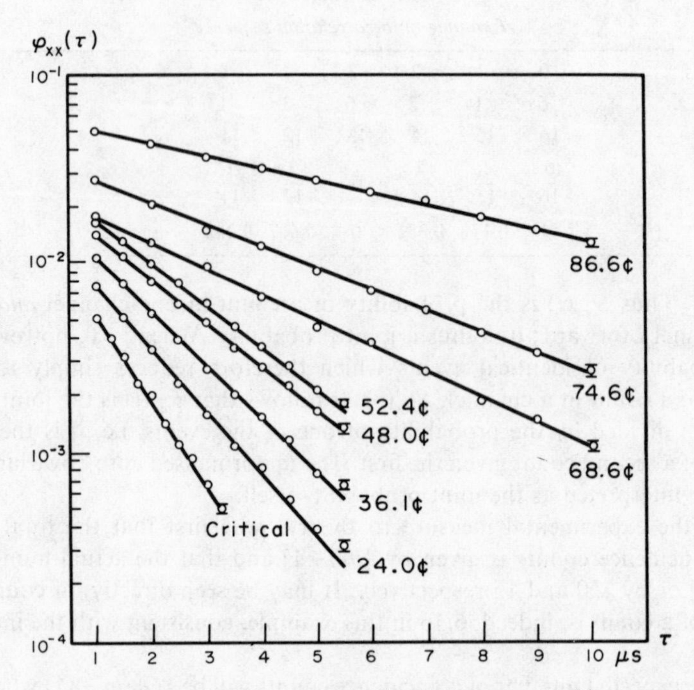

FIG. 6.4. Rossi-α results in the Godiva assembly.[4]

showing log plots of ϕ_{xx} against for different reactivities. These results and eqn. (6.26) illustrate the exponential falling off of $\phi_{xx}(\tau)$ in a Markov system.

Equation (6.26) may be written in terms of the reactor fission rate $F = \langle n \rangle / \tau_f$ to see limitations on the experimental method more clearly. Detector efficiency is not material in distinguishing uncorrelated from correlated terms in this experiment. However, the leading term becomes $[\varepsilon F]^2$ and is evidently the uncorrelated component of the joint PDF $\Gamma(\tau)$. The correlated term has a further factor of $[\frac{1}{2} D / F\Lambda[-\rho]] \exp(\rho\tau/\Lambda)$. Low fission rate favours the correlated term and a fast reactor, where Λ is small, also favours the correlated term. Thus the Rossi-α experiment is generally limited to fast assemblies at low power.

Here, however, it has been a valuable independent method of determining the kinetic parameter α, especially for the case of (delayed) critical when $\alpha_d = \beta/\Lambda$, a particularly important parameter. Reasonable detector efficiency, ε is necessary of course, to secure good statistics in the counting interval. In the results quoted, a fission chamber with 0.1 g ^{235}U was used in an assembly of some 50 kg uranium. Thus ε may be estimated as 2×10^{-6} (and hence the conventional form of definition).

FEYNMANN-α EXPERIMENT

A related coincidence experiment was suggested by Feynmann and may also be used to determine the Rossi-α. It consists of determining the variance-to-mean-ratio (VTMR) of counts m in a neutron detector accumulated over an interval 0 to τ. If such an experiment was repeated for the same interval to τ we could build up an estimate of the mean and variance of $m(\tau)$ and hence of the VTMR as a function of τ. In the previous section the theory for this expression was shown to be

$$\frac{\langle m^2\rangle - \langle m\rangle^2}{\langle m\rangle} = 1 + \frac{\varepsilon D}{[-\rho]^2}\left[1 - \frac{1-e^{-\alpha\tau}}{\alpha\tau}\right] \tag{6.27}$$

This curve may be fitted to an experimental determination of the VTMR as a function of τ, thus determining the fitted parameters such as α.

In practice of course, the counts at τ are not repeated in quite this way. Rather, a long count is gated into many channels as a count rate from which an integration from t to $t+\tau$ gives the VTMR as a function of the correlation interval τ. The reactor is run in the steady state, sub-critical with a source and again VTMR$\to$1.

The final bracketed term, which may be called the Feynmann correction, of eqn. (6.27) is worthy of consideration. If $\tau\to\infty$, the correction tends to 1. If τ tends to zero then the Feynmann correction tends to $\frac{1}{2}\tau\to 0$. Thus the VTMR tends to unity. In the absence of fuel (fissionable material), $\rho\to-\infty$, leaving only the source.

We have, therefore, that the departure from unity VTMR is caused by the fission chain process; in Poisson type fluctuations from a random source the VTMR is known to be unity, as in the analysis of the PSD. This departure is promoted by high efficiency detectors and is large for reactivities close to critical. It is not directly a function of the reactor power or reproduction time so that the experiment may be carried out in thermal reactors efficiently.

It is necessary because of the Feynmann correction, however, to make τ adequately large to the point where the prompt neutron behaviour alone, neglecting delayed neutrons as we have done, may be inadequate.

Figure 6.5 shows a fitting of results in the Ford reactor by Albrecht in which a number

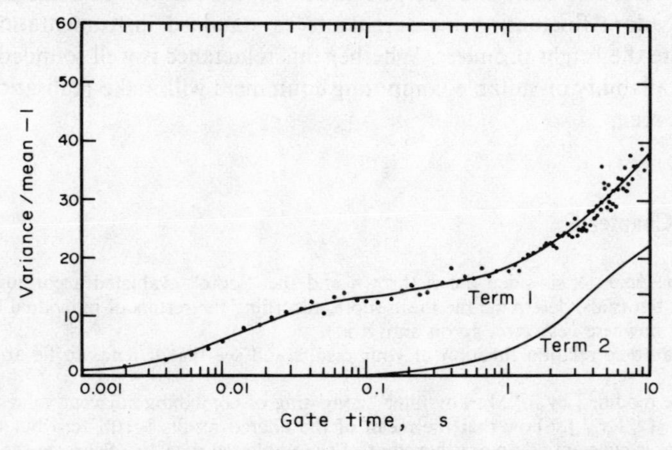

FIG. 6.5. Feynmann-α results in the Ford reactor showing delayed neutron analysis. (After Albrecht.[9])

of delayed neutron terms were taken into account and the necessity for this is evident from the figure. We quote without derivation a one-group of precursors result, that

$$\text{VTMR} = 1 + \frac{\varepsilon D}{[\beta - \rho]^2}\left[1 - \frac{1 - e^{-\alpha\tau}}{\alpha\tau}\right] + \frac{\varepsilon D}{\rho^2}\frac{\beta - \rho}{\beta}\left[1 + \frac{1 - e^{\lambda\rho\tau/\beta-\rho}}{\lambda\rho\tau/[\beta - \rho]}\right] \quad (6.28)$$

Even here there are terms which indicate a difficulty if the experiment is carried out in a critical reactor where $\rho \to 0$. An analysis of the final correction now gives a term proportional to ρ, which cancels only one of the poles arising from the term in $1/\rho^2$. A similar problem arises if we look back at the Rossi-α theory, that there is an unbounded growth in the expression of interest, eqn. (6.26), when the appropriate reactivity is allowed to tend to zero.

There is a straightforward explanation of the difficulty; the whole theory depends on the ergodic, stationary assumption, that amongst other things the variance should, like the mean, tend to a stationary value. This can be demonstrated not to be true in a critical reactor. Put another way, although the probability of having fluctuations in a critical reactor above the mean is small, it is not vanishingly small as far as its effect upon higher moments is concerned. Of course, in practice we do not allow large fluctuations above the mean (or if we did they would invalidate the other assumption of constant properties implying no heating or fuel burn-up); thus it is the constant property model that is naive rather than the mathematical analysis at fault.

Conclusion

There are many other aspects of neutron fluctuations that can be investigated and these are well summarised theoretically by Pacilio *et al.*[7] We remind the reader again that there are other fluctuating state variables of potential interest in the reactor and its plant. It is also worth mentioning that cross-correlation between two or more detectors is valuable as a way of eliminating the background fluctuation of the detectors themselves.

It must be said, however, that despite much experimental work, there has been a reluctance to adopt fluctuation measurements as standard instrumentation in power reactors despite the bright promises. Whether this reluctance is well founded or whether the present availability of suitable computing equipment will make realisation practical, remains to be seen.

Problems for Chapter 6

6.1. *A Markov filter.* A six-sided die is thrown and the "score" evaluated according to the table below. For, say, ten trials, determine the mean score, recording the results of individual events. (If the mean is patently nowhere near zero, go on until it is.)

Evaluate the auto-correlation function of your results and see that it tends to lie around zero for $i \neq 0$.

The scores are modified by a "Markov filter" consisting of combining adjacent values according to the relation $y_i = \frac{1}{2}[x_i + x_{i+1}]$. Show that the mean of the filtered results is still zero but that the auto-correlation of $\{y\}$ is substantially more than that of the unfiltered data $\{x\}$. Estimate the "memory" of the filter and discuss its analytical representation in the frequency transform space.

(This is an amusing blackboard topic for the lecturer willing to submit professional reputation to the vagaries of chance.)

Score table:	die shows:	1	2	3	4	5	6
($\langle \text{score} \rangle = 0$)	score:	1	2	2	−1	−2	−2

6.2. Find the power spectral density for the six-group precursor model.

The following five problems introduce the compact notation of the generating function to problems involving probability equations.

6.3. *Radioactive counting processes.* Show from physical principles that the probability $P_m(t)$ of observing m counts in α-decay in the interval 0 to t satisfies the equation

$$\frac{dP_m}{dt} = a[P_{m-1} - P_m]; \quad P_m(0) = \delta_{no}$$

where a is the expected number of counts per unit time. Show that the solution is a Poisson distribution

$$P_m(t) = \frac{[at]^m}{m!} e^{-at}$$

What is the mean and what is the variance of this distribution?

Following the method around eqn. (6.25), find the auto-covariance of such a source.

6.4. *Distribution of radioactive atoms.* Let $p_n(t)$ be the probability of exactly n atoms existing in α-decay if there were exactly N atoms at time zero and the decay parameter is λ. Show that the p_n satisfy

$$\frac{dp_n}{dt} = \lambda[n+1]p_{n+1} - \lambda n p_n; \quad n = 0, 1, 2, \cdots, N-1$$

$$\frac{dp_N}{dt} = -\lambda N p_N$$

i.e. an $N+1$th order differential system. What are the boundary conditions?

Show that the system can be converted to a single partial differential equation

$$\frac{\partial G}{\partial t} = \lambda[1 - x]\frac{\partial G}{\partial x}; \quad G(x,0) = x^N$$

where the generating function $G(x,t) = \sum_{0}^{\infty} x^n p_n(t)$. Solve, using Lagrange's method for linear first order partial differential equations.

Demonstrate the utility of the generating function by considering

$$\begin{aligned} G(0,t) &= p_0 & G(1,t) &= <1> = 1 \\ G_x(0,t) &= p_1 & G_x(1,t) &= <n> \\ G_{xx}(0,t) &= 2p_2 & G_{xx}(1,t) &= <n(n-1)> \end{aligned}$$

etc., where the notation $G_x(y,t)$ is $\partial G/\partial x|_{x=y}$, etc.

6.5. *Duplet fission.* Consider the probabilistic equations in the lumped model with no delayed neutrons and no detector in the special case of duplet fission, i.e. that $p_\nu = \delta_{2\nu}$; a fission always produces two neutrons exactly. Omitting the source, show that the infinite set of ordinary differential equations can be represented as the single partial differential equation via the generating function of problem 6.4:

$$\frac{\partial G}{\partial t} = \frac{1}{\tau_f}[x-1][x-1+\rho]\frac{\partial G}{\partial x}; \quad G(x,0) = x^N$$

and solve for G. (*Hint.* Take $N=1$ in the first instance and consider $G_N = G_1{}^N$ after using the Lagrange method for linear partial differential equations to solve for G_1). What happens to the solutions in a critical reactor? Find expressions for the mean and variance of n.

6.6. Derive the generating function equations for the general case of fission and source, i.e. convert eqn. (6.2). Obtain from this notation directly the equations for the first and second (factorial) moments, $<m>$, $<n>$ and $<n(n-1)>$, $<nm>$, $<m(m-1)>$.

6.7. Extend the probabilistic equations by considering a second detector. Show that the cross-covariance is given by

$$\Gamma_{m_1,m_2}(\tau) = \frac{\epsilon_1\epsilon_2}{\tau_f^2} <n>^2 \left[1 + \tfrac{1}{2}\frac{D\nu}{S\Lambda}\exp(\rho\tau/\Lambda)\right]$$

Note that the cross-correlation expression is little different from the auto-correlation expression but the use of two detectors eliminates random *detector* fluctuations. (*Hint*: Determine directly the Laplace transform of $\frac{1}{2}d^2 <m_1 m_2>/dt^2$ in the developed solutions.)

6.8. Derive the one-group delayed neutron correction terms for the Rossi-α and Feynman-α experiments. (*Hint*: Use Laplace transforms on the first and second moment equations after making the developed solution assumptions.)

6.9. Derive eqn. (6.23) for the power spectral density relation. Extend to I precursor groups.

6.10. Attempt to take Fourier transforms of the uncorrelated part of the auto-covariance function, eqn. (6.26). Is this made any easier if the function is shifted to refer to a variation about a mean value (where the uncorrelated part must have a delta distribution)?

6.11. *Runs test*. A useful test for randomness in a population which is supposed to be approximately normally distributed, i.e. a test for independence and zero correlation in the sequence of observations, is the Runs test. Successive observations may be compared to the mean and classified as above, 1, or below, 0, and thus represented as a sequence such as 00101011001. Each *change*, from 0 to 1 or 1 to 0, is a break point and the number of runs in the sequence having the same character is 1 plus the number of break points. The more correlation or the more the data comes from more than one population, the more the data will tend to lie together above or below the median. High R measures this tendency of non-randomness, a test that is made quantitative as follows.

For a normal distribution the expected value of R in n observations is $\frac{1}{2}n+1$ (but take n the next lower integer if n odd). The variance is approximately $n(n-2)/4(n-1) \simeq (n-1)/4$. Compare this with the probability of obtaining the observed value of R from the tables of the normal probability integral

$$\text{Prob}\left(Z \geqslant \frac{R - <R>}{\sigma}\right)$$

Test the following sequences for independent randomness

10110011111000010011 and 10010101010001101010 0

Note. These ideas may be extended into polarity correlations where the sign only of signals is observed and greatly eases the problem of computing correlation functions and their transforms (Pacilio).[7]

What could you take as $<R>$ and var (R) if the sequence comes from a table of random numbers, probability distribution function $f(x)=1$, $0 \leqslant x \leqslant 1$.

What test would you apply for the hypothesis of the actual result being given by the assumed distribution?

6.12. As a preliminary to verifying the factorial moment equations derived from eqn. (6.2) show that

$$\sum_{n=0}^{\infty} \sum_{\nu=0}^{n} [n+1-\nu] p_\nu P_{n+1-\nu} = <n>$$

$$\sum_{n=0}^{\infty} \sum_{\nu=0}^{n} n[n+1-\nu] p_\nu P_{n+1-\nu} = <n(n+\nu-1)>$$

$$\sum_{n=0}^{\infty} \sum_{\nu=0}^{n} n[n-1][n+1-\nu] p_\nu P_{n+1-\nu} = <n(n-1)(n+2\nu-2)> + <\nu(\nu-1)><n>$$

where

$$\sum_0^\infty P_n = 1; \qquad \sum_0^\infty nP_n = <n>; \quad \sum_0^\infty n^2 P_n = <n^2>; \qquad \sum_0^\infty n^3 P_n = <n^3>$$

and

$$\sum_0^\infty p_\nu = 1; \qquad \sum_0^\infty \nu p_\nu = <\nu>; \qquad \sum_0^\infty \nu[\nu-1] p_\nu = <\nu(\nu-1)>$$

References for Chapter 6

1. M. M. R. Williams, *Random Processes in Nuclear Reactors*, Pergamon, Oxford, 1974.
2. J. A. Thie, *Reactor Noise*, Rowman & Littlefield, New York, 1963.
3. N. Pacilio, *Reactor-noise Analysis in the Time Domain*, Nat. Bureau Standards, Springfields, USA, 1969.
4. J. D. Orndoff, Prompt neutron periods in metal critical assemblies, *Nucl. Sci. and Engr.* **2**, 450 (1957).
5. M. D. Bartlett, *Introduction to Stochastic Processes*, Cambridge, 1955.
6. J. Lewins, On the interpretation of Markov processes in a nuclear reactor, *Atomkernergie* **23**, 172 (1974).
7. N. Pacilio *et al.*, Theory of reactor noise, *Adv. in Nucl. Sci. and Tech.* **11**, Plenum, New York, 1978.
8. G. E. P. Box and G. M. Jenkins, *Time Series Analysis* (rev. ed.) Holden-Day, San Francisco, 1976.
9. R. W. Albrecht, Measurement of dynamic nuclear reactor parameters, *Trans. Amer. Nucl. Soc.* **4**, 311 (1961).
10. W. Feller, *An Introduction to Probability Theory and Its Applications* (vols. 1 and 2), Wiley, New York, 1966.

CHAPTER 7

Safety and Reliability

Introduction

This chapter deals principally with aspects of reliability and its consequences for safety. In the course of the discussion, general material on control and the instrumentation needed to provide for such control in a nuclear reactor will be developed without, however, going into the design and operation of such instrumentation. To speak of reliability we must speak of the probabilities of failure of elements and how these elements may be combined in the design to reduce the risks associated with their failure to acceptable levels where the occurrences arise stochastically or by chance with a consequent variability. In addition, then, to the mathematics of probabilistic reliability the chapter discusses suitable guidelines and philosophies to be implemented in a safe and reliable control system for a nuclear reactor. Valuable supplementary material will be found in *Nuclear Power Station Operation*(8) and in the US *Light Water Reactor Safety Studies*.(7)

Everyone concerned with the design, construction and operation of a nuclear reactor must give consideration to safety, and this is by no means the sole province for the specialist in reactor kinetics and control. Such a person, however, is likely to be deeply involved in certain safety aspects in view of the contribution to be made by the dynamical study of what might happen in accident situations, almost of necessity time dependent. He will play a leading part, therefore, in specifying the control system to act in a safe fashion and in studying the consequences of malfunctions or even "deliberate accidents". This work cannot be done satisfactorily without some knowledge of the likelihood of malfunction and corresponding reliability theory. Concepts of risks are not only necessary to analyse a given system but are a valuable tool in the synthesis or design of a system *ab initio*.

If any justification of safety were needed it could be given under the main headings of a nuclear engineer's responsibility for:

(a) the safety of the general public: now and future generations;
(b) the safety of the work force engaged in reactor operations, etc.;
(c) the economic safety of the plant itself.

In view of the large capital expenditure in the plant, this last factor alone would be adequate motivation for those responsible to see that high standards are applied to prevent both outage and the consequent loss of revenue and to prevent damage, though most of us would accept the principle that the safety of human life should be at least as important as the direct economic safety criterion.

There are three stages in developing a safety approach to reactors which themselves reflect an historical development:

(a) reactors should be built with certain intrinsic properties promoting safety, e.g. a negative void coefficient of reactivity;
(b) reactors should be built with a containment vessel to hold back the consequences of a malfunction, especially of course the release of radioactive material;
(c) safety against unlikely accidents having nevertheless severe consequences should be engineered with supplementary safeguards, e.g. the ice pressure suppression systems of LWRs and the emergency core cooling systems (ECCS).

Two different approaches to the same goal have been developed and the choice between the two philosophies is still an open one. On the one hand, there is an attractiveness about the analysis of risks in terms of probabilities and consequences as a logical basis for the degree of safety demanded and provided. This approach enables rational comparisons to be made with other risk situations (coal mining, smoking, etc.) and the comparative cost benefits of different policies displayed (see Farmer and Beattie[17] as an introduction). On the other hand, some will have grave doubts about the validity of the end product of the immensely complicated risk analysis called for and the significance of probabilities postulated in the 10^{-7} or 10^{-8} range where no direct frequency evidence is available.

Such critics tend to prefer a more qualitative approach (which might be called engineering design rather than engineering science) where professional judgement is applied to make the damaging consequences of a failure incredible rather than, as it were, design for failure at some low frequency. It is obviously easier to convey in a textbook some ideas about the first, analytical approach from first principles than it is to encapsulate all engineering experience. Good practice, however, must at present be a question of accommodating both points of view.

SAFETY GUIDELINES

Experience suggests a number of guidelines in developing the design of control and safety systems for reactors. The first two of these might even be dignified as philosophies in their own right.

(a) REDUNDANCY

It is generally thought unwise to rely upon one flux-measuring device, say, to indicate a safe level of power operation; a fault may develop in the instrument, giving a misleading reading without the operator or control system being aware of the fault. Such devices should at least be duplicated or triplicated. It is reasonable to suppose that all three instruments will not malfunction at once, so that there is a better reliability for the indication of unsafe operation and consequent corrective action.

(b) "FAIL SAFE" OPERATION

If there should be a failure of a component, however, it is highly desirable that it should "fail safe". That is, a malfunction should promote safety. If this is an instrument malfunction, the likely failure mode should be into a situation indicating danger rather than safety (don't let the needle get stuck in the safe operating zone). If we are considering a safety device, it should operate to shut down the reactor safely in the likely failure mode. A good example of this philosophy is to hold up control absorbers by a magnetic clutch, activated by an electric current. Failure of the electricity supply can allow gravity, which seldom fails, to return the control rod into the safe, shutdown position. (And the designer will take care that the falling control rod is safely brought to rest and does not fall on out of the bottom of the reactor.) Similarly, instrumentation designed to indicate that it is safe to start the reactor up (source and some neutron presence indicates) is interlocked with the magnetic clutch to prevent the control rod being lifted in an unsafe indication.

(c) SPURIOUS SHUTDOWN

As they stand, these two guidelines lead to difficulties over spurious faults which would shut the reactor down.† Spurious faults must be expected; a figure of one per month in a complex instrument is not unreasonable. If every fault shut down the reactor and with many instruments, the time lost would be substantial. A rapid shutdown may of itself promote undesirable consequences, so that there are good reasons for limiting the overriding effect of spurious faults. There has grown up, therefore, a so-called "2 out of 3" philosophy. A control system may be designed so that if 2 of the 3 instruments monitoring an important parameter indicate an unsafe state, the system is corrected or (according to the importance given the parameter) shutdown. If only 1 of the 3 so indicates, however, this is regarded as a spurious fault, and although attention is drawn to the reading, the automatic system disregards it. This enables, therefore, the instrument, etc., to be taken out of commission for repair and maintenance without shutting the reactor down. We shall be considering the design of such 2 out of 3 systems in some detail.

(d) AVAILABLE CONTROL MARGIN

The reactivity of a power reactor changes substantially in passing from cold, clean, critical, through the first power level at operating temperatures, after build-up of xenon and, finally, as the fuel is exposed with changes of isotope concentrations. Enough reactivity has to be provided in the form of excess fuel to meet these different demands and the potential excess reactivity held down at all times with enough absorbing power in the control rods, etc., to provide an adequate margin of safety. It is important to note that some power reactors (e.g. graphite, heavy water) are built several times larger than a critical volume so that there is the possibility of a local region becoming critical or supercritical when control rods are available only in other regions. For safety, an adequate shutdown margin must be available at every possible state of the system and sufficiently well distributed to control local reactivities. Equally, the steady operation of the reactor requires a good spatial distribution of control to "shape" the flux in a way that gives good uniformity to temperatures, control of xenon oscillations, etc.

(e) LIMITED REACTIVITY INSERTION RATES

In an emergency it will be appropriate to decrease the reactivity rapidly. It is generally thought safe practice, however, to limit the rate of increase of reactivity (i.e. control rod *withdrawal* and hence motor speed) so that it meets only the anticipated operational

†The action to initiate such a shutdown is called a "trip" in UK terminology and a "scram" in US terminology.

requirement for reactivity increase. This is based on the idea that if you cannot change things too quickly you cannot get into too much trouble. The rate of increase on xenon poisoning (see Chapter 9) is therefore likely to provide an upper limit to the control rod motor specification.

These various points lead to a need to provide three types of reactivity control, commonly in the form of control rods, to cover: fine control to maintain power and temperature levels (including local shaping); coarse or shim to compensate for gross changes in reactivity including startup procedures; safety control where at all times a margin of reactivity control is available to secure a shutdown of the reactor under all conceivable operating and fault conditions.

Water reactors (BWR and PWR) have made use of soluble boron for shim control as distorting the flux shape less than localised control rods as well as for an alternative safety/shutdown function. Heavy water reactors commonly use moderator height to control reactivity. Fast reactors (where absorbers are less effective) may alternatively remove sections of reflector or blanket or fuel elements.

To secure greater reliability, protection against unsafe conditions and protection against spurious faults, the designer makes heavy use of redundancy, the provision of two or three ways for carrying out one purpose. This is such an important principle that a warning is called for to see that the redundant elements are truly independent. If, for example, all instrumentation is run by one power source, it is of no help to double or triple the instruments if the fault arises from the failure of the power source. If the mathematics for the probability of joint independent events is being taken advantage of, the events must, indeed, be independent. Much skill is needed in the analysis of possible faults to see how they may in fact interfere with apparently independent equipment. It is attractive, therefore, to diversify the type as well as the number of redundant elements with the hope of preventing what might be unforeseeable coincidental faults. Thus three instruments from the same manufacturer may—unknown to the user—all contain a faulty batch of castings; three instruments from different manufacturers, using different physical principles, with different power sources and protected from failure due to a common outside interference such as an impacting aeroplane, will lead to a greater confidence in the advantages of redundancy.

Experience in nuclear power stations to date has shown very few accidents or "near" accidents due to random failure of a single component. There have been serious accidents, however (few, to power stations fortunately, involving loss of life), and these have arisen from a combination of human operator errors and common mode or systematic errors. That is to say, some complete branch of the chain of cause and effect has not been analysed, e.g. the reliability of the digital computer program employed or it has not been recognised that apparently redundant instruments depend on some common factor. For example, a change in manufacture of a type of relay led to an anti-corrosive surface of the relay poles not hardening and hence to all relays of this type sticking up. It is not always sufficiently appreciated that physical separation is desirable so that, for example, a missile thrown out by an accident elsewhere in the plant cannot damage more than one channel or that a localised fire cannot burn up the cable connections of more than one channel. Similarly, in routine operations and maintenance, humans deviate from what they have been expected or even trained to do. There is considerable scope for human engineering (ergonomics) to improve safety and reliability.

References 4, 6, 7 and 11 will give further details of the *fault tree* methods touched upon here of representing interacting failure modes and the careful analysis needed before quantitative results on reliability can be obtained when the independence of the events considered is not clear cut.

Logical and Safety Circuits

REACTOR MONITORING PARAMETERS

The major parameters to be monitored in a reactor are, first, the temperature, measured over suitably many locations, on the grounds that the most likely damage in a reactor is done in the first instance by either temperatures being too high (melting, phase change of fuel, etc.) or the rate of temperature rise or fall being too high (strains of differential expansions, creep, ratcheting, etc.). Efficient use of the reactor will probably dictate that all parts should be run up to the normal operating temperatures, and variations of these temperatures locally are probably an important measure of demand in the automatic control system. Secondly, the neutron levels (or flux) are measured. While these indirectly indicate the power level (with implications for operation), measurements of neutron level and rate of change are important for safety, particularly during startup, where absolute power levels are not easily measured. The kinetics studies of Chapter 3 indicate how a reactor may be supercritical to a dangerous degree if it does not have an adequate source; the neutron level itself is no indicator of the reactivity. In these circumstances the rate of change of neutron level, period or inverse period, is an important parameter. Thus we should inhibit startup if the neutron level is either too low or too high or if the inverse period is too high (short period). Because the neutron level changes by many decades from shutdown to full power, there is a need for overlapping instrumentation to cover the range. All these factors can be implied in Table 7.1, which

TABLE 7.1.
Typical power reactor control parameters (*and see Fig.* 7.1)

Parameter	Protection against	Channels	Designation
Neutron flux	Low power counter	3	F1–F3
	Low level log scale/period	3	F4–F6
	Intermediate level	3	F7–F9
	Intermediate period	3	F10–F12
	High level linear	3	F13–F15
	High level log scale/period	3	F16–F18
Temperature	Sector fuel excess	27	T1–T27
	Axial fuel excess	3	T28–T30
Coolant	Low flow	6	C1–C6
	Excess pressure/rate of fall	3	C7–C9

gives typical parameters and their number to be monitored in a graphite/CO_2 reactor. It would be good practice, incidentally, in monitoring for safety, to see that information on an unsafe state was derived from two different physical sources—not just one.

Although direct readings of temperatures are perhaps more fundamental and better calibrated, they tend to be associated with delays in measurement. It is common, therefore, to look to faster-acting neutron flux measurements as a source of information for control in routine operation and for safety.

Other parameters that may be included, dependent on reactor system and design, might be low water levels in steam generators or reactor vessel, loss of coolant pump power and isolation of reactor from turbine. All reactors will have a manual shutdown facility and a high radiation trip.

The parameters are generally triplicated and provide *channels* of information on the state of the system. Trip circuits are designed to open relays if the related parameters lie outside a designated range. These relays are themselves connected in three *guard lines* which result in three suitably redundant signals which indicate whether (a) the reactor is in a safe operating state (no parameter/channels/guard line tripped), (b) whether there is one spurious fault (one guard line tripped), and (c) whether the reactor is to be shutdown (two or more guard lines tripped). Relays from the three guard lines are connected through a suitable logical circuit to implement this 2 out of 3 philosophy (or some variant) as it holds up the control rods. Figure 7.1, adapted from the CEGB manual on power

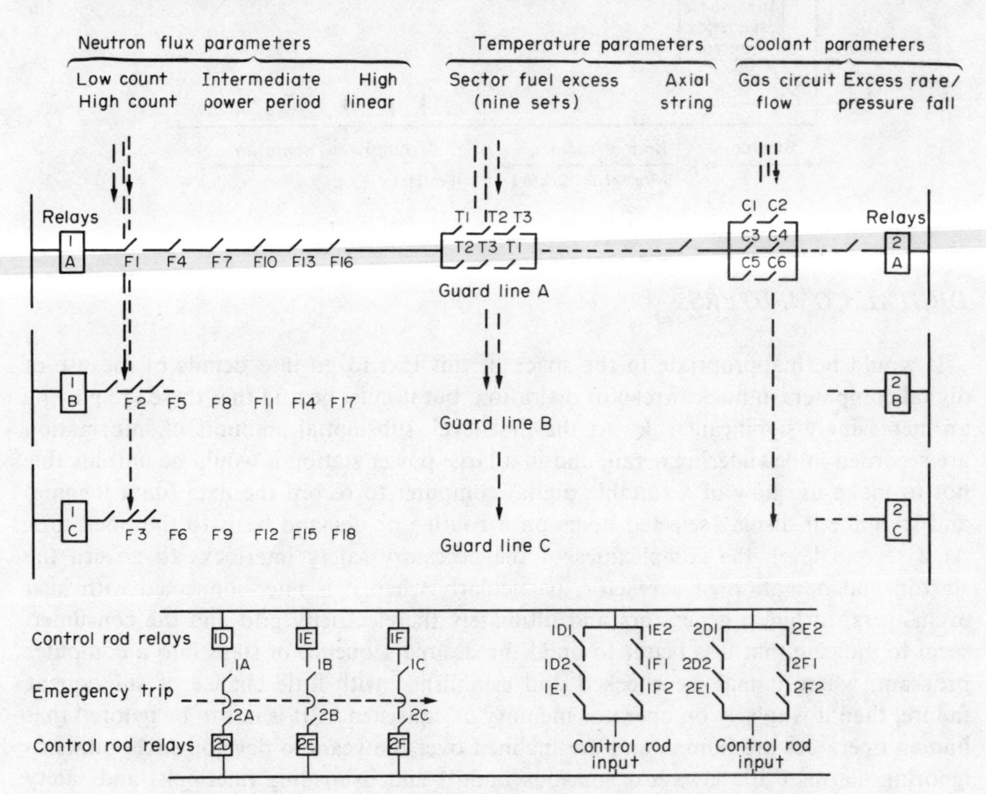

FIG. 7.1. Guard lines connected 2 out of 3 for a Magnox reactor.

stations,[8] shows the schematic for a safety circuit on these lines, using suitable switches and relays as designated in Table 7.1. Magnetic relays of the "laddic" type are now commonly used in preference to less reliable electromagnetic relays.[8, 19]

Figure 7.2 indicates the various regimes for reactor operation and the necessary functions to be provided by the control and instrumentation (after Jervis.[14])

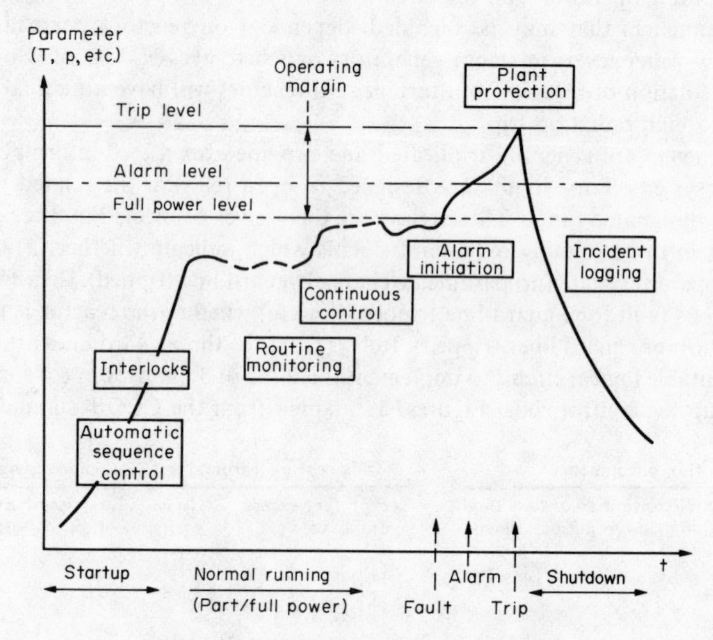

FIG. 7.2. Relation between control functions.

DIGITAL COMPUTERS

It would be inappropriate in the space of this text to go into details of the use of digital computers in nuclear reactor operation, but it must be said that these are playing an increasingly significant role. At the first level, substantial amounts of information are recorded in bewildering detail, and in a large power station it would be unthinkable not to make use now of a suitable digital computer to record the data (data logging) and to make it display selected items on a routine or demand basis to the operators. At a second level, the complexities of the necessary safety interlocks to govern the startup and operation of a reactor, particularly when it is inter-connected with heat exchangers, turbines, generators and ultimately the electricity grid and the consumer, seem to indicate that it is better to build the desired sequence of steps into a computer program, where it may be checked and established with little chance of subsequent failure, than it is to rely on operator memory or consistency. It is not to be ignored that human operators are human and are inclined over the years to develop such habits as ignoring alarms ("it's always a spurious fault") and overriding interlocks and safety devices. Indeed, the more a nuclear reactor behaves well, the less experience the operators have on which to base their reaction to a fault situation.

For all these reasons, digital computers for data-processing purposes play a significant role in the nuclear power station. Hard or permanently wired digital computers can have their functions analysed relatively easily. There should be a proper reluctance to accept *programmable* digital computers for safety control unless care is taken to establish procedures that limit the use of programmes to those that have been thoroughly tested. The danger is not that the computer will fail to operate, but in operating as programmed in unlooked for combination of circumstances, the human programmer will have insufficiently analysed what is required.

Careful thought has to be given to the proper provision of redundancy when control is to be exercised by direct digital control (DDC) though an attractive feature of a digital computer is an adaptability to self-checking features.

In summary, therefore, a digital computer has the characteristics of (a) handling large quantities of digital data, (b) controlling data logging and data display equipment, and (c) undertaking extensive numerical and logical calculations. It is now accepted widely that the first two functions can be implemented in nuclear power stations but there has been some hesitation other than in the UK and Canada to accept DDC. Undoubtedly this will come, and the 1973 material of ref. 13 may be interesting here, as well as the more academic reviews of refs. 14 and 15.

LOGICAL CIRCUITS

The connection of elements in logical fashion can be illustrated very much like an electrical circuit with the following correspondence:

state 1: closed circuit, "working" state
state 0: open circuit, "failed" state

These elements can be connected either in series or in parallel as in Fig. 7.3.

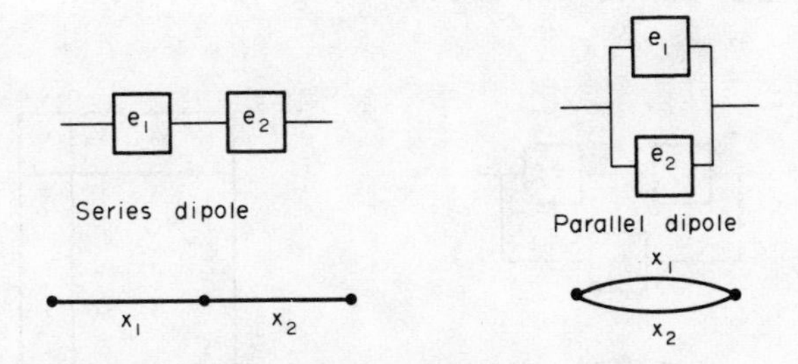

FIG. 7.3. Series and parallel logical circuits and their graphs.

Use e_1, e_2, etc., to refer to the elements themselves and x_1, x_2, etc., to refer to the state of the element, which is also binary, i.e. 1 or 0. The combination of such elements is a logical system which will be either "working" or "failed" with system state $x_s = 1$ or 0.

Evidently if elements are connected logically in series, the failure of any one fails the system; in parallel, all elements must fail to fail the system. The general nature of such circuits can be expressed in a truth table, Table 7.2 for example, which is just an exhaustive enumeration of the possible states of the elements and the consequent system state.

Truth tables become unwieldy for more than a few elements and the *state function*

TABLE 7.2.
Truth table for series and parallel dipole

x_1	x_2	x_s	x_1	x_2	x_s
1	1	1	1	1	1
1	0	0	1	0	1
0	1	0	0	1	1
0	0	0	0	0	0

Series dipole truth table   Parallel dipole truth table

$\psi(x)$ is preferred. What is wanted is a function of the x_i that will take only the binary values 1 and 0 and will take them in accordance with the system state (not every function of x is a state function therefore). It is easily seen that the elementary series and parallel dipoles of Fig. 7.3 have state functions.

$$\left.\begin{aligned}\psi_s(x) &= x_1x_2 && \text{series dipole}\\ \psi_p(x) &= 1 - [1 - x_1][1 - x_2] = [x_1'x_2']' && \text{parallel dipole}\end{aligned}\right\} \tag{7.1}$$

where $x' = 1 - x$, etc., and from them any compound series–parallel logic circuit can be constructed.

When complex circuits are multiplied out in the state function, it is commonly found that powers of particular x_i are present. These may usefully be reduced to simple form by noting that as far as a binary variable is concerned, x_i^n can be replaced by x_i itself. This enables the state function to be simplified; the resulting *reduced state function* $S(x)$ will be made use of for estimates of reliability, and this preliminary simplification or reduction is an essential step for such a purpose.

As a first example, consider the circuit of Fig. 7.4 with state function $\psi_s(x) = [x_3'$

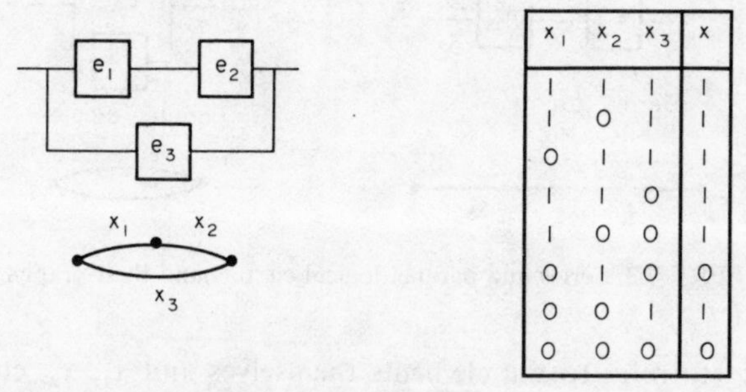

FIG. 7.4. Example logical circuit.

$[x_1x_2]']' = x_1x_2x_3' + x_3$ needing no reduction. The next example, however, is a little more complex. Figure 7.5 illustrates a situation where two independent control rods and their supply systems are thought to need further redundancy of supplies from a standby generator capable of driving either rod.

The logical circuit may be obtained by a device known in electrical theory as the equivalent series–parallel circuit. In this technique the *minimum paths* are first determined where a *path* is a set of elements whose operation (state 1) secures the operation of the

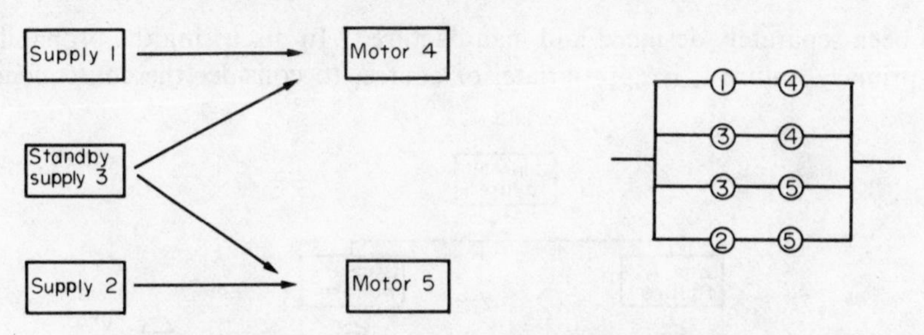

FIG. 7.5. Logical diagram for control rod motors and suppliers.

system ($x_s = 1$) while a *minimum* path means that no element in the set may be changed to state 0 without failing the system ($x_s = 0$). There are four such minimum paths. The equivalent circuit is constructed by putting the elements of a minimum path in series and then connecting all the minimum paths in parallel as shown in the figure.

It is now simple to write a state function for the problem

$$\psi_s(x) = 1-[1-x_1x_4][1-x_2x_5][1-x_3x_4][1-x_3x_5] = [[x_1x_4]'[x_2x_5]'[x_3x_4]'[x_3x_5]']'$$

When this is multiplied out and *reduced* we have

$$S(x) = \begin{array}{l} x_1x_4+x_2x_5+x_3x_4+x_3x_5 \\ -x_3x_4x_5-x_1x_3x_4-x_2x_3x_5 \\ -x_1x_2x_4x_5+x_1x_2x_3x_4x_5 \end{array} \tag{7.2}$$

Logical circuits using this method can be developed also with the help of the truth table for the system, and it may also be noted (see problems) that there is an equivalent method using *minimum cuts* arranged in parallel–series.

FAULT TREE DIAGRAMS

A relationship can sometimes be more usefully represented in the form of event trees and fault trees. The purpose of a fault tree diagram is to show the logical interrelation of the basic events that taken apart or together may lead to a system or device failure, the top fault, using a combination of "and" and "or" symbols. That is, a state may arise if all subsidiary states occur (equivalent to a parallelled circuit)—"and"; a state may arise if any one of a number of subsidiary states arise (series circuit)—"or". It is sometimes necessary to distinguish this use of "or" from the logical "or" where this latter may exclude *both* or *all* events that occur simultaneously (either/or but not both).

Figure 7.6 illustrates how a fault tree may be used in analysing the possible failure modes of the control motor system, leading to a top fault where no control rod motion is available.

Particular care in interpretation has to be given when the same event occurs in several branches of the tree (common failure modes) if the probability of failure is to be correctly expressed. Of course there will be much qualitative skill in knowing what events should be included as the base events initiating a fault, especially "thinking the unthinkable". For example, are both motors liable to destruction by a single missile initiating from an accident within the reactor plant or from outside? If the wiring of a motor fails, is this a generic fault that would be likely to occur simultaneously in the other motor or has this been separately designed and manufactured? In discussing the probability of a serious primary failure it is appropriate, of course, to consider the consequences and

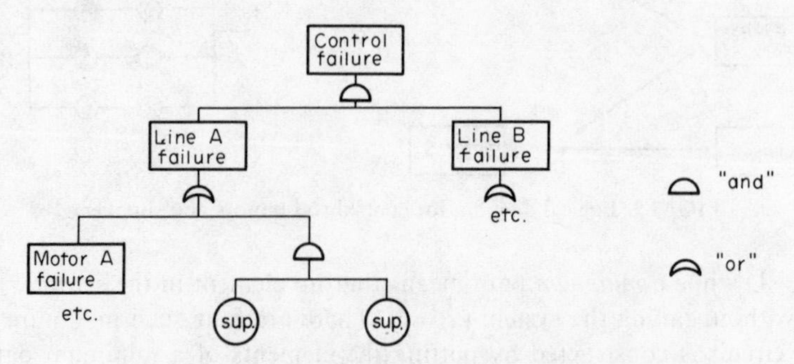

FIG. 7.6. Fault tree diagram for control motor system.

causes as they affect other items of equipment and change the probabilities of what may be called secondary failures.

In the second example, Fig. 7.7, an electrical fuse is represented whose failure may

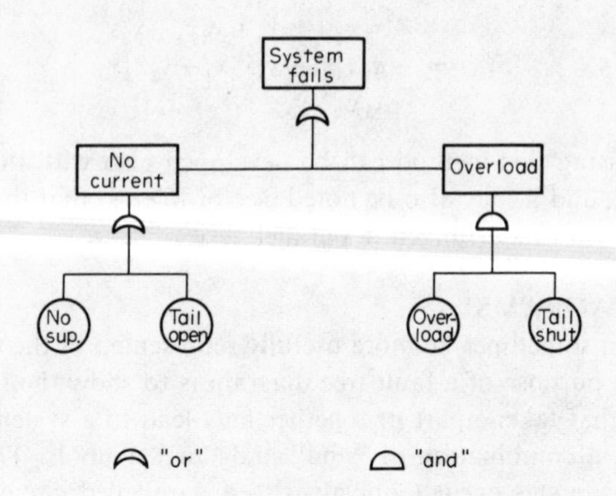

FIG. 7.7. Fault tree diagram for a fused system.

lead to a system failure depending on the failure mode of the fuse and associated conditions. Note the logical "or" in this example since the current cannot at the same time be zero and overloaded, the two possible departures from the normal operating range.

EVENT TREE DIAGRAM

The fault tree of the last section was read from the bottom upwards as a method of determining all those circumstances which together led to a *particular* fault situation. In the event tree, however, the tree is read downwards as leading from an initial event to determine the consequences, usually in terms of success or failure in accordance with some criterion. It is convenient to adopt the convention that success branches left and failure branches right. The tree itself is a useful graphical aid; it can be combined with a recording of the (conditional) probabilities of each step that makes the determination of over-all success and failure probabilities easily traced. Thus the representatiof on probabilities is also facilitated.

Figure 7.8 illustrates an event tree for human operations in the context of the response

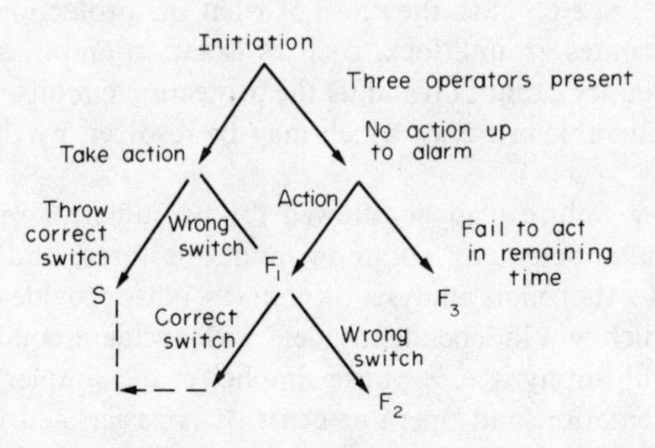

FIG. 7.8. Event tree diagram for operator reaction to emergency.

of operators in a control room emergency situation. Data for such situations is sparse, but has been reviewed in WASH 1400.[7]

The diagram is easily adapted to show recovery from a failure (dotted lines) when some subsequent action, such as an inspection, corrects the fault or failure situation.

INTEGRITY REQUIREMENTS

In two senses reliability costs money. To ensure that a function is performed reliably requires analysis and provision of proper equipment with suitable redundancy. These aspects cost money. On the other hand, if equipment fails there may be a direct cost in terms of unavailability, repairs, damage, etc., quite apart from human safety. To determine the degree of effort appropriate to provide reliability in the safety sense, the following classification with the associated integrity specification is helpful.

Class 1. Essential reactor protection. The system providing for this purpose should be such that no single fault can lead to a loss of the protective function.

Class 2. Information and control items such that significant reactor operation is not possible without continuous availability of the essential functions of the equipment. It is desirable that no single fault can lead to a loss of protective function so that redundant components are indicated.

Class 3. Information or control aids to operating staff which enhance operation. Failure may lead to a degradation of over-all performance, etc. It is generally acceptable for a single fault to cause a partial loss of facility or a total loss in less significant applications, in which case repair and restoration should be possible before limitation of operations is necessary (e.g. spare parts available).

Class 4. Data-logging systems and aids to conventional instrumentation with a facility to check results by other means. A single fault causing total loss of facility is acceptable but the equipment should be designed to give an adequate mean time between failures and for rapid fault detection.

A few further philosophic principles can usefully be borne in mind. If we distinguish between protection and control, the priority must be given to the protection of the plant in the fullest sense. There is something to be said for employing the same parameter measurements as used for protection purposes for the subsidiary purpose of (automatic) control so that operators, etc., have the sense of what the protection instrumentation is providing. This necessitates an interlock, such as isolation amplifiers, to ensure that a fault in the control circuitry cannot invalidate the protection circuits, and so the principle has come in for considerable criticism which may be resolved by the advent of *optical* isolation devices.

Independent random failures can be allowed for by sufficient redundancy. It is far harder to deal with unanticipated "common mode" failures, and here the best protection—in addition to thorough analysis of course—is to provide separate functional protection systems which will independently deal with accident conditions.

Finally, a useful philosophy is to espouse simplicity; the simpler the system the less it is liable to instrumentation and operator error. It is easier said than done, but one realisation of such a philosophy might be to simplify the reaction taken to accident conditions by, for example, taking a common and adequate reaction to a range of situations rather than trying to provide for the identification of too many separate situations and distinct reactions to them.

Failure Expectations

PROBABILITY OF FAILURE

Suppose we may assign, at any given moment, probabilities for the working of each element of a logical circuit with a view to finding the probability that $x_s=1$, i.e. that the system described by the circuit is working. We make the severe assumption that the failure of each element is independent of the remaining elements.

A series dipole with probabilities p_1 and p_2 has the probability $p_s=p_1p_2$ and note immediately that this result could be obtained by substituting p for x-values in the series dipole state function, $\psi_{s2}(x)\rightarrow S_{s2}(p)=p_1p_2$. Similarly, the probability of a parallel dipole system working is $[p_1'p_2']'$ since it only fails when both elements are not working each with probability $p_i'=1-p_i$. Again, this result is equivalent to taking the parallel dipole state function and substituting $x=p$. In general, therefore, the *reduced state function* will also be the *reliability function* for the system subject to the assumption of independence. It is important, however, that the *reduction* of powers of x_i^n has been carried out since the corresponding term wanted in the probability function is just p_i and not p_i^n.

We may apply this idea to the problem shown in Fig. 7.5 whose reduced state function was given in eqn. (7.2). If we assume that each supply has a reliability, the probability of operating at the moment in question, of p_e 0.8 and the control rod motors a corresponding reliability of p_m 0.9, then the reliability of the system is easily determined to be

$$S_s(p) = 4p_ep_m - p_ep_m^2 - 2p_e^2p_m - p_e^2p_m^2 + p_{\text{ə}}^3p_m^3 = 0.9976$$

which, of course, is a noticeable improvement over the case with no standby generator, $S(p)=p_e\,p_m[2-p_e\,p_m]=0.9216$.

Cost and complexity will also need to be considered in deciding whether to adopt the standby.

2 OUT OF 3 CIRCUITS

Figure 7.9 shows a simple 2 out of 3 circuit to secure many of the operating relations

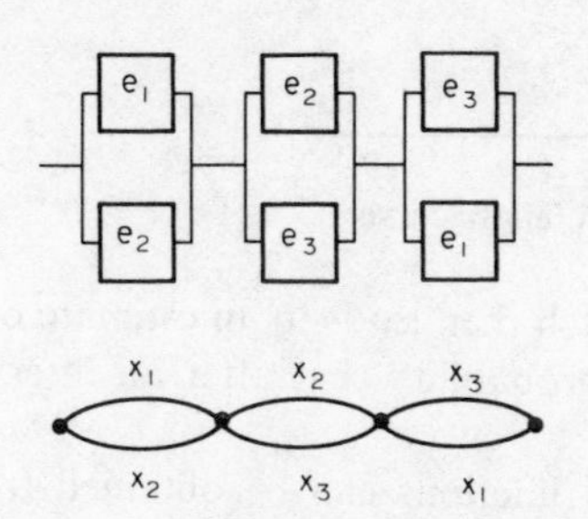

x_1	x_2	x_3	x_s
1	1	1	1
1	1	0	1
1	0	1	1
0	1	1	1
1	0	0	0
0	1	0	0
0	0	1	0
0	0	0	0

FIG. 7.9. A simple 2 out of 3 logical circuit.

we have envisaged. Each signal controls, through relays, etc., as necessary, *two* switches or elements, which may be likened to a two-pole throw switch. The corresponding truth table will verify the claim that if only one element has failed the system is still working.

The state function is $\psi(x)=[1-x_1'x_2'][1-x_2'x_3'][1-x_3'x_1']$ and the reduced state function is

$$S(x) = x_1x_2 + x_2x_3 + x_3x_1 - 2x_1x_2x_3 \tag{7.3}$$

If we assume that each element has a probability of working of 0.9 (related to the frequency of failures and the dead time before a repair is executed), then we see that the system reliability is $S(p)=3p^2-2p^3=0.972$. If, on the other hand, we had adopted a 3 out of 3 philosophy, represented as three elements in series, the reliability would be only $p^3=0.729$. Note also that a 2 out of 2 system would have a reliability of 0.81, so with better reliability, the 2 out of 3 circuit provides the additional ability of dealing with spurious faults.

Reliability alone (or an estimate of its probability) is perhaps insufficient design information. We would also wish to have an estimate of the mean time between spurious faults and, indeed, in the 2 out of 3 case the mean time between two simultaneous spurious faults that will lead to a spurious trip. This requires not only knowledge of the repair down-time but more generally the distribution of element failures as a probability distribution function.

DISTRIBUTION OF FAILURES

We turn to the problem of estimating the reliability probabilities used in the last section and the corresponding mean times to failures. In most cases this can be done on the basis of observed failures, as in Fig. 7.10 where records of failures after construction, manufacture, bringing into service, etc., are kept.

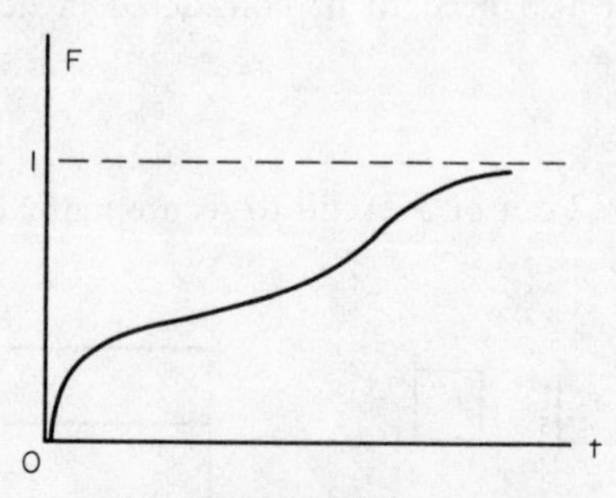

FIG. 7.10. Total failure curve.

Division by the number of items in the batch then leads to an estimate of the cumulative distribution function (CDF) $F(t)$, the probability that after an interval t the unit will have failed.

We recollect that the related probability functions can be obtained from $F(t)$. The *survival function* $R(t)$ is of particular interest:

(a) survival function: $R(t)=F'(t)=1-F(t)$, the probability of surviving to t;
(b) failure distribution function (FDF), $f(t)=-dF/dt$;
(c) age specific–failure rate, $\varphi(t)=-(d/dt)\ln R(t)$;

so that if one is given the others can be found. For example,

$$R(t) = \exp[-\int_0^t \varphi(t')dt'] \tag{7.4}$$

The failure rate $\varphi(t)$ is a measure of the "local" probability of failure.

A direct derivation of the equation involving $\varphi(t)$ may be more helpful. Since $R(t)$ is the survival probability against a failure to time t, the probability of surviving to $t+\delta t$ is $R(t)$ multiplied by the conditional probability of not failing in δt, i.e. $1-\varphi\delta t$. Hence $R(t+\delta t)=R(t)\,[1-\varphi\delta t]$ and we obtain eqn. (7.4) as $\delta t \to 0$ when the boundary condition $R(0)=1$ is utilized.

It is important to note that φ is likely to be time dependent though, of course, the simplest case takes φ to be a constant, μ say. Consider some of these special cases.

1. CONSTANT FAILURE RATE: $\varphi = \mu$

The mathematics are then similar to radioactive decay with φ playing the role of λ. The situation is a fair description perhaps when there are many components. Alternatively, the model is valid if the failure was due to gross misuse, sabotage, etc., brought about at random and having nothing to do with component age, such as an electric fuse. We have

$$R(t) = e^{-\mu t}; \quad F(t) = 1 - e^{-\mu t}; \quad f(t) = \mu e^{-\mu t}$$

This particular simplification leads to an $R(t)$ called the interval distribution. Figure 7.11 illustrates the results for given μ.

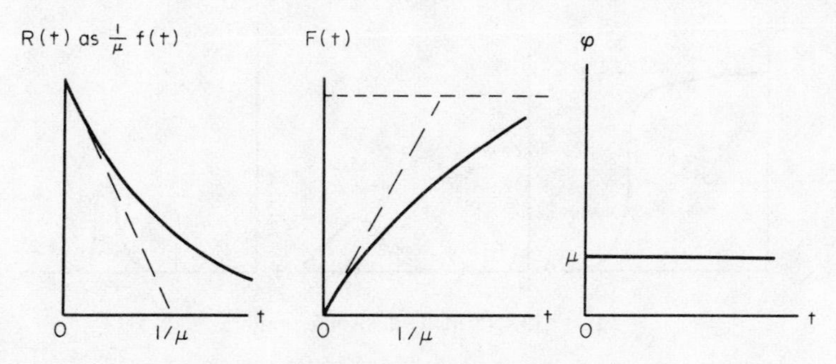

FIG. 7.11. Constant age-specific failure rate.

2. ZERO FAILURE RATE

This is clearly unrealistic.

3. LINEAR AGEING

The next simplest model would be to take $\varphi = kt$, the age-specific failure rate increases linearly with time as the components steadily age. This leads to a reasonable description and a characteristic "bell" curve for the survival curve, see Fig. 7.12.

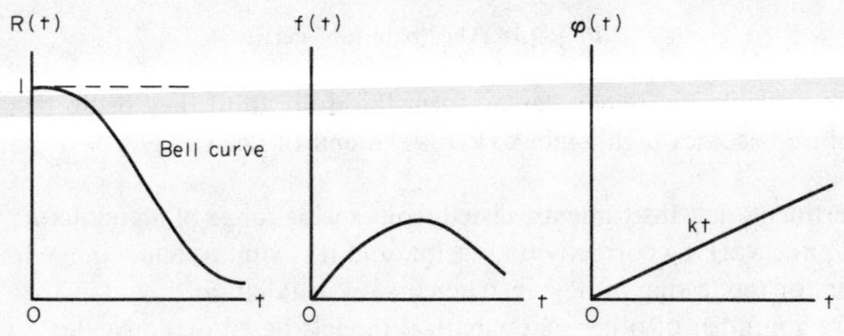

FIG. 7.12. Linear ageing model.

4. FINITE LIFE

If the components are again idealised, this time to work perfectly until the end of their life and then to fail with a precise life τ, we would have $\varphi = \delta(\tau)$. The forms for R and F are then step functions at τ. While this model again represents an oversimplification, it can be improved by supposing that failure is normally distributed about τ with an experimentally determined standard deviation, leading to relations illustrated in Fig. 7.13.

5. THE "BATH TUB" CURVE

A more typical experimental result shows an initial rapid failure (negative ageing) followed by a recovery of reliability in the remaining components and a steady increase in φ (positive ageing). This shows a typical curve in terms of failure distribution $f(t)$, referred to as the "bath tub" (Fig. 7.14).

Thus the output of a car factory may show that a few cars fail very quickly in service

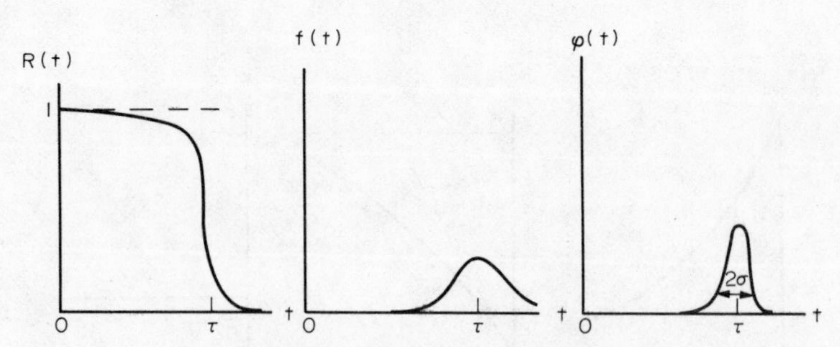

FIG. 7.13. Normally distributed failure.

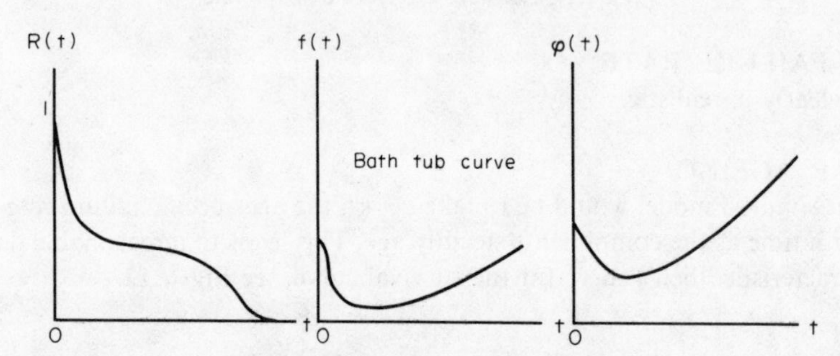

FIG. 7.14. The "bath tub" curves.

(Friday's cars) while the remainder perform faithfully until they begin to age, so that relative failure becomes high again towards the end of what might be a planned obsolescence.

More pertinent, new instruments, tested from a wide range of manufacturers, typically shows 17% not working correctly on receipt, and it is vital to allow in a commissioning programme for the testing of new instruments on installation.

There are a number of other mathematical models based on convenient choices for φ (see Cox,[10] for example). Of the models given here, the first, $\varphi=\mu$, is the most easily handled for complex problems involving *replacement* of faulty parts since it is not necessary to specify the age of the part on replacement. We confine the subsequent example therefore to this idealised model.

MEAN TIMES TO FAILURE

The mean time to failure of a component is given by the integral of time weighted with the PDF, $f(t)$. In the case of the model $\varphi=\mu$ we have

$$<t> = \int_0^\infty t f(t)dt = \int_0^\infty t\mu e^{-\mu t}dt = 1/\mu \tag{7.5}$$

a result that will be familiar from radioactive decay, etc.

For n elements in series, where the system depends against failure from any component, each with $\varphi_i=\mu_i$, we have $R=\exp(-{}_0\int^t \Sigma_i \, \mu_i dt)$:

$$<t_n> = \int_0^\infty t\,[\mu_1 + \mu_2 + \ldots]\,e^{-[\mu_1+\mu_2+\ldots]t}dt = 1/\sum_i \mu_i \tag{7.6}$$

and, in particular, for similar components $\mu_i = \mu$, the mean time to system failure is $1/n$ of the mean time to component failure.

If we have elements in parallel we can expect the system to have a longer mean time to failure than at any rate the shortest-lived component. If this is the problem, consider two elements in parallel first. The probability of both elements failing is given by $F(x) = [1-e^{-\mu_1 t}][1-e^{-\mu_2 t}]$ and the equivalent $f(x)$ PDF for this problem is easily found such that the mean time to failure is

$$<t_p> = \int_0^\infty \mu_1 e^{-\mu 1 t} + \mu_2 e^{-\mu 2 t} - \mu_1\mu_2 e^{-[\mu_1+\mu_2]t}dt = \frac{1}{\mu_1} + \frac{1}{\mu_2} - \frac{1}{\mu_1+\mu_2} \tag{7.7}$$

with the increase anticipated. In particular, for similar elements, the mean time is extended by 50%. However, this is unrealistic in taking no account of *testing* or *replacement/repair*.

EFFECT OF REPLACEMENT AND TESTING

There are important practical modifications to the foregoing results when allowance is made for maintenance, testing and replacement or repair. To develop simple expressions we assume not only the age independent model but also that the failure rate μ is low. Then the probability of failing in time T after being known to work at time zero rises as μt from 0 to T; the mean failure probability in this interval is simply $\frac{1}{2}\mu T$ of course.

Supposing now the device is tested at regular intervals during the interval T. Testing is assumed not to affect the probability of the device breaking down, of course, but it does mean that after every test the device is known to be working, either because it passed the test or because it was then repaired or replaced. The comparative rise in failure probability and corresponding mean probability of failure is shown in Fig. 7.15.

Allowance must be made for the time taken per test and some over-all maintenance time, which may be independent of the test interval and cover repairs or replacements. Of course, if the testing and maintenance can be arranged to fall within over-all system

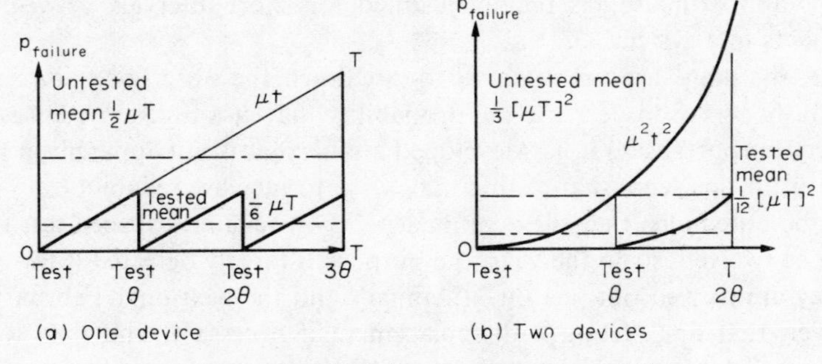

FIG. 7.15. Effect of testing on mean time to failure.

"downtime" when the function of the device is not required, these items can be omitted. More generally, however, the effective downtime is the sum

$$T_{\text{eff}} = \tfrac{1}{2}\frac{1}{n}T + nT_{\text{test}} + T_{\text{maint}} \tag{7.8}$$

where n is the frequency of testing (in the units of time adopted) from which an effective or equivalent probability of failure is obtained. There is, therefore, a trade-off between total testing time and reduction of dead time so that an optimum testing frequency n can be adopted. With this effectively constant device failure probability it is easily seen that the mean time to failure of 2, 3, etc., such elements in parallel is given as in Table 7.3 (no allowance for testing, etc.). The calculation is elementary and illustrated in Fig. 7.15(b).

TABLE 7.3.
Mean times to failure of parallel devices; test interval θ

Number of devices	Failure probability	
	Simultaneous testing	Staggered testing
1	$\frac{1}{2}\mu\theta$	$\frac{1}{2}\mu\theta$
2	$\frac{1}{3}[\mu\theta]^2$	$0.2[\mu\theta]^2$
3	$\frac{1}{4}[\mu\theta]^3$	$0.06[\mu\theta]^3$
4	$\frac{1}{5}[\mu\theta]^4$	$0.004[\mu\theta]^4$

A further improvement, recorded with approximate coefficients in the final column of the table, can be made if the test intervals are uniformly staggered. Thus for two devices tested alternatively, one test every $\frac{1}{2}\theta$,

$$p = \frac{\mu^2}{\frac{1}{2}\theta}\int_0^{\frac{1}{2}\theta} t[t + \tfrac{1}{2}\theta]dt = \frac{5}{24}[\mu\theta]^2 \tag{7.9}$$

These are substantial predicted gains in reliability arising from testing, and a word of caution is advisable. In the first place the theory has assumed constant age-specific failure rates and, in particular, that testing does not in any sense promote the breakdown of the equipment. Those with experience of instruments would have some reservation here. Secondly, for simplicity, failure expectations were averaged over an interval. The effect of this approximation is readily justified for short intervals between tests and several elements in a system.

Of course, the gains are not such as to cut down the observed failure rate in the system but rather serve to decrease the probability that at a time a system is wanted to act (for safety purposes, say) it has developed a fault that is not apparent in the absence of testing. It is in this sense that testing may serve to increase reliability.

It should be noted also that these gains depend on regularly spaced test intervals; if there are to be twelve tests in the year, the purpose is largely defeated if for testing convenience they are carried out one on 31 January and the next on 1 February, etc. It is seen, however, that first testing with replacement if necessary, and then logical combination in parallel, serves to lower expected failure rates dramatically.

Experience shows, however, in many situations, including the control of nuclear power plants, that if a safety device is relied upon then the probability of it failing on demand is closely related to the time interval between tests. Thus it is important to specify regular intervals between tests, on line tests where possible, and that both the fault should be corrected when found and the cause of the fault corrected.

SPURIOUS TRIP IN A 2 OUT OF 3 SYSTEM

A primary advantage of the 2 out of 3 system (or equivalent 2 out of 4, double 2 out of 3, etc.) is that a spurious fault on one parameter in the guard line does not trip the system. Nevertheless, if the original fault is not rectified, a second spurious fault leads in the 2 out of 3 case to a spurious trip. Such trips lead to a waste of the reactor time and are undesirable since they unnecessarily increase the frequency of rapid shutdown and hence rapid temperature transients, etc. If the repair time to identify and rectify the fault is, say, T, we wish to know the benefit in decreased spurious shutdown compared to a single 1 out of 1 system.

Suppose the 1 out of 1 system has a spurious trip frequency of μ in the constant and independent model being studied. The mean time between spurious trips is $1/\mu$. In the 3 guard line system the mean time between trips will be shortened to $1/3\mu$. The probability of a second spurious trip occurring in time T after the first is given by

$$F(T) = \int_0^T 2\mu\, e^{-2\mu t} dt = 1 - e^{-2\mu T} \tag{7.10}$$

Thus the spurious trip rate is decreased according to the probability that the second fault causes a trip because the first has not been repaired. The improvement depends on this probability being less than one-third to offset the factor of 3 increase in the first trip frequency. If T is small, say $\frac{1}{4}$ day compared to the mean life between trips, say $1/\mu = 30$ days, the exponential may be expanded and the lengthening of mean time to spurious shutdown is $1/6\, \mu T$ or, say, by a factor of 20 in this example. Such a gain is well worth the additional complexity of the equipment involved.

This is an important example, of course, but only touches on the whole question of renewal theory which becomes much more difficult if the constant φ model is abandoned and, of course, more difficult again if the independence of failures is an unjustifiable assumption. Cox[10] is a good introduction to the first generalisation with its implications for the economic supply of spare parts in a complex system.

While we are not able here to go on to the closely related topics of the design of instrumentation for detectors and control, it is perhaps worth mentioning that current solid state elements have a big advantage in fast reaction times over their predecessors that makes it possible to test operate them. That is, the elements can be tested to demonstrate working order and then returned to an operational state in times so short that the control element (e.g. electromagnetic release of control rod) is not operated. This facility for routine checking of control systems on-line should be incorporated wherever possible.

In this text we cannot go into the results of fault conditions. However, consideration of the expected consequences and design accordingly is a logical basis for a design procedure of the whole reactor system and is to be commended.

Problems for Chapter 7

"There's a great text in Galatians, once you trip on it entails.
Twenty nine distinct damnations, one sure if another fails"
(Browning.)

7.1. Find the variance in mean life for a set of similar elements with constant age specific failure rate (NO replacement or testing). Will the mean and the variance be the same as those obtained observationally from a sample of the elements in service at a particular time, zero say? How can you best estimate the mean life if the observation is carried only until the fraction $r<1$ fails?

7.2. In the context of Fig. 7.8, determine the over-all failure probability. Data provided includes:

probability of selecting a wrong switch	0.01
probability of failing to act (one operation)	0.01
probability of misunderstanding alarm situation	0.1

Make your own judgement on the degree to which the presence of three operators (one supervisor and two assistants) under the stress of an accident situation may be said to improve on the probability of correct action by one operator on his own.

7.3. A control instrument has a natural mean life expectancy to failure of 10 years (constant age specific failure rate model). Maintenance time per year is estimated at 3 hr while test time is 1 hr per test. The system is required to function with an expected mean failure rate of 10^{-4} per year. Specify the number of such instruments in parallel and the test sequence to meet this need. Comment on the practical advantages of simultaneous versus staggered testing. Is the assumption $p=1-\exp(-\mu t)\simeq\mu t$ a *conservative* one in the circumstances?

7.4. For the supply and control motor system of Fig. 7.5 write out the truth table showing that it has 25 lines. Identify the minimum paths and the minimum cuts. Find the equivalent series–parallel circuit from the minimum cuts and show the resulting reduced state function is the same as given in the text.

7.5. A logical circuit has n elements with the same reliability p. Sketch the relationship of $S(p)$ against n for (a) all elements in series, and (b) all elements in parallel.

7.6. The circuit shown in the diagram has an insufficient reliability and this is to be improved by redundancy. It is contemplated that twice as many elements will be used, but the question arises whether these should be used to duplicate each existing element in parallel or duplicate the whole system in parallel. Determine for sample values of p which is the better approach. Show that it is generally better to duplicate components than systems.

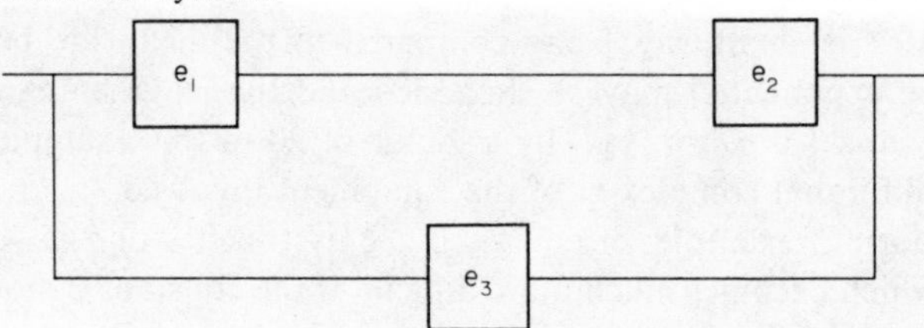

7.7. For the constant age-specific failure rate model $\phi_i=\mu_i$ determine the mean time to failure of parallel circuits with 2, 3, . . ., n elements. (*Hint*: For the general case it will be sufficient to leave in combinatorial form.) Specialise to similar values $\mu_i=\mu$.

7.8. Double 2 out of 3 circuits. Analyse the operation of the double 2 out of 3 circuit diagram shown below. What operational advantages are there?

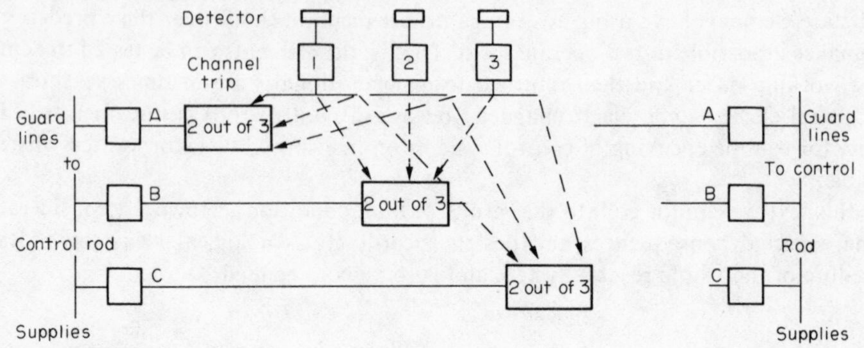

7.9. Design a 2 out of 6 circuit.

7.10. Find a logical equivalent 2 out of 3 circuit in relation to Fig. 7.9 on the concept of equivalents between series and parallel circuits using paths in the truth table.

7.11. Discuss the principles you might wish to impose upon a design using a digital computer for direct reactor control (DDC—direct digital control). How will you make provision for adequate redundancy and protection from spurious shutdown in (a) hardware, and (b) software? What part can automatic self-testing play?

Reference 15, issued in 1976 as a British Standards draft, may be useful here.

References for Chapter 7

1. N. M. Morris, *Logical Circuits*, McGraw-Hill, New York, 1971.
2. *IEEE/JCNPS Guide for Reliability Analysis of Nuclear Power Plant Protection Systems*, 1975.
3. A. Kaufmann, *Reliability*, Transworld, London, 1969.
4. J. B. Fussell, *Fault Tree Analysis—Concepts and Techniques*, NATO Advanced Study Institute, Nordhoff, 1976.
5. W. Feller, *An Introduction to Probability Theory and its Applications* (2nd edn.), Wiley, New York, 1957.
6. D. F. Haasl, Advanced concepts in fault tree analysis, *Systems Safety Symposium Boeing, Seattle*, 1965.
7. N. Rasmussen (ed.), *Reactor Safety Study*, WASH-1400, USNTIC, Washington, 1975.
8. *Modern Power Station Practice: 8, Nuclear Power Generation* (2nd edn.), Pergamon, Oxford, 1971.
9. M. L. Shooman, *Probabalistic Reliability*, McGraw-Hill, New York, 1968.
10. D. R. Cox, *Renewal Theory*, Methuen, London, 1962.
11. R. E. Barlow, J. B. Fussell and N. D. Singpurwall (eds.), *Reliability and Fault Tree Analysis*, SIAM, 1975.
12. T. J. Thompson and J. G. Beckerley, *The Technology of Nuclear Reactor Safety*, MIT, Cambridge, Mass., 1964.
13. *Nuclear Engineering International*, IPC, London, 1973.
14. M. W. Jervis, Digital control of nuclear power stations, *Adv. in Nucl. Sci. and Tech.* **11**, Plenum, New York, 1978.
15. *Application of Digital Computers to Nuclear Reactor Instr. Control and Protection* (IEC 45A 39), BSI draft, London, 1976.
16. Reliability of nuclear power plants, *Proc. Symp. IAEA, Vienna*, 1975.
17. F. R. Farmer and J. R. Beattie, Nuclear power reactors and the evaluation of population hazards, *Adv. in Nucl. Sci. and Tech.* **9**, Academic, New York, 1976.
18. *Draft for Development: DDs* 10–16 (*Engineering Reliability*), BSI, London, 1974–6. See also the work of the International Electrotechnical Commission (IEC) Subject Committee 45A for work on reactor instrumentation.
19. A. H. Weaving and J. Sherlock, Magnetic logic applied to reactor safety circuits, *Jl. BNES* 74, Jan. 1963.
20. A. E. Green and A. J. Bourne, *Reliability Technology*, Wiley, London, 1972.

CHAPTER 8

Nonlinear Systems; Stability and Control†

The analysis of reactors in Chapter 4, using the frequency space and Laplace transforms involved linearising the equations around an operating point. The studies were then necessarily confined to small transients or departures from the operating point and to questions of local stability or stability in the small. Within its context, this approach is valuable. If the predicted transients are small, the approximation is justified; if large (linearly unstable) this information is helpful to a designer denoting something to be avoided.

But there are other situations for study. Some of these are operating conditions with large changes of state as in startup and shutdown. Others concern safety studies where it is desirable to know that the consequences of abnormal excursions can be limited. In this latter case, the emphasis is not so much on tracking the behaviour for individual or particular excursions since safety studies have to provide some guarantee against all possible excursions. The emphasis is therefore more on determining bounds on the behaviour of a class of disturbances. This is just as well because the introduction of nonlinearities into the system state equations produce formidable obstacles to obtaining exact, explicit solutions.

This chapter discusses some techniques for such nonlinear studies centred on the use of the *state space* and its equations in the time domain, in contrast to the *frequency* space of classical control theory. The examples developed will be simple ones, but in addition to illustrating methods they have a number of practical features giving insight into the nature of such systems. In professional practice such studies are supplemented by detailed and laborious computer safety calculations, but it would be naive to enter into such an expensive undertaking for a particular design without appreciating the general nature of nonlinear dynamics and the new situations that can arise in matters of stability and operating performance.

At the same time we can extend our consideration beyond simply controlling a system to questions of controlling it in some "best" way. This aspect of modern control theory is also better served by state space studies. The procedure in this chapter is therefore to tackle a few simple but illustrative problems direct, to make some general statements about stability in the light of the new features so discovered, to introduce Liapunov's method for studying the stability of systems and, finally, give a short account of Pontryagin's optimum control theory and its application to reactor dynamics.

†This chapter contains advanced material which may be omitted on a first reading.

State Space Representations

As the state, x_1, x_2, etc., of a system change, the values can be thought of geometrically in a state space as tracing out a trajectory, the history of the system from its initial point $x(t_i)$ where x is a vector (x_1, x_2). The term phase space is reserved for occasions where x_2 is the rate of change of x_1 but is probably a familiar example of the state space despite being a special case. When we take a region in the state space it is a composite of points describing different states and we may consider the many possible trajectories tracing through various sets of these points. For stability purposes we will want to convince ourselves of the nature of the region for all possible trajectories passing through it.

We shall limit ourselves to situations where the model is described using two dependent or state variables with time as the independent variable. Systems with more than two dimensions become progressively more difficult to handle. The state variables are what the model takes to be a sufficient description of the state of the system as opposed to the coefficients which are presumed known. Suppose, for example, that the neutron level and the reactor temperature are taken as the two (sufficient) descriptions of the state of the reactor as it develops in time from some known initial condition. The neutron balance equation neglecting delayed neutrons and independent sources would be

$$\frac{dn}{dt} = \frac{\rho - \beta}{\Lambda} n \tag{8.1}$$

so that this model is suited to the study of rapid transients in accident conditions.

The equation for the reactor temperature would be derived from an energy balance and depends evidently on the heat removal mechanism considered. In the simplest case there may be no such mechanism and we propose to write a temperature equation,

$$\frac{dT}{dt} = \kappa n \tag{8.2}$$

where κ is the ratio of energy produced at unit neutron level to the reactor heat capacity. Such a model is called adiabatic—no heat removal.

The two state equations are supplemented by an algebraic equation for the dependence of reactivity on temperature. We take this in the present instance to be $\rho = \rho_0 + \alpha T$, where ρ_0 is some initially impressed reactivity and α is the coefficient of reactivity change with temperature. Both ρ_0 and α will be taken as constants though the sign of α may be left undetermined, to be settled by stability considerations arising from the analysis. In this sense, α would be called a parameter.

Introducing this relation we have the so-called Nordheim–Fuchs model with state equations

$$\dot{x} = \begin{pmatrix} \dot{x}_1 \\ \dot{x}_2 \end{pmatrix} = \begin{pmatrix} \dfrac{dn}{dt} \\ \dfrac{dT}{dt} \end{pmatrix} = \begin{pmatrix} \dfrac{\rho_0 - \beta + \alpha T}{\Lambda} n \\ \kappa n \end{pmatrix} = \begin{pmatrix} f_1 \\ f_2 \end{pmatrix} = f \tag{8.3}$$

where $x=(x_1, x_2)$ is a vector of the state variable and $f(x)=(f_1, f_2)$ a vector of the terms in the state equation. Equation (8.3) is supplemented by the initial conditions: $x(0)=(n_0, T_0)$ but we may here choose to measure temperature relative to the initial temperature so that $T_0=0$.

This particular example shows no *explicit* dependence of $f(x)$ on time and so is said to be *autonomous*. That is, if the transient is initiated at a later time the solution is simply shifted in time by this delay.

The natural first step is to investigate possible *equilibrium* points at which each term of x is zero, i.e. solve $f(x)=0$. In this case any temperature combined with a zero neutron level is such a critical point though we notice the special claim of $x=(0, -[\rho_0-\beta]/\alpha)$ in view of the double zero in eqn. (8.1). These particular critical points can be called *shutdown* points in view of the requirement for n to vanish. There is nothing to say, however, so far that any one of these shutdown points is actually *stable* against the introduction of a small neutron, temperature disturbance, and this must be investigated dynamically.

The two-dimensional autonomous case provides an immediate simplification since the two first order equations (equivalent to one second order system) can be reduced to a single first order equation by writing

$$\frac{dn}{dT} = \frac{\rho_0 - \beta + \alpha T}{\Lambda\kappa} \tag{8.4}$$

In the general situation it would be helpful to study the right hand side of eqn. (8.4) by the method of *isoclines* or constant slopes. Putting eqn. (8.4) equal to various constants yields a locus of points in the state or (x_1, x_2) plane of equal slopes. With enough of these determined, various trajectories or solutions of the system can be sketched in.

In this case, however, eqn. (8.4) can be integrated to obtain

$$\kappa\Lambda[n - n_0] = [\rho_0 - \beta]T + \tfrac{1}{2}\alpha T^2 \tag{8.5}$$

and this first integral yields much information. Figure 8.1 shows the solution for a

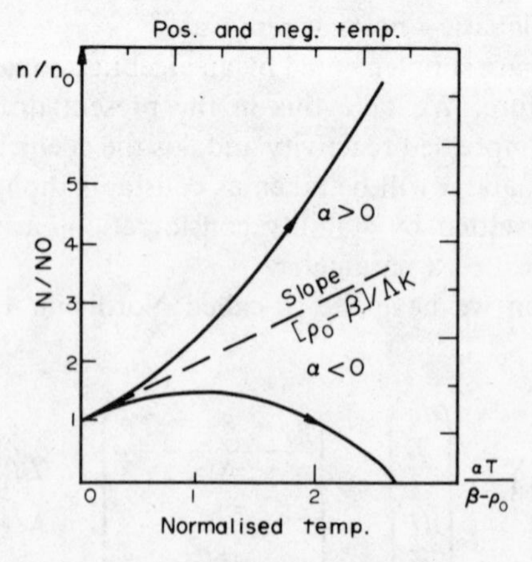

FIG. 8.1. Adiabatic excursions.

particular value of ρ_0 for either $\alpha > 0$ or $\alpha < 0$. The direction of the development of the trajectory in time is seen by reference to eqn. (8.1) which also shows that the state at $n=0$ is left and approached infinitely slowly. The formal solution of eqn. (8.5) is not continued, therefore, into the non-physical fourth quadrant. As we might expect, the nature of the solution depends dramatically on the sign of the coefficient of reactivity. Where α is positive, there is no bound to the excursion in the model used. When negative, the trajectory for $\rho_0 > 0$ passes through a maximum neutron level and steadily increasing temperature until the rise is sufficient to drive the reactivity so far negative as to shut down the reactor. This is seen to occur for a temperature approaching $-2\rho_0/\alpha$ and for small values of n_0 compared to the peak value $n_{\max} = n_0 - [\rho_0 - \beta]^2/2\kappa\Lambda\alpha$, this maximum temperature occurs at $-2g\rho_0/\alpha$, where $g = 1 + \kappa\Lambda\alpha n_0/2[\rho_0 - \beta]^2$. The maximum neutron level occurs at $-\rho_0/\alpha$.

The initial inverse period, $(dn/dt)/n$ initial, is, of course, $[\rho_0 - \beta]/\Lambda \equiv \omega_0$, and since this is a readily measured quantity we note the relation of this inverse period ω_0 to the square of the peak power level through

$$n_{\max} = \frac{\Lambda}{2\kappa\alpha}\,\omega_0^2 + n_0 \tag{8.6}$$

It is tempting to suppose that n_0 is so small in comparison with $n_{\max}$ as to be negligible in this expression and, indeed, there is good experimental support for the observation that for the fast "burst" transients discussed here, the peak power is proportional to the square of the impressed prompt reactivity. Nevertheless, the defects of the model must be noted. If we let $n_0 \to 0$ then the maximum temperature and shutdown point are given by $(0, -2\rho_0/\alpha)$ but, of course, the model without a source and with zero initial condition suggests that it takes an infinite time to initiate this excursion. We have a supercritical system waiting for a stray neutron to start the transient. The reactor would not actually shut down because of the delayed neutron sources ignored in the constant precursor form of eqn. (8.1), and this shows up in the supposition that the reactivity at the maximum temperature $\rho_{\max} = 2[\rho_0 - \beta]$ is still positive if the impressed reactivity $\rho_0 < \beta$. Another defect of the model is that the initial operation, before the transient was initiated by the application of the reactivity ρ_0, was not in steady state unless the power level $n_0 = 0$ since there is no heat removal mechanism. We leave further discussion of this model to the problems therefore and take as an improved model an allowance for constant heat removal that would enable the reactor to be run at an initial level n_0 in the steady state,

$$\frac{dT}{dt} = \kappa[n - n_0]; \quad n(0) = n_0 \tag{8.7}$$

Again this autonomous system enables a first integral to be obtained from

$$\frac{dn}{dT} = \frac{\rho_0 - \beta + \alpha T}{\kappa\Lambda[n - n_0]} \tag{8.8}$$

to yield

$$\kappa\Lambda\left[n - n_0 - n_0\, ln\frac{n}{n_0}\right] = [\rho_0 - \beta]T + \tfrac{1}{2}\alpha T^2 \tag{8.9}$$

In contrast to Fig. 8.1, these are closed figures in the state space, and Fig. 8.2 shows the

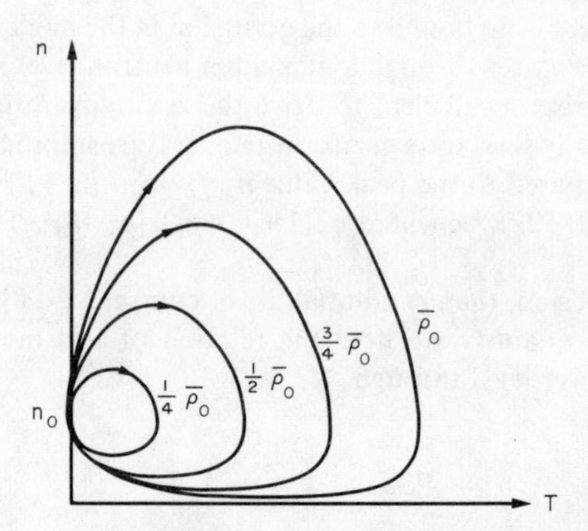

FIG. 8.2. Constant heat removal study.

resulting *orbital* solutions for increasing values of the reactivity disturbance $\rho_0-\beta$. It is seen that small disturbances lead to a circular cyclic behaviour (expand $ln\ n/n_0$) and the larger disturbances are distorted. The equilibrium point from $f(x)=0$ is a *centre* of these cycles.

The temperature at the centre, $T=-[\rho_0-\beta]/\alpha$, also corresponds to points of maximum and minimum power level and a temperature of half the maximum excursion (Nordheim's reflection law).

It is worth noting the physical mechanism producing this orbital cycle. The initial reactivity induces a rise in neutron level and a corresponding increase in energy production, exceeding the energy removal rate. The temperature therefore rises and in so doing decreases the reactivity. Even when $\rho-\beta$ is returned to zero the temperature rise does not stop since the energy production rate is greater than the removal rate. Only increasing temperature and negative reactivity drives the neutron level below its initial value. Energy removal then cools the system until the reactivity increases again and so enables the cycle to be repeated.

The case for $\alpha>0$ is, of course, unstable. Results for improved models will be taken up in the problems.

Stability in Nonlinear Systems

These two examples serve to introduce some of the additional complexities of nonlinear systems, and it is time now to make some general statements about the nature of nonlinear dynamics and stability. These remarks will not be mathematically rigorous, and the reader wishing to pursue the mathematics of the state space is recommended to Barnett[1] in the first instance, and to Smets[3] in relation to nuclear reactors.

We assume autonomous equations (where time does not appear explicitly in the coefficients) after, if necessary, manipulation to achieve this. The interest is also to study the reaction of the system to small disturbances rather than to driving terms such as sources, so we take these terms to be zero in the model.

The first step after formulating the state equations is to investigate the existence of possible *equilibrium* or *critical* points by equating $f(x)=0$. Unlike linear systems, there may be more than one resulting point. Each is examined in turn and we may suppose that a shift of variables leaves the point being studied at the origin. The behaviour of trajectories close to the equilibrium point is determined by linearising the state equations for small departures from the (new) origin and solving the resulting equations for initial conditions near the origin (see problem 8.1). Four of the several possible results are shown in Fig. 8.3.

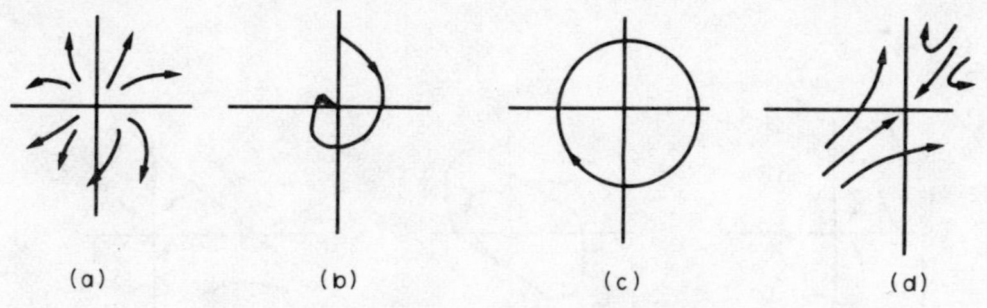

FIG. 8.3. Trajectories close to a critical point.

Case (a) shows all trajectories leaving the origin which is therefore a (local) unstable equilibrium point. Any small disturbance from the origin leads to trajectories which do not return. Case (b) shows a point of stable equilibrium, being the focus of a stable spiral. All small disturbances return (after sufficient time) to the origin. Case (c) indicates the point is a centre of an orbital cycle whilst case (d) indicates a saddle point (which is unstable except for exceptional isolated trajectories, called meta-stable trajectories, which arrive at the point but are clearly not stable against small departures).

Mixed cases may occur where the state space is divided up into zones by surfaces (lines in two-dimensions) called sepatrices. A sepatrix denotes the change from one behaviour to another as in the case of Fig. 8.4 (the viscous damped oscillator of the next section).

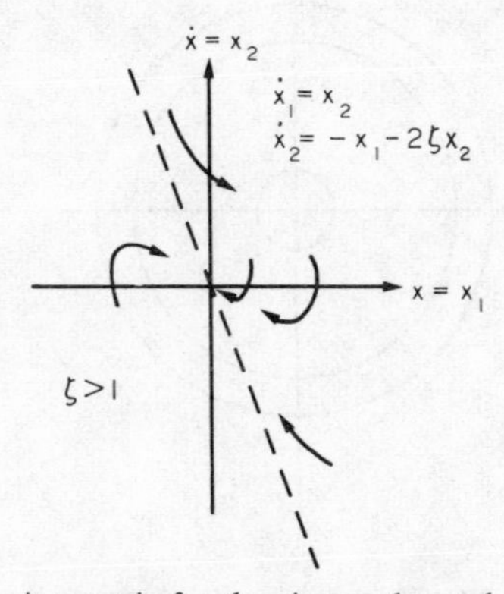

FIG. 8.4. A sepatrix for the viscous damped oscillator.

So far the linearisation has only justified conclusions for small departures from the equilibrium point. On a bigger scale the system may be found to follow patterns such as those illustrated in Fig. 8.5. In (a) the system is locally asymptotically stable, having all trajectories returning to the origin, but this property applies only in a limited region. In (b) the origin is apparently an unstable point but the trajectories are limited to an orbital cycle which in these circumstances is called a limit cycle.

Trajectories outside the limit cycle also move inwards so it is a stable limit cycle. Case (c) shows a limited region of stability around one equilibrium point together with a saddle point. A region where all trajectories move progressively inwards to the critical point is called a region of *attraction* for that point. Obviously it is a major goal of the analysis to find the largest regions of attraction around each critical point. The point is

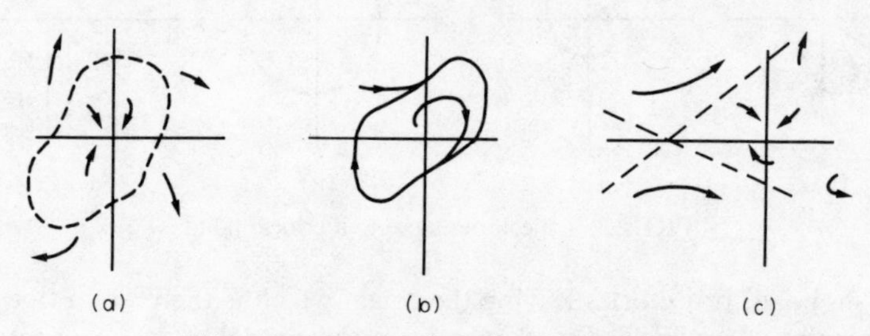

FIG. 8.5. Examples of nonlinear behaviour.

itself *asymptotically* stable for disturbances in that region. Case (b) shows that a linearisation would indicate an unstable system whereas the real situation is bounded and stable. The van der Pol equation examined in Chapter 9 is an example.

We now seek to define stability, which we do graphically, using Fig. 8.6. The critical

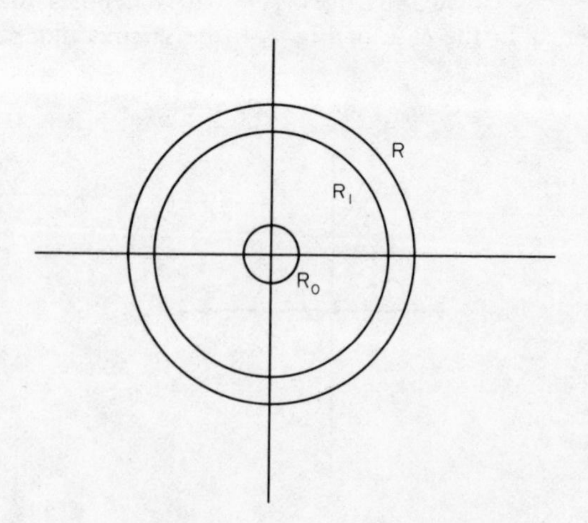

FIG. 8.6. Geometry for stability definitions.

point is at the origin and we define finite concentric regions R, R_1 and R_0 around it. The regions are circular in two-dimensions (spherical in three, etc.). The following definitions follow Liapunov: If for every R there exists an R_1 such that any trajectory starting in R_1 remains in R, the origin is stable. If for every R there exists an R_0 in which any trajectory that starts tends ultimately to the origin, the point is asymptotically stable. If R_1 can be taken as large as we like the point is stable in the large (or globally stable). Otherwise it is locally stable. Because this is true for every R, R can be made small and a nonlinear system is not Liapunov-stable unless the linearised system is stable at the same critical point. Note that R_0 is not itself a region of attraction since the requirement is that the trajectories remain in R not R_0. This raises a practical defect of the definitions: R_1 may exist finite but small, but the trajectories leading from it may travel through the large region R. A designer would generally think it a disadvantage that small disturbances could grow large even though not unbounded.

Lagrange offers an alternative definition that a critical point is stable in the sense of Lagrange if all the trajectories initiating in R are bounded in R.

For our purposes we shall take a more practical concept of seeking to establish regions (not necessarily circular) where the trajectories are bounded and travel *inwards*. If we take successive regions of smaller size where this is true and finally shrink the region onto the centre, all the trajectories from the original region have been traced to the centre, which is therefore asymptotically stable in respect of the original region (at least). We can then ask how big the original region can be and retain this property to investigate stability in the large. Correspondingly, if all trajectories on the surface of a region are directed *outwards*, the critical point is certainly unstable to this region. If the region can again be "shrunk" onto the centre with outward directed trajectories at the surface, the centre is totally unstable.

This approach needs careful understanding of what is meant by "surrounding" the origin. Geometrically, the figure must completely enclose the origin, case (a) of Fig. 8.7,

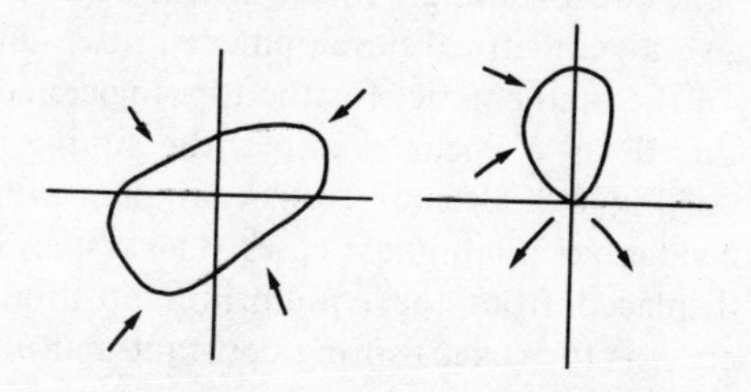

FIG. 8.7. Stability illustration.

and this nature is related later to the positive definiteness of the function describing the figure. Case (b) does not surround the origin and would permit some trajectories to depart from the origin which would therefore be unstable according to Liapunov.

The difference between linear and nonlinear systems can be restated for emphasis. In a linear system an equilibrium point can be either stable (in which case it is stable for large as well as small departures and the trajectories return to the point) or on the verge of stability (in which case trajectories are cyclic) or unstable. More generally in nonlinear systems, however, a system may be locally stable without being asymptotically stable and not necessarily stable in the large; it may lead to orbital stable solutions combined or not with trajectories leading to limit cycles, again in the small and not necessarily in the large. Limit cycles may be stable or unstable; it may be unstable locally (leaving a region) but not necessarily in the large, i.e. there may be other stable areas.

The mathematical definitions of stability do not always accord with the designer's desire to know the bounds of trajectories. It would seem more fruitful to study stability from the point of view of Fig. 8.7(a) with a view to determining a region that *surrounds* the origin for which trajectories do not leave the region. This idea is now developed in the Liapunov direct method.

Liapunov's Direct Stability Study

If the state equations could be solved explicitly, one might attempt to study all possible initial conditions as perturbations around an equilibrium point to determine the nature of the point and the various regions of stability and boundedness. Even if a solution was available it might be tedious to compute for all cases, and generally it will not be available. Liapunov instead studies stability as such and hence directly in terms of a tendency for trajectories to travel inwards. In the mathematical development a positive definite function is proposed which can be shrunk continuously onto the equilibrium point (at the origin). If such a function can be found so that *everywhere* the trajectories crossing its contour surfaces are directed inwards (negative definite), then stability has been demonstrated and the origin is asymptotically stable for at least the largest region covered by the Liapunov function. If the Liapunov function cannot be shrunk as far as the origin, we have still found a bounded region in the sense of Lagrange. Note that the method does not attempt to match up to the more general definition of stability given by Liapunov himself, but fortunately this is of less practical interest.

In the following we give a geometrical development in two-dimensions leaving aside some of the formalities of the mathematics for the topological concepts.

The concept is introduced via a linear example, the simple harmonic oscillator or spring and mass system of Fig. 8.8. Of course this example can be solved explicitly and the method will be more valuable in nonlinear cases. The system shown, without driving term, will oscillate if displaced from the equilibrium position (origin) at a natural frequency $\omega^2 = k/m$, where k is the (fixed) spring constant and m the mass according to a second order equation

$$m\ddot{y} + ky = 0; \quad y(0) = y_0; \quad \dot{y}(0) = \dot{y}_0$$

We start by ignoring the viscous damping pot shown in Fig. 8.8.

These state equations can be written in standard form with suitable normalisation of time as:

$$\dot{x}_1 = x_2; \quad x_1(0) = x_{10}$$
$$\dot{x}_2 = -x_1; \quad x_2(0) = x_{20}$$

A first integral in the state space yields

$$x_1^2 + x_2^2 = C = x_{10}^2 + x_{20}^2 \tag{8.10}$$

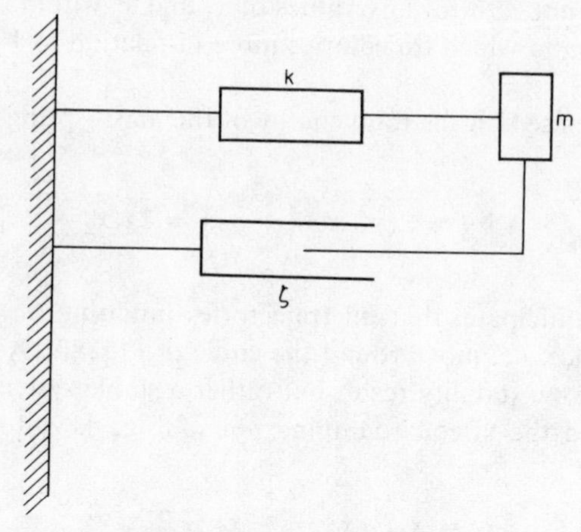

FIG. 8.8. Mass spring system.

and it is convenient to refer to this form of solution as $V(x)=x_1^2+x_2^2=C$. With our normalisation the solution in the state space is seen to be circular with radius dependent on the initial displacement and initial velocity (Fig. 8.9).

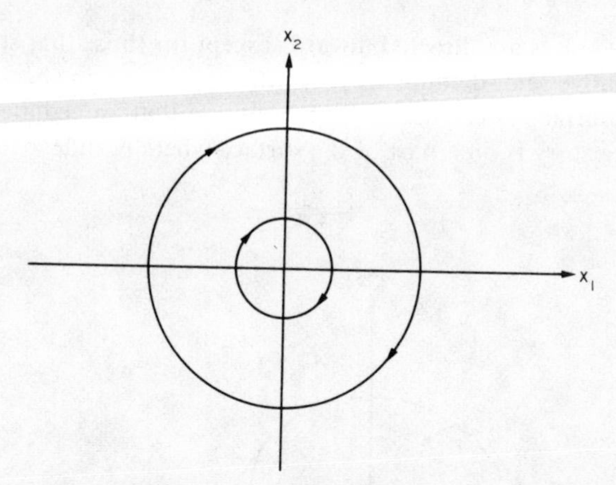

FIG. 8.9. State space solution for mass spring system (no damping).

The close connection of this V-function to the total energy $(\frac{1}{2}m\dot{y}^2+\frac{1}{2}ky^2)$ should be noted and, indeed, $V(x)$ is twice the total of the kinetic and potential energy of this system.

This $V(x)$-function corresponds to a solid figure, a circle in this case, enclosing the origin, the equilibrium point. If a trajectory on a surface $V(x)=C_0$ is directed *inside*, there is no way the trajectory can subsequently leave that V-surface. If the V-function is continuous (and the property of inward direction remains true) and is shrunk down onto the origin, then we have shown that the region bounded by the original $V=C_0$

surface must be a region of asymptotic stability. This evidently requires that $V(x)$ should be positive definite (i.e. not zero for any values of x_1 and x_2 within the region of $V=C_0$) and that $\dot{V}$, the direction in which trajectories more in relation to V, should be negative definite.

Take the V-function based on the total energy of the mass–spring system and we have

$$\dot{V} = \frac{\partial V}{\partial x}\frac{dx}{dt} = \nabla V f = 2(x_1, x_2)\begin{pmatrix} x_2 \\ -x_1 \end{pmatrix} = 2x_1x_2 - 2x_1x_2 = 0 \qquad (8.11)$$

We see, as we might anticipate, that all trajectories initiating on a surface $V=C$ are directed *along* the surface, i.e. move round the circle of Fig. 8.9. As already known, we do not have an asymptotic stability result but rather a stable system in an orbit.

Suppose now we add the viscous damping pot of Fig. 8.8 into consideration. The state equations become

$$\dot{x}_1 = x_2; \quad \dot{x}_2 = -x_1 - 2\zeta x_2$$

where ζ is the damping factor. Again these equations can be solved explicitly but instead use the original V-function with the new state equations to obtain

$$\dot{V} = \nabla V f = 2(x_1, x_2)\begin{pmatrix} x_2 \\ -x_1 - 2\zeta x_2 \end{pmatrix} = -2\zeta x_2^2 \qquad (8.12)$$

We see that all trajectories are directed inwards except for those that start with $x_2=0$. The V-function is negative semi-definite only. A little consideration shows that certainly no trajectory leaves a surface of constant V outwards and that immediately after leaving a point $(x_1, 0)$, the trajectory is on a part of the surface where it, indeed, crosses inwards (Fig. 8.10).

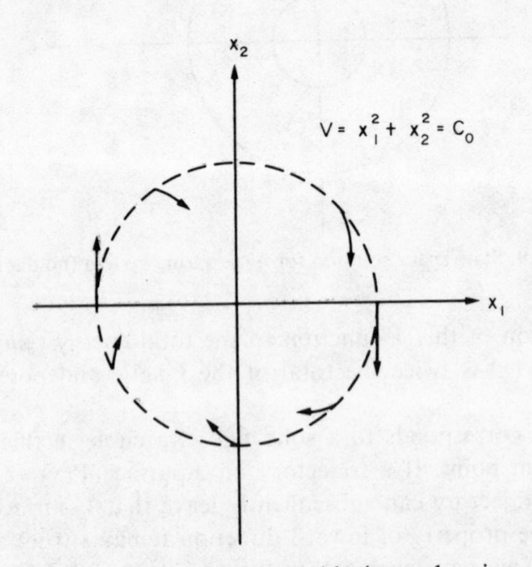

FIG. 8.10. Mass–spring sytem with viscous damping.

With this exceptional point allowed for,† we see that the trajectory for viscous damping necessarily spirals inwards. The V-function is continuous and can be shrunk to the equilibrium point which is demonstrated to be a point of asymptotic stability. Since the original V-function can be taken as large as we like, we have asymptotic stability in the large.

Two points are worth making—an extension and an interpretation. If the original V-function can only be shrunk partially around the origin before either V loses its positive definite character or $\dot{V}$ loses its negative definite character, then we have still demonstrated Lagrange stability; furthermore, we have achieved the valuable practical point of knowing a bound to the transients. Correspondingly, if all trajectories leave the V-surface we have demonstrated instability.

If, of course, our $\dot{V}$-function is not definite—some trajectories leave, some enter—we have proved nothing. (If our V-function is not definite we cannot prove anything; we must take a "solid" figure.)

Therefore we have the philosophic point that using Liapunov's technique is as much an art as a science. It is up to us to propose a suitable V-function and if the first attempt is inconclusive we may try other functions based on experience and the study of the problem at hand. It is comforting to know that if the system is asymptotically stable, a V-function does exist that will demonstrate this property.

DEFINITE QUADRATIC FUNCTIONS

While we may want to use more complicated functions (whose positive definite nature can perhaps be demonstrated geometrically) it is helpful to establish some properties of quadratic functions. First, a function such as $V = ax_1^2 + bx_1x_2 + cx_2^2$ is a general second dimension quadratic function. We wish to know whether for any real value of x_1 and any independent real value of x_2 if the value of V is always positive. There are many ways of testing this but perhaps the simplest in the two-dimensional case is based on the matrix representation of V. We may always write the general quadratic in terms of a 2×2 matrix and the x vector and its transpose:

$$ax_1^2 + 2bx_1x_2 + cx_2^2 = (x_1, x_2)\begin{pmatrix} a, & b \\ b, & c \end{pmatrix}\begin{pmatrix} x_1 \\ x_2 \end{pmatrix} \tag{8.13}$$

Consider the principal minors of the determinant of this matrix, a and $ac - b^2$. If both are positive, V is positive definite. If one is zero and the other positive, V is positive semi-definite. If $-V$ satisfies the same tests, V is negative definite or negative semi-definite. If none of these cases so far is true, i.e. principal minors of opposite sign, V is indefinite.

Quadratic functions are only one example of possibly positive definite functions. However, for linear systems a quadratic function is sufficient for consideration (and serves to derive the various stability criteria in the linear frequency space). Note the connection with Routh's criteria.

Similar tests can be built up for higher than quadratic functions but it may be possible

†Consider further differentials of $\dot{V}$; see also problem 8.6.

to limit their use to cases that are positive definite by inspection (such as $x_1^2+x_2^2$ or $(x_1^2+x_2)^2$).†

LIAPUNOV THEOREM RESTATED

If the system equations are $\dot{x}=f$ and $V(x)$ is a continuous positive definite function over the finite region studied, with $V(\infty)\rightarrow\infty$ and $V(\text{o})=0$, consider $\dot{V}=\nabla V f$. If $\dot{V}$ is negative definite the system is stable in the region. If $\dot{V}(x)$ can be shrunk to $\dot{V}(\text{o})$ retaining this property, the system is asymptotically stable with the origin as equilibrium point. If the region can be enlarged indefinitely, stability has been proved in the large. If $\dot{V}$ is only negative semi-definite, the system is stable but this stability may be of the form of a limit cycle or orbit (which has been found if $\dot{V}$ vanishes). Inverted arguments can be used for positive definite $\dot{V}$ to demonstrate instability. If the V result is indefinite, nothing has been proved. In the above, the shift of axes to the origin is only a convenience and the relations may be tested about any point without shifting the origin.

CONSTRUCTING A LIAPUNOV V-FUNCTION

The functions used so far have either been taken from the energy function of a dynamical system or based on the fortuitous availability of a positive definite (or at least space-enclosing) function obtained as the solution of a simpler problem. One could, of course, guess a succession of functions, but it is as well to tackle such an approach systematically. The variable gradient method is one such systemization of guessing Liapunov functions.

The idea here is not to start from a known function but rather to take the simplest *gradient*, ensure that it can come from a function (i.e. that it can be integrated) and simultaneously study V and $\dot{V}$ for definiteness. If the simplest case fails, take the next simplest.

In two dimensions the gradient can be written as

$$\nabla V = \begin{pmatrix} a_{11}x_1 + a_{12}x_2 \\ \\ a_{21}x_1 + a_{22}x_2 \end{pmatrix} \tag{8.14}$$

where the a_{ij} are in principle functions of x. We can start with constant a_{ij}, then proceed to such forms as $c_{ij}x_1+d_{ij}x_2$ and go on complicating the representation until we either find satisfactory functions or give up.

The condition that the gradient has come from a scalar function is the integrability condition (or vanishing of the curl of ∇V) which in the two-dimensional case is simply

$$\frac{\partial[a_{11}x_1 + a_{12}x_2]}{\partial x_2} = \frac{\partial[a_{21}x_1 + a_{22}x_2]}{\partial x_1} \tag{8.15}$$

If we use constant coefficients, this condition reduces simply to the symmetry property $a_{12} = a_{21}$.

†Positive definite in two but not, of course, in three dimensions.

As an example, we try to construct a V-function for the simple harmonic oscillator (mass–spring) with no damping.

Using the simplest case to start with, the integration of the symmetric gradient gives

$$V = \tfrac{1}{2}a_{11}x_1^2 + a_{12}x_1x_2 + \tfrac{1}{2}a_{22}x_2^2 \tag{8.16}$$

with principal minors $a_{11}>0$ and $a_{11}a_{22}-a_{12}^2>0$ in order to be positive definite. Hence $a_{22}<0$. The V-function, however, is

$$\dot{V} = \nabla V f = [a_{11}-a_{22}]x_1x_2 + a_{12}[x_2^2-x_1^2] \tag{8.17}$$

The best that can be made of the conflicting requirement that some firm conclusion is to be drawn is to take $a_{12}=0$ and $a_{11}=a_{22}$. We recover the energy V-function (or a scalar multiple) and the conclusion that $\dot{V}=0$. This demonstrates that the system has an orbit for any displacement with the obtained V-function actually being the solution of the state equations and the description of the periodic solution.

In the problems the variable gradient method is used to develop a rigorous Liapunov function for the damped oscillator. The coefficients a_{ij} are not limited to constants or even polynomials in x_1, x_2, so the method is a systematic device for trying possible V-functions. Further applications to reactor studies are developed in the problems.

Application to Reactor Studies

In the problems we extend the model with no delayed neutrons to encompass arbitrary cooling mechanisms. Here we use the prompt jump approximation to re-introduce delayed neutrons while retaining the order of the system unchanged. Again take one (constant and negative) temperature coefficient of reactivity and the constant energy removal model. State equations are therefore

$$\left.\begin{aligned}\frac{dc}{dt} &= \frac{\lambda\rho}{\beta-\rho}c = \frac{\lambda\alpha T}{\beta-\alpha T}c; \quad c(\mathrm{o}) = c_0\\ \frac{dT}{dt} &= \kappa[n-n_0] = \frac{\kappa\lambda\Lambda}{\beta-\alpha T}\left[c - \frac{\beta-\alpha T}{\beta}c_0\right]; \quad T(\mathrm{o}) = T_0\end{aligned}\right\} \tag{8.18}$$

There is no imposed reactivity, ρ_0 and the disturbance consists of an imposed temperature disturbance T_0.

The only equilibrium point is at $c=c_0$, $T=0$. It is not essential to shift the origin in applying the method. How shall we select a Liapunov function to study this case? The idea is to solve a simplified version and use this as a guide.

The obvious approximation is to take $\beta-\rho\to\beta$ in the denominator of the reduced state equation

$$\frac{dc}{dT} = \frac{\alpha}{\kappa\Lambda}\frac{Tc}{\left[c - \dfrac{\beta-\rho}{\beta}c_0\right]} \simeq \frac{\alpha}{\kappa\Lambda}\frac{Tc}{c-c_0} \tag{8.19}$$

Certainly the prompt jump approximation itself will fail if ρ approaches β so this should be reasonable. The approximation can now be integrated to give

$$-\frac{\alpha}{2\kappa\Lambda}[T^2 - T_0^2] - c_0 \ln\frac{c}{c_0} + c - c_0 = 0 \tag{8.20}$$

or

$$V(c, T) = -\frac{\alpha}{2\kappa\Lambda}T^2 + c - c_0 - c_0 \ln\frac{c}{c_0} = -\frac{\alpha}{2\kappa\Lambda}T_0^2; \quad \text{constant}$$

which is seen (Fig. 8.11) to indicate orbit cycles in the state place for negative reactivity

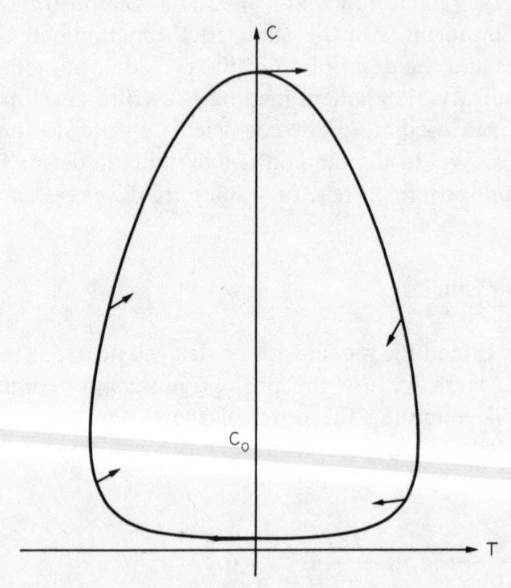

FIG. 8.11. Prompt jump model: Liapunov study.

coefficient. Let us suppose this to be the case but seek to improve on the approximation just made.

Take eqn. (8.20) to provide a Liapunov function. It is positive definite and can be shrunk continuously on the critical point $(c_0, 0)$. The resulting value of $\dot{V}$ is

$$\dot{V} = \left(1 - c_0/c, \quad -\frac{\alpha T}{\kappa\Lambda}\right) f = -\frac{\lambda\alpha^2 T^2 c_0}{\beta[\beta - \alpha T]} \tag{8.21}$$

The resulting function is then negative semi-definite (vanishes at $T=0$) for $\rho = \alpha T < \beta$ We have demonstrated with the full prompt jump approximation that the system is bounded for $\alpha < 0$ having at worst a limit cycle or orbit. Again, second order consideration of what happens to the trajectory initiated at $T=0$ shows that *all* trajectories are inwardly directed and the system is actually asymptotically stable with $(c_0, 0)$ as the stable equilibrium point, the centre of the approximate study. Chapter 9 indicates methods by which this problem can be studied on an analogue computer. It is also instructive to look for a rigorous Liapunov function (i.e. having $\dot{V}$ negative definite) along the lines of problem 8.7.

Smets classic reactor study extends these ideas into the three-dimensional (n, c, T) state space.[3]

Optimum Control Theory

Suppose that a system is to be brought from one state to another, perhaps in the form of a correction to a substantial disturbance. We suppose that there are control variables that can be changed to achieve this aim—the system is controllable and can in principle reach the desired state with any conditions imposed. These control variables, written as a vector $u=(u_1, u_2, \ldots)$ will appear in the coefficients of the state equations. It is likely that there will be a range of choices open amongst the possible control patterns achieving the specification so that the possibility exists of choosing one particular control pattern as being in some way better than all others.

This desire for a "best" control calls for a precise definition of what might be called a cost function $\bar{C}$ which is to be minimised (or a profit function to be maximised). One simple cost function might be the time taken to achieve the return to the desired state. In this case $\bar{C}$ is simply $\int_{t_i}^{t_f} dt'=\tau$, where $t_i=0$ is the initial time, at which the control problem starts, t_f the final time at which it finishes and $\tau=t_f-t_i$ the so far unknown control period. More generally the cost function will depend on both the state vector x and the control vector u, and can be written

$$\bar{C} = \int_{t_i}^{t_f} f_0(x, u)dt' \tag{8.22}$$

For example: if $f_0=1$ we have the original minimum time problem;
if $f_0=u$ we have a problem based on minimising the control effort, e.g. the fuel used;
if $f_0=x_1^2$ we have a problem related to the mean value of the square error of the first state variable, to be minimised.

Note that the choice of cost function is independent of the state description and equations; different cost functions will usually lead to different "best" control choices, each achieving the specified change of state.

It would be naive to try and solve the state equations explicitly for every possible choice of control pattern and then compare the cost functions direct in seeking a minimum cost. The need is to solve for *one* choice of u and the resulting x and then to find the *change* in $\bar{C}$ about this value in some way that does not require solving the state equations over again. We then note that if the first choice of u was indeed optimum, all possible small changes around it must lead to an *increase* in $\bar{C}$ or at worst no change. This approach[7] is in the spirit of the perturbation formulae using adjoint functions introduced in Chapter 2.

Thus the sequence of ideas in this presentation of the blind mathematician Pontryagin's optimal control theory is as follows:

(a) Construct an expression for the cost function that is not sensitive to small changes in the *state* variable. This will involve an adjoint equation and adjoint variable.
(b) From this stationary function, find the changes in the cost for small changes in the *control* variable direct without having to find the effect on $\bar{C}$ of the corresponding changes in the state variable.
(c) Seek that value u^* such that all small changes of u around u^* lead to an increased cost or at worst no increase.

To follow this procedure we set up a functional expression L which is to yield the cost function $\bar{C}$ while being stationary to errors or changes in x, the state variable. We take

$$L = \int_{t_i}^{t_f} [H - px]dt = \bar{C} + \int_{t_i}^{t_f} p\left[f(x, u) - \frac{dx}{dt}\right] dt \tag{8.23}$$

Here p is the adjoint vector (sometimes called the co-state vector), $p=(p_1, p_2, \ldots)$ corresponding to $x=(x_1, x_2, \ldots)$. H (called the Hamilton density) is a convenient expression for $H=f_0+pf$. So far p is undetermined. The choice of L is governed by the observation that if the value inserted for x is exact (solution of the state equation for x), then the term in brackets vanishes and leaves only $\bar{C}$, again exactly evaluated. What we need now is that if we make small errors in giving a value to x in working out L, that the change in L should be negligible, leaving the correct value of $\bar{C}$.

Recollect that $\bar{C}$ itself depends on x and on making a change δx around the correct value we may after integrating by parts write

$$\begin{aligned}\delta L(\delta x) &= \delta\bar{C}(\delta x) + \int_{t_i}^{t_f} \left[p\frac{\partial f}{\partial x}\,\delta x - \frac{d}{dt}[\delta x]\right] dt \\ &= \int_{t_i}^{t_f} \delta x \left[\frac{\partial f_0}{\partial x} + \frac{\partial}{\partial x}[fp]\right] dt + \int_{t_i}^{t_f} \delta x \frac{dp}{dt}\,dt - [p\delta x]_{t_i}^{t_f}\end{aligned} \tag{8.24}$$

where $fp=f_1p_1+f_2p_2+\cdots$.† We want $\delta L=0$ and so we choose to take

$$\int_{t_i}^{t_f} \delta x \left\{\frac{dp}{dt} + \frac{\partial}{\partial x}[fp] + \frac{\partial f_0}{\partial x}\right\} dt = 0 \quad \text{and} \quad [p\delta x]_{t_i}^{t_f} = 0 \tag{8.25}$$

Consider the first of these conditions, where δx is unknown. We can, nevertheless, make the expression vanish for arbitrary $\delta x(t)$ by requiring p to satisfy the adjoint equation

$$-\frac{dp}{dt} = \frac{\partial}{\partial x}[fp] + \frac{\partial f_0}{\partial x} = \frac{\partial H}{\partial x} \tag{8.26}$$

Here we see the term $\partial[fp]/\partial x=\nabla[fp]$ as a coefficient for the terms linear in p while ∇f_0 is an adjoint source term, depending on the choice of cost function.

†See problem 8.11 for clarification of the vector notation.

Consider the second of these conditions. Presumably at t_i we know x so that $\delta x_i = 0$; the expression vanishes at its lower limit. We probably do not know every detail of $\delta x(t_f)$ and so we choose $p(t_f) = 0$ to satisfy the requirement. We have added adjoint boundary conditions at the *final* time to the adjoint equation.

EXAMPLE: THE RACING CAR PROBLEM

The usual simple example is based on a second order servo system with inertia but no friction driven by a controllable torque with a view to returning it from a given displacement to the required position, taken to be at the origin. With a different interpretation this could be the racing driver who has brake and accelerator at his disposal in an attempt to travel in the shortest time between rest at one traffic light before returning to rest at another traffic light.† The problem then is one of minimum time where the state equations have the form

$$\dot{x}_1 = x_2; \quad \dot{x}_2 = u \tag{8.27}$$

with $u(t)$ the (single) available control, $u = F/m$ where the torque or controlling force F can be varied in time, and m is the inertia or mass.

It is a practical observation that controls can usually only be varied between finite limits dependent on availability, safety, etc., and we may in this example suppose $-1 < u(\tau) < 1$. The condition for the termination of this control problem specified $\delta x_1(t_f) = \delta x_2(t_f) = 0$, so that no imposition is made on the adjoint variable at t_f. We have $f(x, u) = (x_2, u)$ so that $fp = x_2 p_1 + u p_2$ and the adjoint equations are

$$\frac{\partial H}{\partial x_1} = -\dot{p}_1 = 0; \quad \frac{\partial H}{\partial x_2} = -\dot{p}_2 = p_1 \tag{8.28}$$

The adjoint solutions may therefore be obtained in general form dependent on two arbitrary constants

$$p_1(t) = C_1; \quad p_2(t) = C_1 t + C_2 \tag{8.29}$$

We consider the change in $\bar{C}$ brought about by a change in u applied to the functional $\bar{L}$. We have

$$\delta \bar{L}(\delta u) = \int_0^{t_f} \delta u \frac{\partial H}{\partial u} dt = \int_0^{t_f} \delta u \frac{\partial}{\partial u} [pf] dt = \int_0^{t_f} \delta u p_2 dt \tag{8.30}$$

Now δu is essentially arbitrary as long as overall the system terminates at the origin. If we thought of the change as made up of many delta function components we would see that wherever the term $\partial H/\partial u = p_2$ was positive, a positive addition of u would add to the cost/time. Wherever $\partial H/\partial u = p_2$ is negative, however, an addition to u would *decrease* the cost. The expression $\partial H/\partial u$ is called the *switching function* since clearly when it changes sign it is advantageous to switch from adding to subtracting elements

†With a suitable "controller" this makes an excellent analogue problem to demonstrate optimum control in the form of a challenge to the user to find the optimum control pattern. All the following examples can usefully be set up this way and tackled heuristically.

to u. This process of improving the result by changing u in accordance with the sign of the switching function has to stop, however, when u reaches its bounds and no more can be added (subtracted). We may conclude in this case that the optimum pattern in u consists of segments with u either at $u_{max}(=1)$ or $u_{min}(=-1)$ according to the *sign* of the *switching function*.

We next note that when the general solution of the switching function is examined there is at most one occasion (dependent on the ratio of the two constants of the second order system) when p_2 vanishes in between t_i and t_f. There can at most be two segments in the optimum control, one at $u=1$ and one at $u=-1$. This final degree of freedom, choice of C_1/C_2, allows us to adapt the optimum solution to arbitrary initial conditions, $x(\text{o})$.

Figure 8.12 shows the general solution, built up of parabolas at one extreme value of

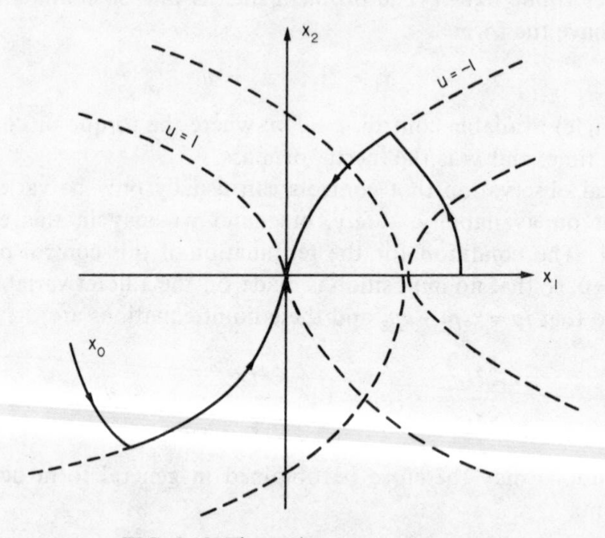

FIG. 8.12. The optimum servo system.

u to take the state of the system towards the origin as fast as possible, followed by a segment at the other extreme of u to kill the speed as the system is brought to rest at the origin in the shortest time. Exceptionally the system starts on a switching point as it were, and only one control segment is required.

This was a simple example and a few more points need explanation before attempting further examples or the problems. The first concerns the situation when instead of a switch it is possible for the switching function to remain at zero for a finite time, which would admit a segment of the optimal control at that value of u which serves to keep $d/dt\ [\partial H/\partial u]=0$. In the last example it may be shown by differentiating that there is no singular solution that keeps the switching function zero for a finite time.

Secondly, the terminal conditions may be more complicated. Rather than specifying $x(t_f)$ fully it may be sufficient to require some relationship between the elements of x to hold and thus finish the control period. Instead of a desire to carry x to some specific point, we need only bring it to some *target curve*. The expression $p\delta x|_{t_f}$, however, is still to vanish and this will replace the number of specifications in $x(t_f)$ that have been discarded with exactly the number of specifications in $p(t_f)$ to restore the exact "solubility" of the system of state and adjoint equations. In particular, if $x_i(t_f)$ is entirely free, the $p_i(t_f)$ must be taken to be zero. If the end time t_f is known rather than being free as in the example, we will find that this imposes an extra condition on the choice of controls available to again compensate for the change in degrees of freedom.

Thirdly, it will have been seen that the example was autonomous; time did not appear explicitly in the state equations. If time should appear there is a simple trick to make the system look autonomous (at the expense of some complication) and thus have these results applicable: we define time t in the coefficients of the state equation as a further state variable to be added to x (and hence require a further adjoint variable added to p). We now have an explicitly autonomous system.

Finally, all the examples are linear state equations. Nevertheless, the *control* may appear as a multiplier of a state variable, and from this point of view we loosely, perhaps, speak of a nonlinear system. The next two examples are nonlinear in this control sense while the last example was linear. In the "linear" system only it may be shown that optimum control is composed just of segments at extreme values, and if the order of the system is, say, n (2 in the example) the maximum number of segments in each optimum control is n. For nonlinear control systems we have the possibility of singular optimum controls that make the switching function vanish identically.

Optimum Reactor Control

REACTOR STARTUP

The additional adjoint equations, particularly that they introduce the need to solve at *two* points (t_i and t_f) simultaneously, bring substantial difficulties in all but simple cases. It is possible to set up digital or analogue computer programs for some general cases, though it is commonly found necessary to limit the state equations to two or three dimensions. We give here two examples from reactor theory which are additionally simplified by having only a brief discussion of the role of bounds on the *state* variables.[7]

The first example concerns starting up a reactor from low power and raising it as quickly as possible to a predetermined level with no subsequent overshoot of the neutron population. This is not simply a matter of inserting the largest reactivity available (even if this were operationally sound and safe) and withdrawing it instantly when the power level reached the desired level since the delay in the production of neutrons from precursors would then lead to a subsequent increase in the neutron level to be offset by further reactivity changes which must be taken account of before the control time can be said to have terminated.

We grossly simplify the example by taking one group of precursors, assuming no source term and ignoring all feedback effects. The state equations are thus

$$\left.\begin{aligned} \frac{dn}{dt} &= \frac{\rho - \beta}{\Lambda} n + \lambda c; \quad n(0) = n_0; \quad n(t_f) = n_f \\ \frac{dc}{dt} &= \frac{\beta}{\Lambda} n - \lambda c; \quad c(0) = \frac{\beta}{\Lambda\lambda} n_0; \quad c(t_f) = \frac{\beta}{\Lambda\lambda} n_f \end{aligned}\right\} \tag{8.31}$$

and we assume the initial state is in equilibrium, as also the final state, with a change of neutron level from n_0 to n_f being required. As a control system, this is nonlinear because of the product ρn. We have

$$H = 1 + p_1\left[\frac{\rho - \beta}{\Lambda} n + \lambda c\right] + p_2\left[\frac{\beta}{\Lambda} n - \lambda c\right]$$

The adjoint equations are therefore

$$\left.\begin{aligned} -\dot{p}_1 &= \frac{\rho - \beta}{\Lambda} p_1 + \frac{\beta}{\Lambda} p_2; \quad -\dot{p}_2 = \lambda p_1 - \lambda p_2 \\ \frac{dp_1}{dp_2} &= \frac{[\rho - \beta]p_1 + \beta p_2}{\Lambda\lambda[p_1 - p_2]} \end{aligned}\right\} \tag{8.32}$$

and the switching function $\partial H/\partial \rho = p_1 n/\Lambda$, whose sign depends upon p_1. A little analysis in the state plane and in the adjoint plane is helpful. We first note, Fig. 8.13(a), for

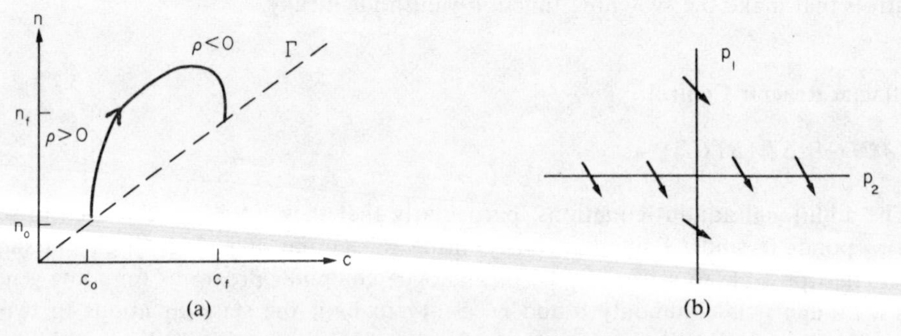

FIG. 8.13. State and adjoint spaces for fastest startup.

the state trajectory to rise above the line Γ, $n = \lambda\Lambda c/\beta$, requires a positive reactivity. To return to Γ in finite time from above requires a negative reactivity. We next observe from (b) that the axes of the adjoint plane serve to determine the direction of the adjoint trajectory so that the adjoint variable p_1 of the switching function may change sign at most once. It must change sign at least once; it cannot support a finite period at which the switching function is zero, so we have done enough to completely synthesise the required optimal solution: apply the greatest available reactivity at the start and switch to the smallest available reactivity at that time that will bring the trajectory to the required value n_f on the Γ line in the state space.

A better model including restrictions on the rate of reactivity insert is given in ref (8).

For the next example we need an extension of theory. Consider the Hamilton density in an autonomous system, $H = f_0 +$ so that

$$\frac{dH}{dt} = \frac{\partial H}{\partial x}\frac{dx}{dt} + \frac{\partial H}{\partial p}\frac{dp}{dt} + \frac{\partial H}{\partial u}\frac{du}{dt}$$

The first two terms on the right always cancel in view of the state and adjoint equations, $\partial H/\partial p = dx/dt$ and $-\partial H/\partial x = dp/dt$. In an optimum autonomous system, either $\partial H/\partial u = 0$ (the singular solution) or $du/dt = 0$ (the "bang-bang" solution) where u is at one of its (constant) limits, $u_{\min}$ or $u_{\max}$. Thus $dH/dt = 0$ and $H(t) = \text{const}$. Now consider the free-end time problem so that δt_f may have either sign. The variation is $\delta \bar{L}(\delta t_f) = H(t_f)\delta t_f \to \delta\tau$, and for this to vanish, $H(t_f) = 0$, so that H in an autonomous optimum free end time system is identically zero.

The singular solutions obtained from considering the switching function to be identically zero over a finite period, correspond to classical variational solutions. The recognition of "bang-bang" solutions, i.e. periods when the controller is switched as rapidly as possible between its extreme values, arose later in variational studies though control engineers had heuristically recognised the practical value of such controls in many cases before a consistent theory was developed.

In more complicated cases it is necessary to solve the adjoint equations to determine the switching function and by monitoring this as the solution develops, switch optimally. The next example, however, while considering the possibility of a singular solution and introducing a fuel optimum example, again is sufficiently simple for an optimum control to be synthesised completely.

XENON SHUTDOWN

As a second example, consider the problem of xenon poisoning.[9] On shutting down from a steady operating point, the xenon density grows hugely by iodine decay to ^{135}Xe, no longer being burnt out by the operating neutron level. This will continue until such time as decay of xenon removes the poisoning which may well prevent the reactor from being started up for up to 48 hr—a matter of some cost and embarrassment to reactor operators. Perhaps, instead of shutting down outright, we can try a control period at an intermediate flux, enough to keep the xenon level below that at which the reactor cannot be given enough reactivity to overcome the poisoning effect. This intermediate neutron level is then maintained until such time as the free system would tend to decrease in xenon, a situation in which we seek to carry the xenon, iodine state trajectory towards a target curve Ω, on and below which it is known that no further poisoning out can occur (Fig. 8.14).

There are many such possible ways of using the neutron level (or flux) as a control parameter to achieve the desired controlled shutdown and we may therefore impose an additional restriction to shut down in some "best" way. One such choice is to minimise the amount of fuel used up in the control period, i.e. to minimise the integral of the flux over the control period, with $f_0 = \varphi$, $\bar{C} = \int_{t_i}^{t_f} \varphi\, dt$.

The state equations are now (neglecting direct yield)

$$\frac{dI}{dt} = \gamma\Sigma_f\varphi - \lambda_i I; \quad \frac{dX}{dt} = \lambda_i I - \lambda_x X - \varphi\sigma X \tag{8.33}$$

and the equation for the target curve can be written as $\Omega(X, I) = 0$. The adjoint equations become

$$\frac{dp_1}{dt} = \lambda_i[p_1 - p_2]; \quad \frac{dp_2}{dt} = [\lambda_x + \varphi\sigma]p_2 \tag{8.34}$$

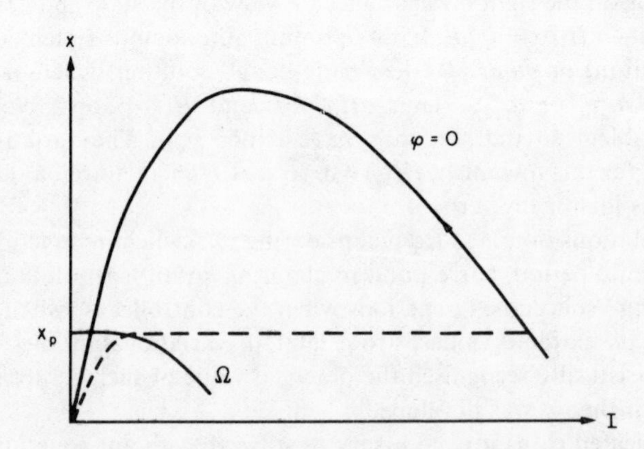

FIG. 8.14. The xenon shutdown uncontrolled and target curves.

subject to the boundary conditions obtained from the target curve as acceptable variations move along Ω, interconnecting δI_f and δX_f:

$$p\delta x|_{t_f} = 0 = \frac{\partial\Omega}{\partial x}\,\delta x|_{t_f} \to (p_1, p_2)_{t_f} = \left(\frac{\partial\Omega}{\partial I}, \frac{\partial\Omega}{\partial X}\right)_{t_f} \quad (8.35)$$

The Hamilton density for this problem should be zero:

$$H = \varphi + p_1[\gamma\Sigma^f\varphi - \lambda_i I] + p_2[\lambda_i I - \lambda_x X - \varphi\sigma X] = 0 \quad (8.36)$$

and the switching function is

$$\frac{\partial H}{\partial\varphi} = 1 + p_1\gamma\Sigma^f - p_2\sigma X \quad (8.37)$$

We check for the possibility of a singular solution where $\partial H/\partial\varphi$ and its rate of change as well as H are also zero. These three simultaneous conditions lead to an equation for I, X of the form

$$I^2 + \frac{\lambda_x}{\lambda_i}\frac{\gamma\Sigma^f}{\sigma}X = 0 \quad (8.38)$$

It is evident, however, that this lies in the third and fourth quadrant and is not physically realisable. Thus singular solutions do not appear.

We next note that unless the shutdown trajectory is from a sufficiently high power as to tend to exceed the available reactivity and lead to a xenon value higher than the target curve, there is no problem. However, we are obliged to prevent $X(t)$ rising above this value X_p and this imposes a state restraint for which we have developed no theory. While such a theory can be given [7] the physical situation will here suffice.

Since all switching points lie above the continuum curve for $\partial H/\partial\varphi = d/dt[\partial H/\partial\varphi] = 0$, then switching can only be in the direction allowed by $d/dt[\partial H/\partial\varphi]$. As the trajectory leaves the operating point it is at $\varphi_{min} = 0$ and rising. It can only, according to the unrestrained problem, switch to φ_{max} once, and initially this would not bring the trajectory to the target curve; the only trajectory at φ_{min} to reach the target curve approaches in the horizontal portion to the left of and including the xenon peak X_p of the target curve. Thus when the trajectory reaches the restraint, $X = X_p$, a new flux, sufficient to restrain $X(t)$ to X_p, must be employed. The question remains: Does the trajectory remain on the restraint until reaching the target peak or does it at some point switch to φ_{max}, where this segment would lead it to the target curve?

To determine this question, consider the solution in reverse time leaving the target curve away from the peak and therefore away from the restraint and necessarily at φ_{max}. The adjoint boundary conditions derived from the target curve are to be satisfied simultaneously with the vanishing H condition:

$$\varphi_{max} + \frac{\partial\Omega}{\partial X}\frac{dI}{dt}\bigg|_{\varphi_{max}} \left[\frac{dX}{dI}\bigg|_{\Omega} - \frac{dX}{dI}\bigg|_{\varphi_{max}}\right] = 0 \tag{8.39}$$

These requirements are generally incompatible except for the special case that the trajectory intersects the target curve at its peak, at which time the slope is zero and the necessary flux to bring the trajectory along the restraint dwindles also to zero. Figure 8.15 shows the optimum trajectory for this problem. Further examples amongst the problems for this chapter will show how a change in the specified cost function alters the nature of the optimum control.

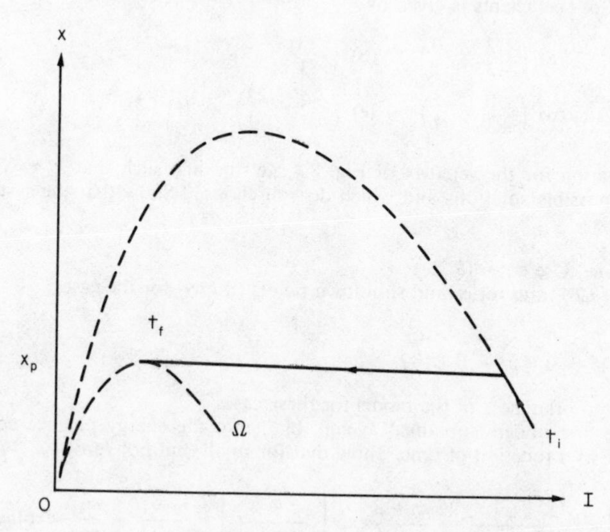

FIG. 8.15. Minimum fuel usage xenon shutdown.

Conclusion

There are several other interesting developments in modern control theory which space precludes. Methods of optimisation are well reviewed by Terney and Wade[11] in a general context. Control theory based on the extension of classical mechanics in the state space admits the treatment of uncertainties in some of the state vectors (e.g. refs. 12 and 13), while an interesting development suitable for direct digital control (DDC) lies in the region of adaptive control.[14]

It is fair criticism of these methods that when they are applied to problems of any complexity instead of the specially simple examples used here, then a considerable burden of computation arises. It is likely, therefore, that their use in practice will not arise until DDC itself is accepted as safe and reliable in the operation of nuclear reactors. If we assume that this very reasonable expectation will be met, then we can foresee a developing use of state space methods in DDC since they provide a more logical and consistent way to optimise behaviour than the empirical optimisation of analogue control elements.

Problems for Chapter 8

SOLUTION OF DYNAMIC EQUATIONS

8.1. *Equilibrium points.* The following equations were obtained by linearising the system equations around an equilibrium point. Investigate the nature of each point by solving the equations for the general solution of the system dynamics in this linear approximation and categorise the points as stable or unstable. Sketch the nature of the trajectories around each point.

$$x = f = Mx$$

where M, the matrix of coefficients is given by:

$$\text{(a)} \begin{pmatrix} 2 & 0 \\ -3 & 1 \end{pmatrix} \quad \text{(b)} \begin{pmatrix} 3 & 2 \\ 1 & 3 \end{pmatrix} \quad \text{(c)} \begin{pmatrix} 1 & 2 \\ 1 & 3 \end{pmatrix}$$

$$\text{(d)} \begin{pmatrix} 0 & 2 \\ -3 & 1 \end{pmatrix} \quad \text{(e)} \begin{pmatrix} -2 & -3 \\ 5 & 1 \end{pmatrix} \quad \text{(f)} \begin{pmatrix} 4 & 2 \\ 2 & 1 \end{pmatrix}$$

8.2. Find the equation for the sepatrix of Fig. 8.4, i.e. the line such that $x_2 = \gamma x_1$ with γ constant Why are there two possible solutions and which do you choose? Show that the system must be over-damped for a sepatrix.

8.3. *Adiabatic model.* Use eqns. (8.3).

(a) What are the (n,T) trajectories and shutdown points (if any) for the cases:

(i) $\alpha < 0$; $\rho_0 - \beta > 0$, $\rho_0 - \beta < 0$;
(ii) $\alpha > 0$; $\rho_0 - \beta > 0$, $\rho_0 - \beta < 0$?

Comment on the appropriateness of the model for these cases.

(b) Substitute the first integral obtained at eqn. (8.5) into the energy balance equation and hence find the temperature as a function of time. Show that for small (but not zero) n_0,

$$\frac{1}{T(t)} \to \frac{\alpha}{2[\rho_0 - \beta]} - \frac{\rho_0 - \beta}{\alpha\kappa n_0} \exp\left[\frac{\rho_0 - \beta}{\Lambda} t\right] = \frac{\alpha}{2\omega_0\Lambda} - \frac{\omega_0\Lambda}{\alpha\kappa n_0} \exp[\omega_0 t]$$

where $\omega_0 = [\rho_0 - \beta]/\Lambda$, when $\alpha < 0$. Find the corresponding function for $n(t)$. (*Hint*: Expand the quadratic solution in powers of n_0.)

(c) (i) What is the total energy release in the excursion?
(ii) What is the maximum rate of energy release and when does it occur?
(iii) What is the energy release in the major part of the excursion between half the maximum neutron level and the moment when the system returns to $\frac{1}{2}n_{max}$ on its way to a shutdown point (i.e. the energy release in the half width)?

(d) Show that the answers to (i) and (iii) above are similar and so the time taken in (iii) is a reasonable expression of the duration of the excursion. Why should we not use the full duration? Show on the basis of (i) that the excursion duration is about four times the *initial* period.

(e) Having solved the source free equations, we might expect to be able to solve the equations with a (constant) source in the neutron balance. Investigate this possibility. Will success add materially to the usefulness of the model?

(f) Try this problem for smaller reactivities in the prompt jump model. Investigate the three-dimensional system with one group of precursors. (*Hint*: Try linearising the equations to study departures in the small.)

8.4. *Constant energy removal.* Suppose $\rho = \rho_0 + \alpha T + \mu t$ with $\alpha < 0$, $\mu > 0$, both constants, so that the system is being subjected to a fast ramp reactivity input (thus helping to justify the neglect of delayed neutrons) where this reactivity has a negative feedback. Use the constant heat removal model.

In order to eliminate the explicit dependence on time, solve the reduced equation in the $(n, \rho - \beta)$ state space. Show that this equation is

$$[\rho - \beta] \frac{d[\rho - \beta]}{dn} = \frac{\Lambda \alpha \kappa}{n} \left[n + \frac{\mu}{\alpha \kappa} - n_0 \right]$$

where, again,

$$[\rho - \beta]/\Lambda = \omega(t) = \frac{d}{dt} \ln n$$

Sketch the resulting trajectories in the (n, ρ) space and the time dependent trajectories $\rho(t)$, $n(t)$ and $T(t)$. Relate ω^2 at the beginning of a cycle to the peak temperature.

STABILITY STUDIES

8.5. *Nonlinear spring.* If the undamped mass–spring system has a nonlinear spring such that the restoring force for a displacement x_1 from rest is proportional to $g(x_1) = \int_0^{x_1} k(x_1')x_1' dx_1'$, show that the total energy is

$$\tfrac{1}{2}mx_2^2 + \int_{x_0}^{x_1} g(x_1')dx_1'$$

Use this as a Liapunov function to show that the system moves in stable orbits. What restriction, if any, should be placed on $g(x_1)$ in this conclusion and is this restriction physically reasonable?

Extend the consideration to include viscous damping. Is there an analogous development for the arbitrary reactivity coefficient such that $\partial\rho/\partial T = \alpha(T)$ in the adiabatic model?

8.6. *Variable gradient method.* We have been a little loose in our use of Liapunov functions. To prove asymptotic stability strictly it is desirable that the V-function is positive definite and its derivative $\dot{V}$ is strictly negative definite and not the semi-definite used on occasions in the text. These exceptional points (since they are isolated) can be dealt with directly, but it would be attractive to find a rigorous Liapunov function for the damped oscillator problem.

Using the variable gradient method, it will be sufficient to consider constant coefficients a_{ij} with

$$\nabla V = \begin{pmatrix} a_{11}x_1 + a_{12}x_2 \\ a_{21}x_1 + a_{22}x_2 \end{pmatrix}$$

Impose the integrability condition and write $V(x)$ in terms of the remaining coefficients. Find $\dot{V}$ from ∇Vf, where $f = (x_2, -x_1 - 2\zeta x_2)$.

Express the condition for definiteness in terms of the principal minors for V and $\dot{V}$ and demonstrate the relationships to be satisfied by the coefficients. Show that for $\zeta = 1$, $V = x_1^2 + 2x_1x_2 + 3x_2^2$ meets these conditions and that both V and $\dot{V}$ have the desired form. Hence the system is asymptotically stable in the large. Sketch $V = C$ in the state plane.

8.7. *Newton's law of cooling.* Suppose an energy removal term is added to our prompt reactivity model to give state equations

$$\frac{dn}{dt} = \frac{\alpha T}{\Lambda} n; \quad \frac{dT}{dt} = \kappa[n - n_0] - \mu T$$

where $\mu \geqslant 0$ and describes the transfer of energy proportional to the difference in temperature. Write as a function around the centre.

What is the equilibrium point? (If there is more than one, comment). Suppose an initial disturbance to n, T occurs. Find the resulting trajectories for $\mu=0$.

Use this result as a Liapunov V-function for the general case $\mu>0$. Is $\dot{V}$ negative definite? If not, can you still draw useful conclusions?

Compare this problem with that of adding a step reactivity $\rho_0-\beta$. Would the same approach work? Give physical reasons why the trajectories with $\mu>0$ may cross outwards the $\mu=0$ trajectory in this case.

Can you find a rigorous (negative definite) V-function? (*Hint*: Find a transformation of T to eliminate n_0 from the second state equation. Considering (n_0, T_0) as centre for $\mu=0$, try

$$V = n - n_0 - n_0 \ln\frac{n}{n_0} - \frac{1}{2}\frac{\Lambda}{\alpha\kappa}\left[\frac{\alpha}{\Lambda}[T - T_0] + \mu \ln\frac{n}{n_0}\right]^2$$

For what n, n_0, T_0 is this positive definite? What restriction do you then put upon α and μ?

8.8. *General laws of cooling.* Suppose the Newton's law of problem 8.7 is replaced with the term $-g(T-T_0)$. Using geometrical arguments, discuss the necessary form of the function $g(x)$ for stability.

OPTIMUM CONTROL

8.9. *Variable viscous damping.*[10] Suppose ζ in the viscous damping system described by

$$\ddot{y} + 2\zeta\dot{y} + y = 0$$

is a controllable function of time. Show that for a process time minimum problem $\bar{C}=\int_{t_i}^{t_f} dt$ there are no singular optimum trajectories so that the optimum trajectory consists of segments at alternate minimum and maximum damping. If $\zeta_{\min}=0$ and $\zeta_{\max}=\infty$, sketch the optimum trajectory in the state plane and find the time to return the system to rest after a displacement. Compare this result with the time taken at *fixed* ζ when the value is selected to minimise the time taken to bring the system within 5% of its original displacement (the 1/20 settling time).

8.10. *Ramp startup.* Suppose the rate of introduction of reactivity is limited by the control rod motor, etc. (for sound reasons of safety). By taking $u=\dot{\rho}=\mu(t)$ and treating ρ as an additional state variable, analyse the problem of raising the power without overshoot in the minimum time to a desired level. Show that the optimum control process starts and finishes with a segment at maximum reactivity rate insertion.

8.11. *Xenon override.* Analyse the xenon shutdown/override problem for two further cost functions:

(a) minimum process time: $\bar{C} = \int_{t_i}^{t_f} dt$

(b) minimum wasted use: $\bar{C} = \int_{t_i}^{t_f} 1 - \varphi(t)/\varphi_{\max}\, dt$

(*Hints*: Look for the existence of singular solutions and at the target or terminal boundary conditions as they affect the adjoint function.)

8.12. *Nomenclature.* Transposes of vectors have not been indicated as such since only the *inner* product of vectors is used. Transposition may be assumed in relation to the pair $pf=fp$. Matrices, including the matrix generated by the vector gradient of a vector, must have their transposes indicated, of course, e.g. $\partial f/\partial u$ and $\partial f^T/\partial u$.

Taking a 2×2 example (Wigner's acid test for any bright new theory) show that

(a) $$\nabla f = \frac{\partial f}{\partial x} = \begin{pmatrix} \frac{\partial f_1}{\partial x_1}, & \frac{\partial f_1}{\partial x_2} \\ \frac{\partial f_2}{\partial x_1}, & \frac{\partial f_2}{\partial x_2} \end{pmatrix}$$

(b) $$\nabla [fp] = \frac{\partial}{\partial x}[fp] = \begin{pmatrix} \frac{\partial}{\partial x_1}[f_1 p_1 + f_2 p_2] \\ \frac{\partial}{\partial x_2}[f_1 p_1 + f_2 p_2] \end{pmatrix}$$

(c) If p is not a function of x,

$$p \frac{\partial f}{\partial x} \delta x = \left[\frac{\partial f}{\partial x} \delta x \right]^T p = \delta x \frac{\partial f}{\partial x} p = \delta x \frac{\partial}{\partial x}[fp]$$

References for Chapter 8

1. S. Barnett, *Introduction to Mathematical Control Theory*, Clarendon Press, Oxford, 1975.
2. D. L. Hetrick, *Dynamics of Nuclear Reactors*, University of Chicago Press, Chicago, 1971.
3. H. B. Smets, *Problems in Nuclear Power Reactor Stability*, University of Brussels, Brussels, 1962.
4. E. P. Gyftopoulos, Lagrange stability by Liapunov's direct method, *Symposium Reactor Kinetics and Control, USAEC TIE, Washington*, 1964.
5. Z. Akcasu, G. S. Lellouche and L. M. Shotkin, *Mathematical Methods in Nuclear Reactor Dynamics*, Academic, New York, 1971.
6. L. S. Pontryagin *et al.*, *Mathematical Theory of Optimal Processes*, Wiley, New York, 1962.
7. J. Lewins and A. L. Babb, in *Adv. in Nucl. Sci. and Tech.* **4**, Academic, New York, 1968.
8. R. R. Mohler and C. N. Shen, *Optimal Control of Nuclear Reactors*, Academic, New York, 1970.
9. S. Salo, On the use of the maximum principle, *Nucl. Sci. and Engr.* **50**, 46 (1973).
10. S. M. Fadilah, University of London PhD (1974); *Annals Nucl. Energy* **2**, 443 (1975).
11. W. B. Terney and D. C. Wade, Optimization methods, *Adv. in Nucl. Sci. and Tech.* **10** (1977).
12. J. T. Tou, *Modern Control Theory*, McGraw-Hill, 1964.
13. J. S. Meditch, *Stochastic Optimal Linear Estimator and Control*, McGraw-Hill, 1969.
14. J. L. Macdonald and B. V. Koen, Application of artificial intelligence techniques to digital computer control of nuclear reactors, *Nucl. Sci. Engr.* **56**, 142 (1975).
15. J. D. Cummins and M. H. Butterfield, Applications of modern control theory in nuclear power, IAEA-SM-168/B-8, *Nuclear Power Plant Control and Instrumentation, Vienna*, 1973.
16. W. D. Goriatchanko, Methods of stability in the dynamics of nuclear reactors (in Russian) ATOMIZDAT, Moscow, 1971.

CHAPTER 9

Analogue Computing†

Introduction

SCOPE OF CHAPTER

This chapter introduces the modern electric analogue computer and derives the patching or programming rules whereby it can be made to find solutions to problems of interest by virtue of the analogous behaviour of electric circuits. Then the chapter develops a number of examples of intrinsic interest in nuclear engineering showing the programming involved and typical results, including the slower transients of xenon poisoning and fuel burn-up. The greater part of this chapter is written to be independent of any particular machine; correspondingly we will not travel far into the more sophisticated uses of analogue computers combined with logical and digital elements. In so far as the book refers to a particular machine it is the EAI-180, a general purpose transistorised teaching computer of wide availability having, in its standard form, six integrator elements and six further summing elements as well as four multiplying elements. All the examples given may be run on machines of this size. A professional laboratory studying a reactor control system may well employ over 1000 amplifiers and a small digital computer to automate the "setting up" as well as keep track of the output runs. References 1–3 are examples of many texts on analogue computing.

PURPOSE OF ANALOGUE COMPUTING

The solution of differential equations beyond second order with constant coefficients by analytical means gets tedious and in some cases essentially impossible. Digital computers have tremendous power (particularly in handling masses of data) but have some limitations in solving differential equations; they are better suited to differencing than integration.

Considerable analytical skill is needed to choose a solution method if satisfactory answers are to be relied upon. The almost unlimited accuracy that is available to the skilful numerical analyst using a digital computer is purchased with increasing computing time and cost.

Analogue computers offer a way of studying systems, their response and their stability, characterised by relative ease of programming, a need for little analytical skill, by great

†This chapter may be omitted if the reader is familiar with analogue computers except that the examples towards the end are interesting to nuclear engineers in their own right.

flexibility in altering programs and coefficient settings during the investigating and by easily understood visual or graphical output. This flexibility means that the analogue computer is particularly useful in design studies where many parameters have to be varied and perhaps optimised. This flexibility is purchased at the cost of limited accuracy, perhaps 1% or 0.1% depending on the machine and complexity of the problem; such accuracy is usually good enough (and can be increased in skilled hands).

Undoubtedly the *electric* analogue computer is the most convenient analogue machine, based on an electric circuit satisfying the same form of equations as the model under consideration. Voltages at different points in the analogue circuit will then vary in analogy with the state variable of the system to be studied. In particular, the electric analogue using the modern high gain amplifier is an accurate integrator and lends itself to rapid adjustment of initial values and coefficients through variable resistances (potentiometers or "pots"). Also the modern solid state devices lend themselves to producing voltages proportional to the product of two voltages so that multiplication and its variations are available to permit the representation of variable coefficients in state equations.

We mention, but pursue no further, that other analogies can be used (the slide rule being perhaps the most common mechanical analogy for logarithms), e.g. pneumatic and hydraulic elements with special nonlinear properties, wax models or resistance paper for problems in distributed systems, electro-chemical analogies for heat transfer, etc. These should be followed up in the specialist literature of their applications. For the remainder of this chapter the analogue computer is taken to be synonymous with the electric analogue computer using high gain (operational) amplifiers.

Used in real time, the analogue computer may be called a simulator and may then be built into the testing or control of a system. However, it is also most convenient that the time scale may be changed in the computer from the real system. On the one hand, it may then be applied to problems whose natural time constants are very short, such as prompt neutron behaviour, voids or bubble growth. On the other hand, it makes it feasible to study long term changes such as fuel burn-up in a convenient time span. Correspondingly, the ability to prepare, check and "debug" the program in slow time while being able to run and record several cases in fast time may be advantageous as well as the flexibility of relating the output to an oscilloscope (in fast time) or to a pen recorder (more slowly).

The ability to integrate accurately from specified initial conditions makes the analogue computer particularly suitable, as in this text, for open-ended equations of the initial value type and less adaptable to equations with two-point (fully or partially specified) boundary conditions. It is beyond our scope to discuss the high speed two-mode "fly back" analogue computers developed for this situation (though the EAI example has this capability to some extent). The analogue computer is a "natural" when the system to be modelled is itself governed by time dependent ordinary and algebraic equations.

MODELLING

The size of computer puts a limit to the complexity of the model of any system. Thus before being able to program the computer, the user must go through the stage of *modelling* the system to obtain adequate but tractable equations. It is likely that a distributed system will have elements represented by "lumped" approximations or, at best, two or three terms in an expansion representation. Other approximations would be a linearisation valid only for a restricted range of the variables, the neglect of elements with very long or very short characteristic times, etc. In a new situation this preliminary work of modelling the system with acceptable simplifications is likely to be quite the hardest and most sophisticated part of the over-all solution process and cannot be exhaustively dealt with here.

We distinguish particularly between that time T in the model which is to represent real system time and the time t that passes during the analogue computation and is perhaps more "real" than the model time. The notation T and t should keep the distinction clear. Time as represented in the model is given by $T = Kt$, where K is a scaling factor that can be adjusted, as shown later.

Linear Analogue Elements

THE HIGH GAIN, OPERATIONAL AMPLIFIER

The central element in the analogue computer is the operational amplifier having two important characteristics:

(1) *High gain:* between 10^6 and 10^8.
(2) *Low current drain:* negligible in proper use.

It is convenient to represent the amplifier and other standard elements graphically, as

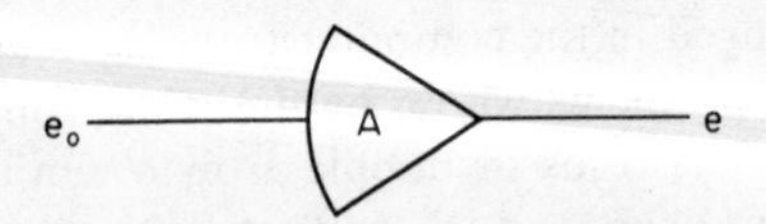

FIG. 9.1. The operational amplifier.

in Fig. 9.1, by standard symbols and a shorthand to be developed. The conventional circuit diagram is then the "program" for the problem and the patching guide to realise the program on the computer.

We use e_0 for the input voltage, e for the output voltage. If the gain is A in magnitude and taking note of the usual sign inversion in electronic amplifiers,

$$e = -Ae_0 \tag{9.1}$$

DYNAMIC MULTIPLICATION

Using the amplifier, we may arrange for a dynamic multiplication by a constant, either larger or smaller than unity in magnitude as shown in Fig. 9.2.

Applying Kirchhoff's laws to the lead/shunt circuit, we have for the current i:

$$i = \frac{e_1 - e_0}{R_1} = \frac{e_0 - e}{R} \tag{9.2}$$

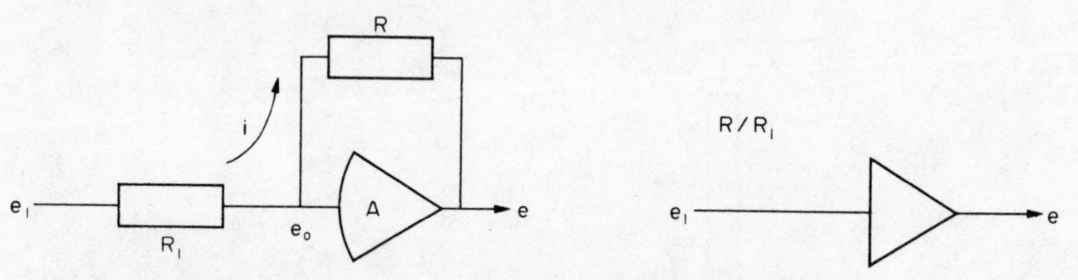

FIG. 9.2. Dynamic multiplication and conventional symbol.

Because A is large and the output is limited to the machine voltage (10 V in the EAI-180), e_0 is very small and, indeed, essentially at the earth or zero voltage (virtual earth amplifier). Then

$$e = -\frac{R}{R_1} e_1 \tag{9.3}$$

which shows we have a circuit to multiply by a constant R_1/R, bigger or smaller than unity, but accompanied by a change of sign.

STATIC OR POT MULTIPLICATION

We may also use a pot to alter a voltage and thus effectively multiply by a constant, though only of magnitude smaller than unity. As shown in Fig. 9.3, if the tapping draws

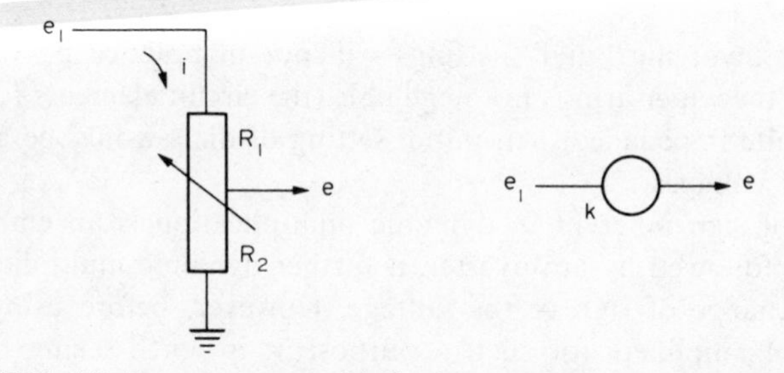

FIG. 9.3. Potentiometer multiplication and conventional symbol.

negligible current (e.g. followed by a high resistance lead into an amplifier), then pot multiplication follows the relation

$$e = \frac{R_2}{R_1 + R_2} e_1 \equiv ke_1 \tag{9.4}$$

thus defining k, which is necessarily in the range $0 \leqslant k \leqslant 1$. Potentiometers may be marked with k-values on the setting dials for ease of use.

Figure 9.3 shows the pot grounded to zero voltage. Alternatively, a double-ended pot can be used as an "averager" as shown in Fig. 9.4.

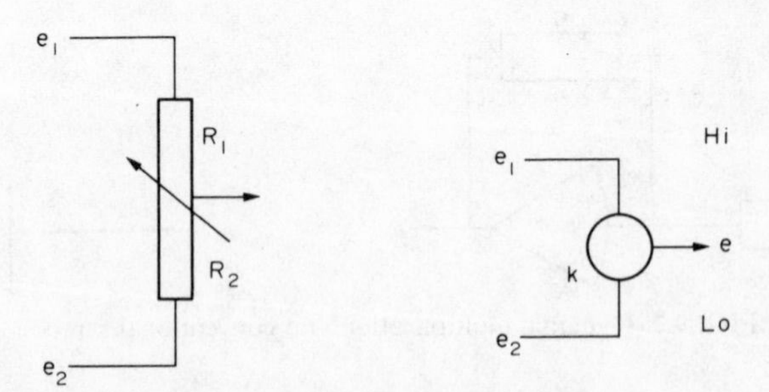

FIG. 9.4. Ungrounded or Hi–Lo pot with conventional symbol.

The usual analysis of the current flowing in the pot leads to

$$e = \frac{R_1 e_2 + R_2 e_1}{R_1 + R_2} = ke_1 + (1 - k)e_2 \tag{9.5}$$

The difference between dynamic and static multiplication in operation, therefore, is between multiplying with a possible increase in magnitude but a change of sign, compared to reduction (attenuation) but with no change of sign. The amplifier is usually provided with a limited range of standard resistors (such as 1 MΩ, 10^5 Ω, etc.) which are convenient to patch and can produce R/R_1 values such as 0.1, 1 and 10. The wire-wound or carbon granule pots available elsewhere in the computer give k-values anywhere (to within 0.1% say) between 0 and 1. It is likely, therefore, that an arbitrary coefficient will require the combination of dynamic and static multiplication to represent it.

These relations involving "dial" settings will not in practice be valid because the current drawn by the wiper arm is not negligible (the circuit elements following the arm do not show infinite impedance). An initial setting of dials would be adjusted with an on-load setting of voltages.

If the change of sign inherent in dynamic multiplication is an embarrassment, the multiplier can be followed by an inverter, a further dynamic multiplier with $R/R_1 = 1$, giving merely a change of sign to the voltage. However, before using up one of the limited number of amplifiers for such a purpose, it is worth seeing how the over-all program can be adjusted to make use of the negative voltage—perhaps it will be inverted in a subsequent stage anyway.

SUMMATION OF MULTIPLIERS

To sum several voltages, each being multiplied by a separate constant, we use the circuit of Fig. 9.5.

Using the same properties of negligible e_0 and amplifier current summation, we have

$$i = \frac{e_1 - e_0}{R_1} + \frac{e_2 - e_0}{R_2} + \cdots \frac{e_n - e_0}{R_n} = \frac{e_0 - e}{R} \tag{9.6}$$

or

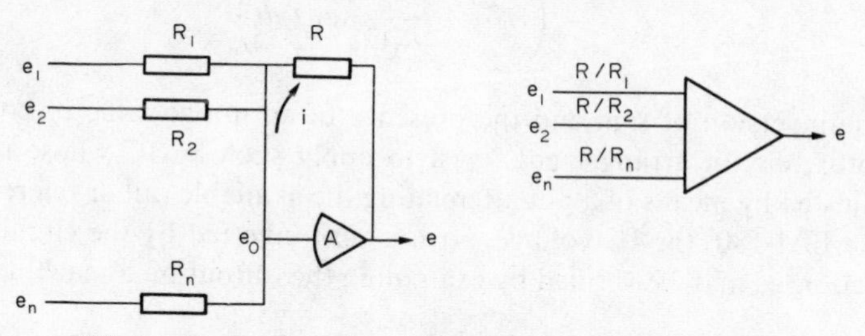

FIG. 9.5. Summation of multipliers and conventional symbol.

$$e = -\left(\frac{R}{R_1}e_1 + \frac{R}{R_2}e_2 + \cdots \frac{R}{R_n}e_n\right) \tag{9.7}$$

thus summing the weighted input voltages, with over-all sign change.

Note the general significance of our assumptions about e_0 and amplifier current being negligible (or wiper arm from a pot). If the resistance values or any additional circuitry allow too great a current to be drawn, then the postulated relationships will fail and the equations will be in error. Correspondingly, the amplifier output (e) may be driven too high and reach saturation, generally some 10 to 20% above the nominal machine voltage. In some cases, amplifier overload lights will give warning to the operator of this condition, which clearly renders the computation nugatory.

INTEGRATION

Integration makes use of a capacitor as shown in Fig. 9.6.

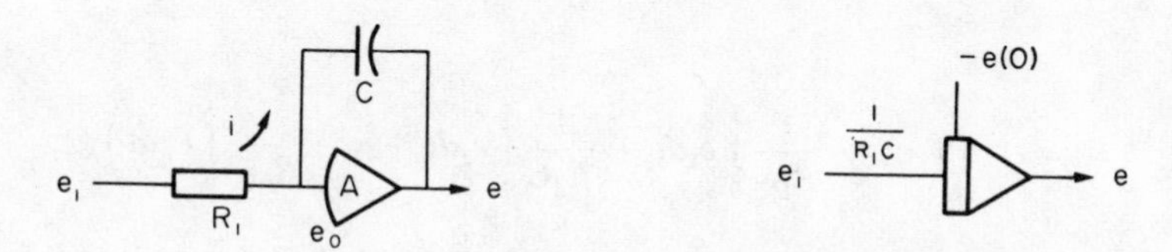

FIG. 9.6. The integrator and conventional symbol.

The charge on the capacitor will be the integral of the current flowing into it, from some initial value, and

$$\int_0^t i\,dt = q(t) - q_0 = (e - e_0)C \tag{9.8}$$

which combined with the expression through the resistor gives

$$-C\frac{de}{dt} = i = \frac{e_1 - e_0}{R_1} \simeq \frac{e_1}{R_1} \tag{9.9}$$

whence

$$e = -\left(e(o) + \frac{1}{R_1C}\int_0^t e_1(t)dt\right) \tag{9.10}$$

Note again the inversion of sign and the presence of an initial value or condition (IC) on the capacitor; circuit arrangements exist to apply such an IC whose magnitude is commonly adjusted by means of a pot attenuating the available rail or reference machine voltage. In the EAI-180, the IC voltage, $e(o)$, is *also* inverted by the circuitry applying it to the capacitor, as may be verified by examining the output on a machine if the point is in doubt.

The R_1C value is, of course, the time constant for this circuit. It is usual to have a choice of capacitors available to each integrator (1 μF and 0.01 μF in the EAI-180) to provide for easy change of time constants. A simple plug change on all integrators therefore varies the time constant from, say, 1 s ($R_1 = 1$ MΩ, $C = 1$ μF) to 0.001 s ($R_1 = 10^5$ Ω, C$=0.01$ μF). Summation and integration are shown in Fig. 9.7.

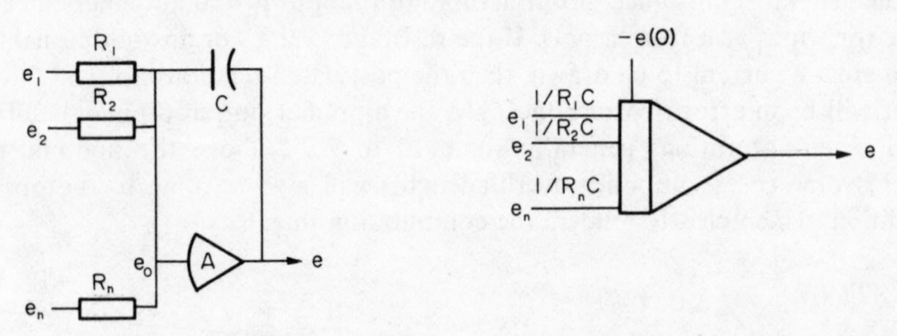

FIG. 9.7. Summation and integration with conventional symbol.

From a similar analysis as before (and assuming again the sign inversion of the initial condition),

$$e = -\left(e(o) + \frac{1}{R_1C}\int_0^t e_1dt + \frac{1}{R_2C}\int_0^t e_2dt + \cdots \frac{1}{R_nC}\int_0^t e_ndt\right) \tag{9.11}$$

Errors due to substandard components, for example, will show up in integrators as an increasing drift away from "correct" values.

Elementary Programs and Output

The question of scaling the system so that voltages represent the model quantitatively will be put off until the next section. In this section we give some elementary example and discuss measuring and recording the output.

TIME AS A COMPUTER VOLTAGE

Perhaps the simplest problem calls for a voltage to vary in proportion to time in the model, i.e. we seek a steadily increasing voltage obtained by integrating a constant.

(Of course, the computer may have a built-in time generator or ramp function.) The equation for our model is then

$$T = \int_0^T dT \tag{9.12}$$

so we wish to program the analogue equation

$$e = -\frac{k}{R_1C}\int_0^t dt = -\frac{k}{R_1C}t \tag{9.13}$$

which is done in Fig. 9.8. Clearly we take zero initial condition voltage (open circuit

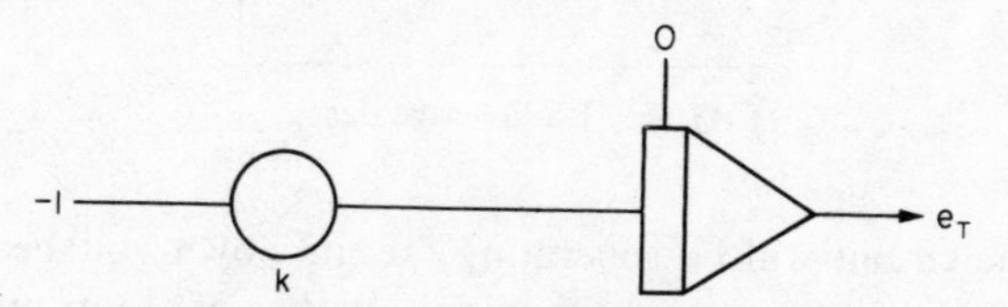

FIG. 9.8. Time base integrator.

IC) so that the analogue voltage starts from zero. Some R_1 and C-values will be selected and a pot with a k-value to be chosen from any subsequent rescaling of time. The constant input is derived from a reference voltage available in the machine and a negative reference is selected in view of the inversion of sign so that the output voltage is positive to correspond with positive T. Reference voltages are conventionally shown as ± 1 whatever the true volts; we speak of one machine unit of voltage, 1 muv. Most machines are designed so that the open circuit as used here for the IC is earthed.

RADIOACTIVE DECAY

The decay of a single radioactive substance has the well known exponential solution. This can be obtained easily in an analogue to satisfy the decay law

$$\frac{dN}{dT} = -\lambda N; \quad N(\text{o}) = N_0 \tag{9.14}$$

with λ a fixed positive constant. The sign inversion makes it more convenient, indeed, to solve for systems tending to decay than systems of growing first order equations. Figure 9.9 shows the simple circuit of program required. If the time scaling is suitably chosen, the decay rate λ will lie in the range $0 < \lambda < 1$ and can be represented by a pot as shown. If not, a suitable R_1C value must be selected—the equivalent of rescaling time in the system.

Note also the feedback or "bootstrap" philosophy of analogue programming. We do not initially have a voltage available to represent N—but have faith, assume we do. Use this voltage around the circuit according to the relationships of the equation and,

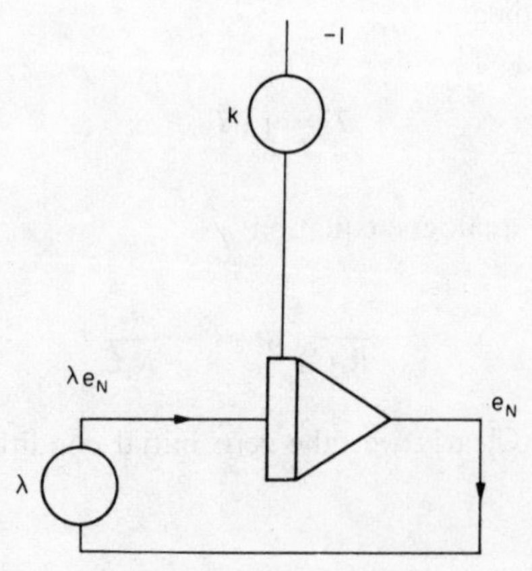

FIG. 9.9. Radioactive decay.

when completed, the circuit will be generating the unknown voltage required. We add a suitable initial condition to represent the initial number of atoms present, $N(o)$. When switched from IC to operate (OP) the voltage e will decrease exponentially and represent the decay $N(t)/N(0)$ of the population of radioactive atoms.

SINE OSCILLATOR

Voltages varying sinusoidally and cosinusoidally with time are useful in checking other components, especially from the viewpoint of the transfer function as the response of a system to a sinusoidal input. Since $X = \sin \omega t$ is known to be a solution of the equation

$$\frac{d^2X}{dT^2} + \omega^2 X = 0; \quad X(o) = X_0; \quad \dot{X}(o) = \dot{X}_0 \tag{9.15}$$

we may generate an analogue voltage through the program illustrated in Fig. 9.10.

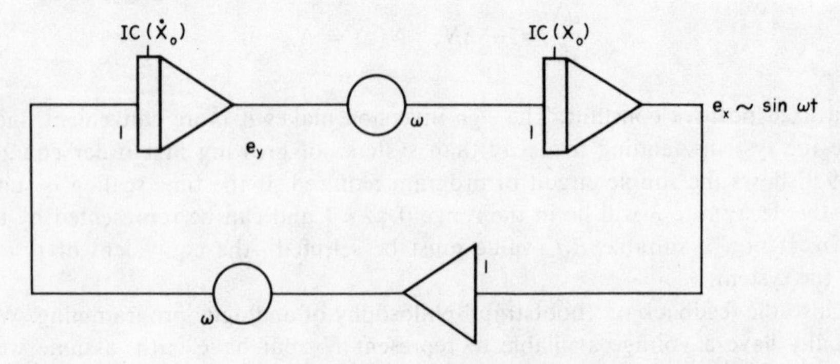

FIG. 9.10. A sine oscillator program.

If an intermediate step in the analysis is needed to understand this program, note the technique of writing the second order system as two first order equations (because we know how to integrate first order equations) and using the "bootstrap" philosophy on

$$\left.\begin{aligned} \frac{dX}{dT} &= -\omega Y; \quad X(o) = X_0 \\ \frac{dY}{dT} &= \omega X; \quad Y(o) = X_0 \end{aligned}\right\} \tag{9.16}$$

where the initial conditions are specified that will determine the magnitude and phase of the output voltage.

Note the use in Fig. 9.10 of the inverter and two pots to represent the frequency ω. If these pots were short-circuited ($k=1$) then with the RC values shown we generate a sine wave with a natural frequency of 1 radian/s. Other RC values can be introduced; a factor of ten on *both* integrators makes a factor of ten change in the natural frequency. If pot settings are introduced, the frequency can be decreased. It is certainly worth setting up this circuit as an exercise to see how the voltages e_x and e_y, representing X and Y, change with pot setting and in the alternative program where only one pot is used, representing ω^2.

OUTPUT AND RECORDING

The interesting result is usually the output of a number of the amplifiers, and most machines will have trunking arrangements to lead such signals to a suitable recording point. This may be for examination by a digital or analogue voltmeter, for observation on an oscilloscope or for recording on a pen recorder. Probably the same voltmeter will be used in setting the pots (pot set) during which time the reference voltage is connected across the pot and the k-setting read off as a fraction of a machine unit voltage. The use of a voltmeter may be adequate therefore during the setting up and debugging process.

For rapidly fluctuating signals, either setting up or in subsequent runs, an oscilloscope is evidently to be preferred and can usefully be combined with repetitive solution so that the light spot retraces the solution mode so frequently as to give to the eye a sustained image. The oscilloscope may have its x-input connected to a voltage varying with time (from the analogue itself or the time base of the oscilloscope) and one or more y-signals, thus giving the time dependent picture of the system behaviour. In many cases it is helpful to plot two other variables in x and y mode so that the spot travels along this state plane representation in accordance with time and a complete picture of the system behaviour emerges. This may usefully be tried with the sine oscillator of the last example. The oscilloscope will have to be adapted in its voltage scaling to be compatible with the analogue output, of course.

It is possible to photograph the output of an oscilloscope but probably it is more convenient to record output permanently by means of a pen recorder. Like the oscilloscope, it is necessary to match the voltages of the analogue to the input of the recorder, and, again, time dependent graphs and state space graphs are both useful. The pen recorder will need a reasonable interval between, say, 5 s and 50 s to record efficiently without being painfully slow, and there is likely, therefore, to be a need to change the time scale in the analogue computer between oscilloscope and pen-recorded output. Most computers have a means of initiating the rise and fall of the pen automatically in accordance with their OP and IC modes.

For either oscilloscope or pen recorder, it is important to note the significance of the scale factors. A simple way to do this is to apply in turn a static zero and 1 muv signal and mark the corresponding interval on the recorder, in x- and y-directions.

Computer Scaling

THEORY OF SCALING

A scaling factor or constant of proportionality must be set for every voltage that is to represent a system variable if, firstly, the computer is to be a true analogue model (by virtue of satisfying the same form of equations with the same values of the coefficients) and, secondly, if the measurement of the voltage outputs is to be given quantitative meaning in the model.

The selection of scaling constants, denoted as K with suitable subscripts, is to a certain extent arbitrary, however, with the setting of the analogue coefficients taking the K-values into account. The degree over which we can choose a K-value is governed by two extremes:

(a) the resulting voltages must not be so small that inaccuracies are proportionally too great;
(b) the resulting voltages must not be so large as to overload the amplifiers or otherwise invalidate the assumed equations governing the computers.

As to (a), every element of the computer will have some uncertainty or inaccuracy associated with it. Thus wired pots will have a certain amount of "jump" between turns and cannot really be set to *any* value between 0 and 1, nor could the VM produce a perfect reading in setting them. Capacitors can be cut to, say, 0.25%, and resistors at least matched in a set so that ratios are correct to 0.25%. All this leads to certain inherent errors and the lower the signal voltages the higher the percentage error in the results. Ideally, therefore, voltages should swing over the full range that is available.

As to (b), the voltage must not go much beyond the nominal range of the reference voltage (1 muv) or various elements in the computer will saturate. The machine will usually have an overload indicator that will operate when the "virtual earth" ($e_0 \sim 0$) condition is violated.

The steps in scaling a program are therefore as follows:

Step one. Estimate the likely range of all variables in the system and select scale factors for each that will cause the voltages to move over a suitable range. Note for each variable

$$K = \frac{\text{system units}}{\text{machine units}}$$

The scale factor K will be used to interpret the results of the measured analogue voltages in the form of machine unit voltage back into system variables.

Step two. Introduce the scale factors into the machine equations and these, with the coefficients values, determine the machine coefficient settings for pots, etc.

In step one, only a rough estimate of the variable range is necessary because if it turns out in running the analogue that the first guess is unacceptable, we will have a guide from the results to improve the scaling factors. Do not forget the corresponding adjustment of step two, however.

TIME SCALING

There is a principal distinction between scaling the state variables of a problem and varying the actual time taken in the computer to develop the analogue solution. Changes in either of these relations will, of course, change the equations programmed on the computer, the machine equations. The time taken to produce the analogue solution, in laboratory time t as opposed to model time T, is governed by the need to produce results without undue delay, on the one hand, and, on the other, to allow the recording equipment adequate time to respond to varying signals. Thus a pen recorder might need at least 10 s to draw graphical results without showing erroneous transients, and it would probably be tedious to have it taken more than a couple of minutes. It will be almost the exception, therefore, to run in "real" time with $T=t$.

Write $dT/dt=K$, a scaling coefficient without a subscript, to denote the special role of the independent variable, time in this context, and time scaling involves in principle a change of operators from d/dT in the system equation to $(d/dt)/K$ in the machine equations. In practice this is readily done by making a uniform change in the RC value in every integrator in the program, varying the capacitor or the lead resistance or both. Computers are generally designed so that this may be conveniently and flexibly implemented by factors of 10, 100 and 1000.

We have spoken of machine units of voltage (muv) throughout, whether the computer is actually a 10 V reference transistorised machine or an older 100 V reference valve computer. It is convenient in scaling to refer to a unit of machine volts, the available reference voltages as 1 muv. Thus the voltmeter (VM) scaled between 0 and 1 muv reads a k-value of a pot directly when reference voltage is applied at the top and the bottom is grounded. The scaled equations are independent of machine voltage and can be transferred to another machine more freely. Finally, nonlinear elements such as multipliers are more easily handled if this convention is adopted (for clearly two 10 V signals multiplied together in a 10 V machine are going to be arranged to give 10 V output, not 100 V, and machine scaling avoids having to correct for this).

SCALING THE TIME INTEGRATOR (Fig. 9.8)

The following examples refer to amplitude scaling of the voltages representing state variable in the model. It is convenient to use a square bracket notation for the voltage representing a corresponding variable, i.e. to write [T] for e_T, the output of the integrator used in Fig. 9.8 to produce a voltage varying with time. We now seek to determine the amplitude scale factor so that a measured computer voltage can be interpreted as time T in the model.

Suppose the time base is being prepared in a problem where system time stretches over 10 years—a fuel cycle study, for example. Note that we are not dealing with time scaling as discussed in the last section but with the scaling of a voltage $[T]$ representing T. Clearly we cannot operate the computer in system time so time scaling is also called for as well as amplitude scaling. To record the results we may wish to have a system time T of 10 years occupy real computer time t of 20 s.

Step one. Time T varies from 0 to 10 years. The voltage $e_T = [T]$ should therefore vary between 0 and 1 muv. Introducing this amplitude scale factor

$$T = K_T e_T = K_T[T] \tag{9.17}$$

and we chose

$$K_T = 10 \frac{\text{system units}}{\text{machine units}} = 10 \frac{\text{years}}{\text{muv}} \tag{9.18}$$

Step two. We showed that the analogue program produced a voltage satisfying $e = kt$ with units of t determined by the R_1C values. If $R_1C = 1$ s (e.g. 1 MΩ and 1 μF) the voltage would grow from 0 to 1 muv in 1 s. The pot setting k can now be chosen to stretch this time to 20 s so that when $t = 20$

$$20k = e = [T] = 1 \text{ muv} \tag{9.19}$$

and so we choose $k = 1/20$. Thus with this pot setting of 0.050 and an R_1C value of 1 s, the analogue computer will generate a voltage $[T]$ proportional to time T according to K_T model units years to 1 muv and such that full scale will be reached after 20 s, equivalent to 10 years in the model.

With the scaling of e_T established, the process of computation may be speeded up or slowed down by making a simultaneous change of the RC values of all integrators, including the time base integrators, of course. In this example, a change from 1 μF to 0.01 μF would speed up the time taken in the computer so that full scale on the time integrator would be reached in 0.2 s, perhaps more suitable for an oscilloscope display. The significance of K_T as an *amplitude* scale factor is unchanged if $[T]$ has had the same change in the time constant of its integrator.

SCALING FOR RADIOACTIVE DECAY (Fig. 9.9)

Figure 9.9 gave a patching to represent the radioactive decay of an isotope. Suppose we have a system with 10^6 active atoms present initially and a decay rate of 10^{-3} s^{-1} and we are to use a pen recorder.

Step one: range of variables. The population of atoms can only decrease but cannot become negative. Therefore 10^6 atoms is the maximum population and can usefully be represented by 1 muv, $N = K_N[N]$ with

$$K_N = 10^6 \frac{\text{system units}}{\text{machine units}} = 10^6 \frac{\text{atoms}}{\text{muv}} \tag{9.20}$$

Step two: coefficient setting. In this problem, real time simulation is possible with the time scaling conversion $K=1$; from the system equation we go to the machine equation:

$$\frac{dN}{dT} = -\lambda N \rightarrow \frac{1}{K}\frac{d}{dt}K_N[N] = -\lambda K_N[N] \rightarrow \frac{d}{dt}[N] = -\lambda[N] \qquad (9.21)$$

and the coefficient in the machine equation is the same as λ, i.e. 0.001.

However, as this value is down to the limit of accuracy of the pot scale, such a setting would not be accurate enough. An alternative way to represent the necessary attenuation of 0.001 should be found. We might use dynamic multipliers, but due to the inversion of sign it would be necessary to employ two amplifiers, which seems wasteful. Another way would be to use two static pots in series with an over-all reduction of 0.001, e.g. one at 0.1 and one at 0.01, both these being in a reasonable part of the range.

Yet another way is to change the time scaling of the problem so that we are no longer simulating system time but propose perhaps to compute 10 s of system time in 1 s of computer time. We can put this formally by taking note of the time scaling factor K, but the simplest interpretation in this case is to note that in units of 10 s, the model equations is governed by a λ-value of 0.01 (10 s^{-1}). Then the pot setting in the computer can be made 0.01 to give the setting accuracy wanted, with the results being speeded up and interpreted through K=system time/computer time =10.

SCALING THE MASS–SPRING OSCILLATOR

A mass–spring oscillator with no damping satisfies the sine oscillator equations, and thus the patching of Fig. 9.10 can be employed. The natural frequency is such that $\omega^2 = k/m$, where k is the spring constant and m the mass, in the usual mechanical system notation. Suppose values to be represented in the system are:

mass	2 kg
spring constant	8 N/m
initial displacement	1 m
initial velocity	0 m/s

Step one. Since the initial speed is zero, the maximum displacement of the mass is equal to its initial displacement in magnitude but will swing positive and negative. A little theory shows that the maximum speed, as the mass passes through the equilibrium position, at $X=0$, is $\dot{X}_{max} = \omega X_{max} = \omega X_0$. Since ω is 2 s^{-1} the maximum speed is therefore 2 m/s. It then seems at first sight convenient to take

$$K_x = 1\,\frac{\text{metre}}{\text{muv}}; \quad K_{\dot{x}} = K_y = 2.0\,\frac{\text{m/s}}{\text{muv}} \qquad (9.22)$$

Step two. The natural frequency is 2 radians/s and this is perhaps a little fast for a pen recorder. We may choose a time scaling factor $K=0.1$, say, for a suitably slowed down output. Then the system equations reduce to machine equations as follows:

$$\left.\begin{aligned}
\frac{dX}{dT} &= \dot{X} = Y & \frac{dY}{dT} &= -\omega^2 X \\
X &= K_x[X] & Y &= K_y[Y] \\
\frac{K_x}{K}\frac{d[X]}{dt} &= Ky[Y] & \frac{K_y}{K}\frac{d[Y]}{dt} &= -\omega^2 K_x[X] \\
\frac{d[X]}{dt} &= K\frac{K_y}{K_x}[Y] & \frac{d[Y]}{dt} &= -\omega^2 K\frac{K_x}{K_y}[X] \\
&= 2K[Y] & &= -2K[X]
\end{aligned}\right\} \tag{9.23}$$

with a final version of the machine equations as

$$\frac{d[X]}{dt} = 0.200[X]; \quad \frac{d[Y]}{dt} = -0.200[Y] \tag{9.24}$$

(NB: This is on the basis of a modification to Fig. 9.10 using only one pot to represent ω^2 rather than two pots each representing ω.)

If these were not convenient for implementation (although they are), another value of K can be selected. We are now able to set initial conditions on the basis that the IC for Y should be open circuit and the IC for the X integrator is to represent 1 m and so requires a voltage of $1/K_x = 1$ muv. The program with its pot settings at this stage is shown in Fig. 9.11.

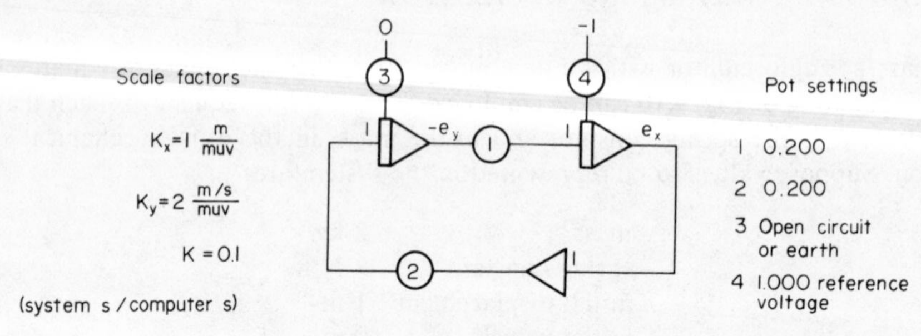

FIG. 9.11. Mass–spring oscillator and pot settings: stage 1.

Having set the problem up as required, we are ambitious and wish to study the effect of different initial conditions. A problem arises because if the initial speed is no longer zero, the maximum values of displacement and speed will change and the system may well overload at the selected scale factors. Some more theory shows that the maximum displacement subsequent to arbitrary initial conditions is $\sqrt{X_0^2+(\dot{X}_0/\omega)^2}$ while the maximum speed is $\omega\sqrt{X_0^2+(\dot{X}_0/\omega)^2}$. We therefore should change the scale factor so that with general initial conditions the computer is not overloaded. This may be done by taking $K_x = 2.0$ and $K_y = 4.0$ for a reasonable range of initial conditions around the first values contemplated. Note that this does not affect the pot settings 1 and 2 but only the pot settings for the initial conditions themselves, 3 and 4.

We may also wish to vary the natural frequency, and this, too, will have consequences on the magnitude of e_y in particular and may in the event call for further changes of the scale factors. The change in frequency itself can be introduced by simultaneously changing the time constants of the two integrators (either through the *RC* values or through the lead resistances of the pots 1 and 2). It might be convenient, however, to alter only one pot to provide a frequency change proportional to ω^2; that is, pot 2 can be varied around the initial value 0.200 in proportion to $\omega'^2/\omega^2=\omega'^2/4$; so giving a maximum frequency for study of some 4.5 radians/s. This example should be made the subject of a practical exercise by the newcomer to analogue computing.

CHECKING AND RUNNING

The last example was treated in some detail appropriate to the first time an analogue computer is used, since the detail of scaling and rescaling seems to be the most difficult point for the tyro to learn. A few further machine independent remarks may also be helpful.

The preparation of the models of the systems and corresponding equations having been done, one can sketch the analogue program readily. The analysis of likely range of values, suitable time scale, etc., should be extended to calculation of the various initial values of the state variables in the system so that when scaling factors are selected, a series of initial voltages to be expected are known.

When a simple problem is patched, it may be sufficient to record the pot settings, scale factors, etc., on the graphical program, though clearly in more complicated cases a systematic recording of settings, etc., is essential, together with a record of the scaling of recording devices and identification of run numbers. When first set up, the patching will need checking against gross and minor errors, and the separately calculated initial voltages are helpful here. Thus a static check should be run of the output of the various amplifiers to see that in the initial condition mode the anticipated voltages are in fact realised. If there is a gross error, the patching is suspect and must be rechecked. If there is a minor error, it probably arises from the difference between the loaded condition of the amplifiers and the initial setting of pots in the pot set mode. The real loading will tend to invalidate the simple relations assumed between input and output. This may be corrected by adjusting the pots on load, in the IC mode, until the various voltages agree with the precalculated values within the accuracy acceptable for the computation.

It is also desirable to perform a dynamic check by recording the various amplifier outputs part way through the run and comparing them (on the basis of a special test case if necessary) with separately established expected values. If these are wildly out, it indicates a gross error that did not show up in the static test. Small discrepancies at least indicate the degree of accuracy that can be expected of the computation and may enable finer adjustments to be made to improve this accuracy.

It should be mentioned that just as numerical methods have their weaknesses and are subject to instability, so this occurs on occasion in analogue computing. This shows up, perhaps fortunately, more in using them to solve simultaneous *algebraic* equations than in the differential equations considered here, and care must be taken in the formation of such a problem. The combination of static and dynamic tests should at least indicate whether the run is introducing such an instability and consequent inaccuracy (see a subsequent example on Newton's law of cooling).

Additional Elements

VARIABLE MULTIPLIER

A variety of elements can be manufactured to give over a suitable operating range an output voltage proportional to the square of the input voltage. By suitable circuitry these multiplier elements can provide a unit capable of not only multiplying two variables, squaring one variable, but also dividing two variables and taking the square root of a varying voltage. Such elements greatly add to the power of the analogue computer. Figure 9.12 shows the patching for these different purposes on the EAI-180.

FIG. 9.12. EAI-180 multiplier patching.

Special care is needed in division to see that overloading due to too small a magnitude of e_x is avoided.

OTHER SPECIAL ELEMENTS

Diode function generators are available to approximate a general (nonlinear) function from a series of straight lines whose slopes and break points can be set as required. Comparators are useful to produce signals according to the sign of the sum of two voltages, $\text{sgn}(e_1 + e_2)$. The output can then be used to operate switches and function relays. Other nonlinear elements to operate as flip-flops, to represent saturation, hysteresis, etc., can be built up from these components.

TRANSFER FUNCTION ELEMENTS

A general relationship for impedance combinations can be given for a shunt impedance Z and n series impedances Z_1, etc.:

$$e = -\left(\frac{e_1}{Z_1} + \frac{e_2}{Z_2} + \cdots \frac{e_n}{Z_n}\right) = -\sum_i^n \frac{e_i}{Z_i} \tag{9.25}$$

This relation can be made the source of a series of transfer function program modules and some of the simpler of these are given explicitly in Table 9.1. In the table $\tau_i = R_i C$, etc.

TABLE 9.1.

Standard transfer functions and program modules

Circuit	Transfer function
R, C, R_1, e_1, e	$-\dfrac{1}{1+p\tau}$
R, C, R_1, e_1, e	$-\dfrac{p\tau}{1+p\tau}$
C, R, R_1, e_1, e	$-\dfrac{1+p\tau}{p\tau_1}$
C, R, R_1, e_1, e	$-(1+p\tau)$
R, C_1, C, R_1, e_1, e	$-\left(\dfrac{1+p\tau_1}{1+p\tau}\right)\dfrac{R}{R_1}$

It is also good practice to follow a convention in drawing patch diagrams or programs as developed here, as summarised in Fig. 9.13 and including the arbitrary function generator.

One suitable convention to follow over numbering, which may be the same as the numbering of the elements on the computer itself if this has been thoughtfully designed, is to use the corresponding number within the conventional symbol. Actual settings can be shown at the side or separately tabulated on the records of the analogue calculations. Some of the elements of Table 9.1 may require special capacitors, etc., to realise. In addition, the modern analogue computer will have logic elements that are

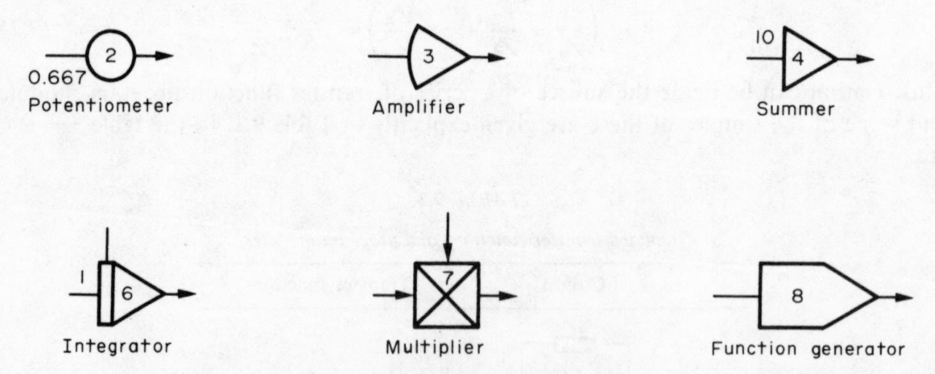

FIG. 9.13. Conventional patch symbols.

well suited to represent exactly such elements as were approximated by describing functions at Fig. 4.30 greatly extending the power of the analogue computer into non-linear models.

Sample Programs

LOW POWER KINETICS

The low power equations using the single precursor group approximation of Chapter 3 are readily adapted to analogue computing. A real time simulation can be considered. In the first sample program, step variations of reactivity between $\pm .01$ can be studied. The initial condition is taken to be a steady state situation with a source maintaining a constant neutron population $n_0 = 10^7$ neutrons at a shutdown reactivity of -0.01. Introduction of the various step additions of reactivity demonstrate the solutions already quoted in Chapter 3. The program with scaling factors and pot settings, suitable for implementation on the EAI-180, is given in Fig. 9.14. Note the inverter needed for $\rho > 0$.

Major features that can be studied include the prompt jump in $n(t)$ after the step change, the rapidly increasing neutron level when the reactivity exceeds or even approaches prompt critical ($\rho = \beta$) and the different steady state levels for negative reactivities. With repetitive solution, the prompt jump itself can be studied on an oscilloscope.

This program is modified in Fig. 9.15 to accept a ramp reactivity increase, linear with time, again from the initial conditions given above. Sample results were also quoted in Chapter 3. Practical studies should include taking note of the danger of fast rates of addition of reactivity leading to a prompt critical system before a significant neutron level has been established, and hence the importance of having a strong enough source to support a good neutron level even in the shutdown condition. This is particularly important on first commissioning because of uncertainties in critical mass and hence reactivities and also because the reactor having been run at power is likely to have enough radioactivity to sustain an acceptable source from photofission.

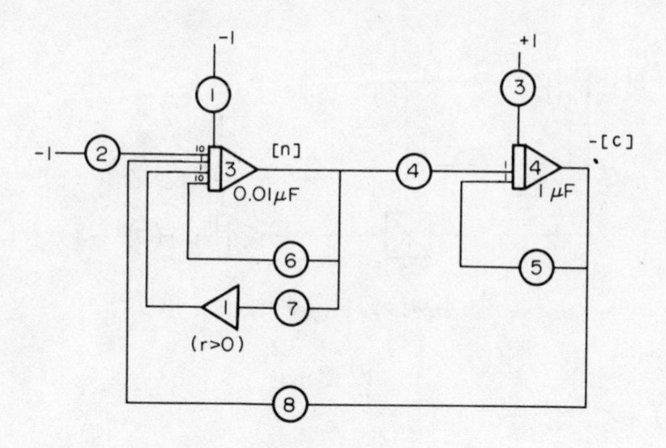

System equations

$$\frac{dn}{dt} = \frac{0.01r - \beta}{\Lambda} n + \lambda c + s$$

$$\frac{dc}{dt} = \frac{\beta}{\Lambda} n - \lambda c$$

$\Lambda = 10^{-4}$ s; s$=10^9$ neutrons/s; $\beta = 0.007$; $\lambda = 0.07865$; $n_0 = 10^7$; $c_0 = 8.9 \times 10^9$; $r = 100\rho$

Machine equations

$$\frac{1}{100}\frac{d[n]}{dt} = (r - 0.7)[n] + 0.7865[c] + 0.1; \quad n_0 = 0.1$$

$$\frac{d[-c]}{dt} = 0.07[-n] - 0.07865[-c]; \quad c_0 = 0.089$$

Pot settings

Pot	Meaning	Setting
1	n_0/K_n	0.100
2	$s/100K_n/10$	0.010
3	c_0/K_c	0.089
4	$K_n\beta/\Lambda K_c$	0.070
5	Λ	0.079
6	$\beta/100\Lambda/10$	0.070
7	$\lvert r\rvert = 100\lvert\rho\rvert$	variable
8	$\lambda K_c/100K_n$	0.787

Scale factors

$K = 1 \quad K_n = 10^8$ neutrons/muv
$K_c = 10^{11}$ precursors/muv

FIG. 9.14. Analogue program for step reactivity changes.

XENON SHUTDOWN EFFECTS

The equations governing the production and removal of ^{135}Xe and its precursor ^{135}I were quoted in Chapter 4 and explored for the linearised form for a contribution to the reactor system transfer function. When there are larger changes in the flux ϕ such that the linearised equations no longer serve, an analogue computation is a convenient way

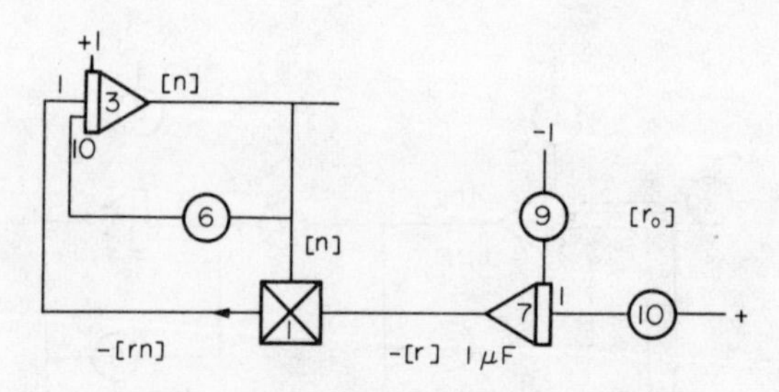

System equation

$$\rho = \rho_0 + \gamma t; \quad \rho_0 = -0.01; \quad r = 100\rho$$

Machine equation

$$\frac{d[r]}{dt} = 100$$

Pot setting

Pot	Meaning	Setting
7	Replaced	not used
9	$-100\rho_0$	1.0 (short circuit)
10	100γ	Variable

Scaling factors

$K = 0.01$; suitable values: 0 to 1 $\rho\%\ s^{-1}$

FIG. 9.15. Ramp reactivity analogue program modification.

to study the substantial effects of xenon poisoning, especially the massive rise after first shutdown with the flux going to zero.

The equations including the small amount of direct ^{135}Xe yield from fission are

$$\left.\begin{aligned} \frac{dI}{dt} &= \gamma_i \Sigma \phi - \lambda_i I \\ \frac{dX}{dt} &= \lambda_i I - \lambda_x X + \gamma_x \Sigma^f \phi - \phi \sigma X \end{aligned}\right\} \tag{9.26}$$

Values given in the *Reactor Physics Handbook*[9] for ^{235}U are repeated for convenience in Table 9.2.

TABLE 9.2.

Xenon Iodine data

$\gamma_i = 0.61$	$\lambda_i = 2.89 \times 10^{-5}\ s^{-1}$ $\sigma_0 = 272 \times 10^{-24}\ m^2$
$\gamma_x = 0.003$	$\lambda_x = 2.09 \times 10^{-5}\ s^{-1}$ (2200 m/s)
i.e. $\lambda_i = 0.0752\ hr^{-1}$	NB: $\sigma = \sigma_0 g$, where g is Westcott's non-1/v correction factor, dependent on effective flux temperature, and fluxes are quoted in nominal 2200 m/s-fluxes.
$\lambda_x = 0.1040\ hr^{-1}$	
$\sigma \simeq 300 \times 10^{-24}\ m^2$	

However, spatial effects should be taken into account via an effectiveness factor for the final term, and the appropriate spatial factor will depend upon the details of the reactor system. Probably the most convenient formulation is to normalise the I- and X-concentrations to the values they take after long exposure to the normal (and presumably maximum) operating flux. These equilibrium or steady state values are

$$I_{ss} \equiv \gamma_i \Sigma^f \phi_0 / \lambda_i; \quad X_{ss} = \frac{(\gamma_i + \gamma_x)\Sigma^f \phi_0}{\lambda_x + \phi_0 \sigma} \tag{9.27}$$

where φ_0 is the normal flux. It is convenient to put

$$r = \phi(t)/\phi_0 \quad i = I(t)/I_{ss} \quad \text{and} \quad x = X(t)/X_{ss}$$

The resulting non-dimensional xenon concentration $x(t)$ measures the relative poisoning or reactivity loss during the transient as compared to the loss in steady operation at the nominal flux. It also means that r and i are known to vary only between 0 and 1. It may be established that for fluxes below 10^{17} neutrons/m² s, the transient is negligible while at 2×10^{18} neutrons/m² s, x varies between 0 and 10. Interesting transients cover about 40 h system time.

Figure 9.16 gives a patching program, pot settings and scale factors for a simplified

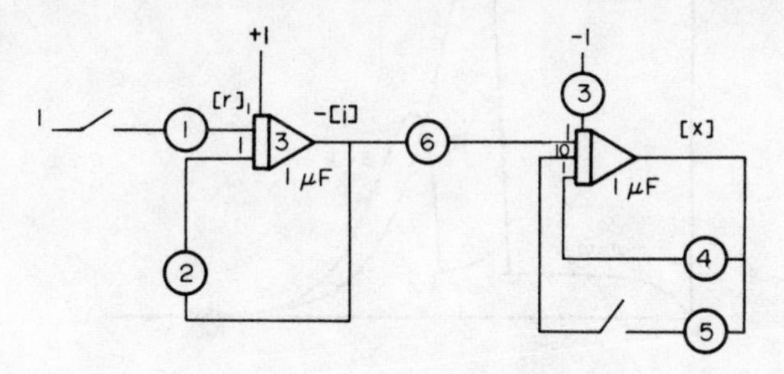

System equations

$$\frac{di}{dt} = \lambda_i r - \lambda_i i$$

$$\frac{dx}{dt} = (\lambda_x + \phi_0 \sigma) i - \lambda_x x - r\phi_0 \sigma x$$

Machine equations

$$\frac{1}{K}\frac{d[-i]}{dt} = -\lambda_i[-i] + \lambda_i[-r]K_r/K_i$$

$$\frac{1}{K}\frac{d[x]}{dt} = \frac{\lambda_x + \phi_0 \sigma}{K_x} K_i[i] - \lambda_x[x] - \phi_0 \sigma K_r[rx]$$

Pot settings

Pot	Meaning	Setting ($\phi = 17.5 \times 10^{18}$ neutrons/m² s)
1	$\lambda_i K$	0.414
2	$\lambda_i K$	0.414
3	$x(0)/K_x$	0.100
4	$\lambda_x K$	0.301
5	$\phi_0 \sigma K/10$	0.762
6	$(\lambda_x + \phi_0 \sigma) K K_i / K_x$	0.792

Scaling factors

$$K_i = 1; \quad K_x = 10; \quad K_r = 1; \quad K = 4\,\frac{\text{system hours}}{\text{machine seconds}}$$

Note use of two switches: open circuit: $\phi = 0$; $r = 0$
closed circuit: $\phi = \phi_0$; $r = 1$

FIG. 9.16. On–Off xenon poisoning analogue computation.

case where we neglect the direct yield of ^{135}Xe and suppose that the flux is either xero or at nominal power (thus $r \to 0$ or 1 only). Switches in the program are closed for $r=1$, open for $r=0$. Figure 9.17 gives sample results based on $\phi\sigma$ value of 1.905 (hr^{-1}) which is equivalent to a flux ϕ of 17.5×10^{18} neutrons/m² s=and $\sigma = 300 \times 10^{-24}$ m².

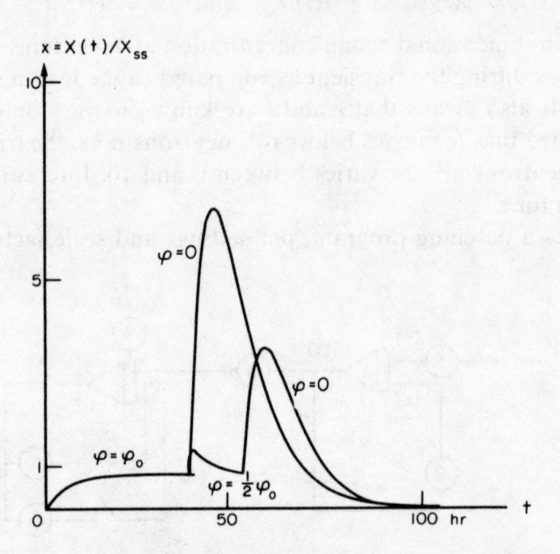

FIG. 9.17. Sample output: xenon poisoning versus time.

Figure 9.18 gives a generalisation where an arbitrary flux ratio function $r(t)$ is introduced. The machine equations and pot settings are essentially the same as before and we merely generate any intermediate flux ratio via pot number 7. Both programs are

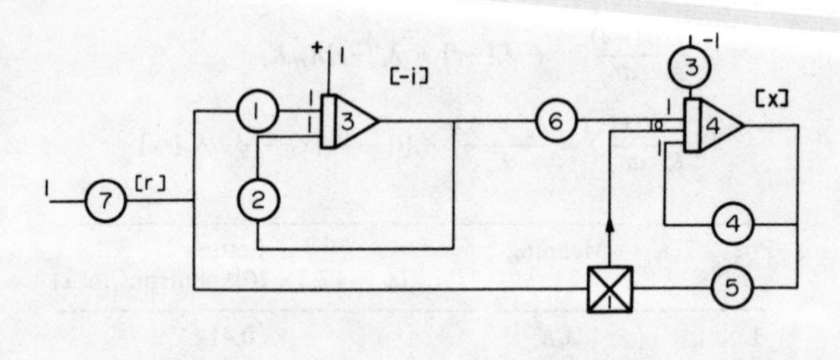

Pot setting

Pot	Meaning	Setting
7	$r = \phi/\phi_0$	variable 0–1

FIG. 9.18. Xenon poisoning modification for variable flux.

provided with initial conditions corresponding to the equilibrium I- and X-values, but equally well these values are established by operating the computer at $r=1$ for sufficient time. Figure 9.17 also shows the effect of such an intermediate flux followed by a subsequent shutdown

Note that the effects of fluxes *lower* than the assumed value of 17.5×10^{18} neutrons/m² s can always be studied with these settings by utilising a corresponding $r=\phi/\phi_0$ value and allowing the equilibrium value to build up accordingly.

The results shown in Fig. 9.17 illustrate the build-up of xenon during the steady operation and the transient increase in xenon—and hence reactivity loss—after shutdown, when the flux no longer provides a removal mechanism and removal is dependent solely on radioactive decay. Clearly there is a likelihood that for periods up to 40 h the transient xenon poisoning may prevent the restart of the reactor unless there is a substantial excess fuel loading to provide a restart capacity. This implies an investment in excess fuel and a substantial economic penalty for spurious shutdown in a high flux reactor. In Chapter 4 the relation of xenon concentration to reactivity loss was discussed and given for the steady state loss as $\rho_{ss}=\gamma/\nu\, \phi_0\sigma/(\lambda_x+\phi_0\sigma)$, and on this basis the normalised xenon concentration $x(t)$ is also a normalised reactivity loss.

In Chapter 8 the question of operating the reactor at some intermediate flux to limit this poisoning was investigated (and can usefully be studied with the modified analogue program just given). Another important area to study is the rate of growth of xenon poisoning during startup from a clean condition in terms of both the total reactivity loss ρ_{ss} (for which extra fuel and compensating reactivity must be provided) and also the rate of change of this reactivity loss, since this is likely to be the most severe case which in turn would govern the maximum rate of withdrawal of shim rods, soluble poison, etc., holding down the excess reactivity.

It is fairly readily shown in this model that the maximum *rate of change* of xenon after shutdown from a steady operating situation occurs immediately with a value $\phi_0\sigma X$ and with a little more care that the maximum rate of change after a clean startup occurs after an interval t given as the solution of

$$\lambda_i \exp(-\lambda_i t) = (\lambda_x+\phi_0\sigma) \exp[-(\lambda_x+\phi_0\sigma)t] \qquad (9.28)$$

The analogue program is quite suitable in determining values for required reactivity addition rates to meet these situations. Plots of $x(t)$ versus $i(t)$, i.e. the state space, are also informative (see Chapter 8).

BUILD-UP OF TRANSURANIC ISOTOPES

The build-up of transuranic isotopes, particularly of plutonium, leads to reactivity effects on a relatively long scale and is significant in itself, of course, for fuel cycle studies. When these studies take reloading and recycling into account they can be complex indeed, even without allowance for spatial distribution in the reactor. Benedict and Pigford, etc., may be consulted [10–12]; here we give a simple model for the build-up of the plutonium isotopes ^{239}Pu, ^{240}Pu and ^{241}Pu in a natural (or low enrichment) thermal reactor. There is a facility for altering the design value of $\varepsilon p'=(1-p)\varepsilon$ (the fast fission and the resonance escape probability) to study the effect upon plutonium production. The build-up of these isotopes can be combined with the estimation of long term fission product build-up to compute variations in long term reactivity and the consequent need for refuelling or excess fuel, changes in temperature coefficients of reactivity, etc. (see problems).

The system equations are here quoted without further justification save that they are derived from the usual mass balance of production less removal; constant ^{238}U concentration is assumed:

$$^{235}\text{U balance:} \quad \frac{dN_5}{dT} = -N_5\sigma_5\phi$$

$$^{239}\text{Pu balance:} \quad \frac{dN_9}{dT} = N_8\sigma_8\phi + (\eta_5\sigma_5 N_5 + \eta_9\sigma_9 N_9 + \eta_1\sigma_1 N_1)\phi\varepsilon p \cdot$$

$$^{240}\text{Pu balance:} \quad \frac{dN_0}{dT} = (\sigma_9 - \sigma_9^f)N_9\phi - N_0\sigma_0\phi \qquad (9.28)$$

$$^{241}\text{Pu balance:} \quad \frac{dN_1}{dT} = N_0\sigma_0\phi - N_1\sigma_1\phi$$

Here η is the yield of neutrons per absorption, and σ the microscopic absorption cross-section and σ^f the fission cross-section.

At a flux of 10^{19} neutrons/m^2 s it is known that interesting developments of isotope production take between 10 and 100 days. If other fluxes are of interest, the time scale is proportionately varied so that the product of flux time is the independent parameter. Here we take a nominal flux of 10^{19} neutrons/m^2 s.

Table 9.3 gives the atomic data appropriate to the model and Table 9.4 gives the necessary nuclear data.

The program for this model, Fig. 9.19, has been kept elementary to enable the solution scheme to be easily seen. Accuracy can be improved with more complicated arrange-

TABLE 9.3.

Initial uranium concentrations

Enrichment (%)	^{238}U (atoms/m^3)	^{235}U (atoms/m^3)
0.713 (nat.)	4.69×10^{28}	3.37×10^{26}
1.0	4.67×10^{28}	4.7×10^{26}
2.0	4.63×10^{28}	9.4×10^{26}
3.0	4.58×10^{28}	14.1×10^{26}

TABLE 9.4.

Nuclear data for burn-up

Isotope	Absorption [a] cross-section	$\phi\sigma$ [b]	Fission cross-section [a]	$\phi\sigma$ [b]	η-values
^{235}U	683	0.0540	582	0.0503	2.08
^{238}U	2.71	0.000234	—	—	—
^{239}Pu	1028	0.0888	742	0.0641	2.08
^{240}Pu	286	0.0247	—	—	—
^{241}Pu	1400	0.121	1010	0.0873	2.21

[a] In barns, 10^{-28} m^2. [b] Nominal flux 10^{19} neutrons/m^2 s.

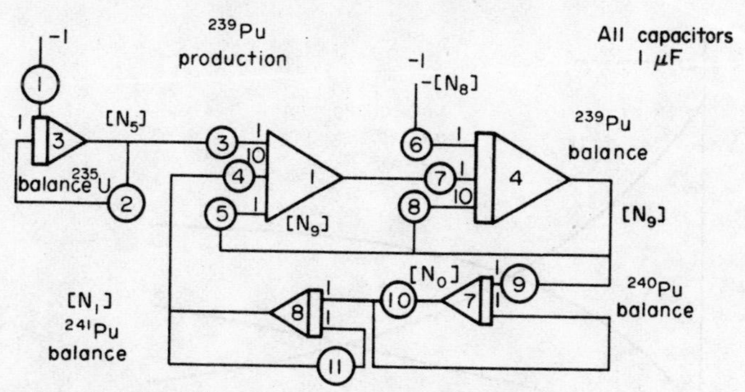

Machine equations (nominal flux 10^{19} neutrons/m² s) natural enrichment

$$\frac{d[N_5]}{dt} = -0.059[N_5]; \qquad \frac{d[N_9]}{dt} = 0.0234[N_8] - 0.0888[N_9]$$

$$\frac{d[N_0]}{dt} = .00247[N_9] - .0247[N_0]; \qquad + \{0.1227[N_5]$$

$$\frac{d[N_1]}{dt} = .00247[N_0] - .121[N_1]; \qquad + 0.2673[N_1] + 0.1847[N_9]\}\epsilon p'$$

Scale factors

$$K_8 = 5 \times 10^{28} \frac{\text{atoms/m}^3}{\text{muv}}; \quad K_5 = K_9 = K_0 = K_1 = 5 \times 10^{26} \frac{\text{atoms/m}^3}{\text{muv}}$$

$$K = 1 \frac{\text{system days zero}}{\text{machine seconds}}$$

Pot settings

Pot	Meaning	Setting	Pot	Meaning	Setting
1	$N_5(\text{o})/K_5$	0.674	7	$(1-p)\epsilon$ say	0.150
2	$\sigma_5\phi K$	0.059	8	$\sigma_9\phi K/10$	0.009
3	$\eta_5\sigma_5\phi K$	0.123	9	$\sigma_{f9}\phi K$	0.025
4	$\eta_1\sigma_{1f}\phi K/10$	0.027	10	$\sigma_0\phi K$	0.025
5	$\eta_9\sigma_9\phi K$	0.185	11	$\sigma_1\phi K$	0.121
6	$\sigma_8\phi K/K_8$	0.219			

FIG. 9.19. Transuranic isotope build-up program.

ments to allow pots to be set in a better range. Integrators set using 1 μF capacitors are suitable for X-Y plotting on a pen recorder and a change to 0.01 μF would be suitable for a repetitive calculation for oscilloscope display.

Figure 9.20 gives a typical output and the cross-over where the ^{235}U concentration is overtaken by the ^{239}Pu concentration is seen. It is also clear that the build-up of the higher isotopes are second and third order effects.

NONLINEAR KINETICS

In Chapter 8 a model of a power reactor with a single temperature coefficient of reactivity (α, assumed constant) was studied for stability in a situation where closed

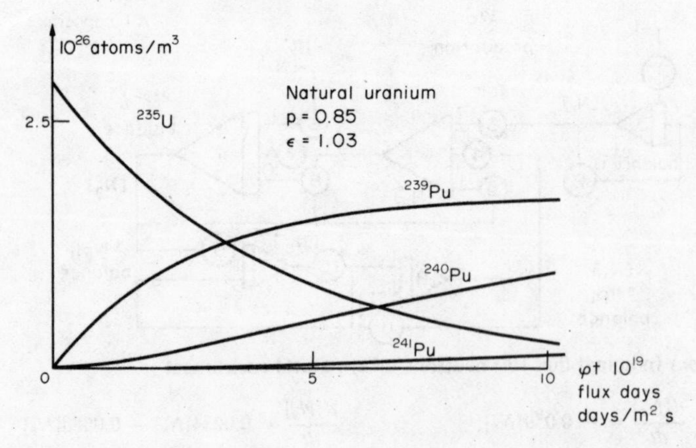

FIG. 9.20. Transuranic isotope build-up and ^{235}U burnup—sample results.

analytic solutions are not readily available. The following analogue program can realise this solution readily, for $\rho_0 = \rho - \beta$, the prompt reactivity.

The state equations for neutron level n and temperature T are:†

$$\left.\begin{aligned} \frac{dn}{dt'} &= \frac{\rho_0 + \alpha T}{\Lambda} n; \quad n(\mathrm{o}) = n_0 \\ \frac{dT}{dt'} &= (n - n_0)\kappa - \mu T; \quad T(\mathrm{o}) = 0 \end{aligned}\right\} \tag{9.30}$$

where κ is the power/heat capacity factor and the term in μ represents Newton's law of cooling, transfer of heat proportional to temperature differences (Λ, κ, $\mu > 0$).

Knowledge of the solution with $\mu = 0$ suggests that for $\alpha < 0$, $\rho_0 > 0$, stable cycles exist, and if $\rho_0 > \frac{1}{2}\sqrt{|\alpha|\kappa\Lambda n_0}$ the maximum values in the cycle are $n_{max} \sim 10n_0$, $T_{\max}$ $10\sqrt{\Lambda\kappa n_0/|\alpha|}$. The system equations can then be transformed to machine equations as

$$\left.\begin{aligned} \frac{d[n]}{dt} &= \frac{K\rho_0}{\Lambda}[n] + \frac{KK_T|\alpha|}{\Lambda}\,\mathrm{sgn}(\alpha)[nT]; \quad [n(\mathrm{o})] = n_0/K_n \\ \frac{d[T]}{dt} &= \frac{KK_n\kappa}{K}[n - n_0] - \mu K[T]; \quad [T(\mathrm{o})] = 0 \end{aligned}\right\} \tag{9.31}$$

It is therefore convenient to take the time scaling factor $K = \Lambda/K_T|\alpha|$ as well as $K_n = 10n_0$, $K_T = 10\sqrt{\kappa\Lambda n_0/|\alpha|}$ and to put the reactivity as a fraction of this largest acceptable value: $\zeta = 2\rho_0/\sqrt{|\alpha|\Lambda\kappa n_0}$. Figure 9.21 shows a suitable program for a range of coefficient settings to illustrate the behaviour of the system; note the use of the multiplier with assumed inversion of sign and the use of two inverters. For the normal case of interest, inverter No. 1 is used (for $\rho_0 > 0$) and No. 2 is used (for $\alpha < 0$).

†T for temperature in this section, not system time.

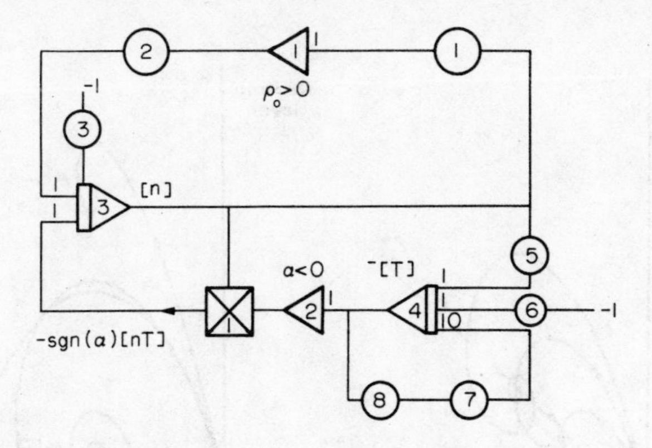

Machine equations

$$\frac{d[n]}{dt} = 0.2\zeta[n] + \text{sgn}(\alpha)[nT]; \quad [n_0] = 0.1$$

$$\frac{d[T]}{dt} = 0.1[n - n_0] - 0.1\mu\sqrt{\Lambda/|\alpha|\kappa n_0}\,[T]; \quad [T_0] = 0$$

Scale factors

$$K = 0.1\sqrt{\Lambda/\kappa|\alpha|n_0}; \quad K_n = 10n_0; \quad K_T = 10\sqrt{\Lambda\kappa n_0/|\alpha|}$$

Pot settings

Pot	1	2	3	5	6	7	8
Meaning		ζ	n_0/n_{max}	$n_0/10n_{\text{max}}$	$n_0/10n_{\text{maa}}$	$0.1\sqrt{\Lambda/|\alpha|\kappa n_0}$	μ
Range or setting	0.200	0–1	0.100	0.100	0.010	0.050	0–1

FIG. 9.21. Temperature coefficient of reactivity study.

It may be demonstrated, for example, that with $\mu > 0$ the trajectories in the (n, T) space move to a steady operating point (if $\rho_0 > 0$, $\alpha < 0$) given by $\hat{n} = n_0 - \mu\rho_0/\kappa\alpha$, $\hat{T} = -\rho_0/\alpha$ so that for given ρ_0, as μ is increased, the stable operating centres lie on a straight line through $(n_0, 0)$. Other aspects worth study include:

(a) effect of μ;
(b) effect of $\rho_0 < 0$;
(c) effect of $\alpha > 0$;
(d) dependence of n_{max}, T_{max} on ρ_0; $\rho_0 > 0$, $\alpha < 0$.

It should also be noted by practical demonstration that for $\mu = 0$, as $\zeta \to 1$, the analogue computer will lose accuracy in the final section where it is called upon to integrate the product of two small currents $[nT]$. The analytical treatment of Chapter 8 shows that the trajectories for $\mu = 0$ should be closed cycles. The stability studies of that chapter may usefully be illustrated by computing a series of trajectories at $\mu > 0$, starting at different locations on one of the closed cycles, where $\mu = 0$, to show which trajectories with Newton's law of cooling are inwardly directed to the stable operating point. Figure 9.22 illustrates such a study.

The state equations may also be "linearised" and the analogue computer used to

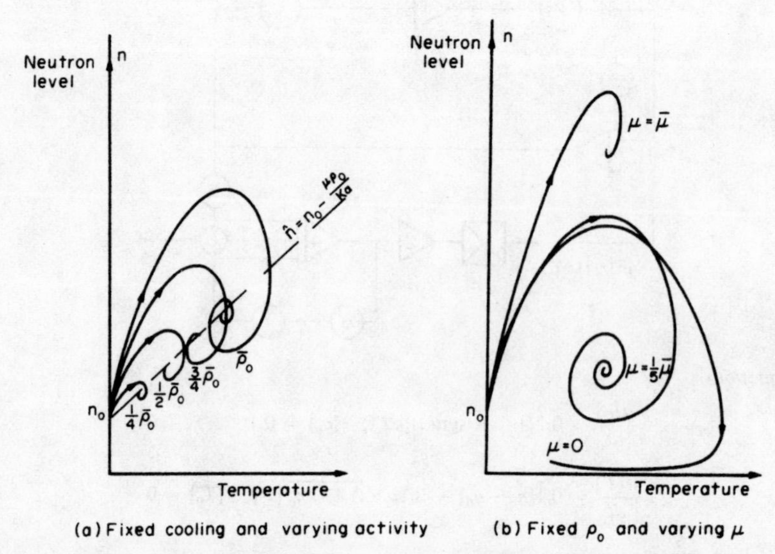

FIG. 9.22. Newton's law of cooling study with varying reactivity and cooling.

study this approximation and compare the results with Fig. 9.22 for small and large departures from the stable operating point.

Figure 9.23, on the other hand, shows trajectories around an equilibrium point (together with the orbit cycle for $\mu=0$) illustrating the Liapunov stability study of Chapter 8.

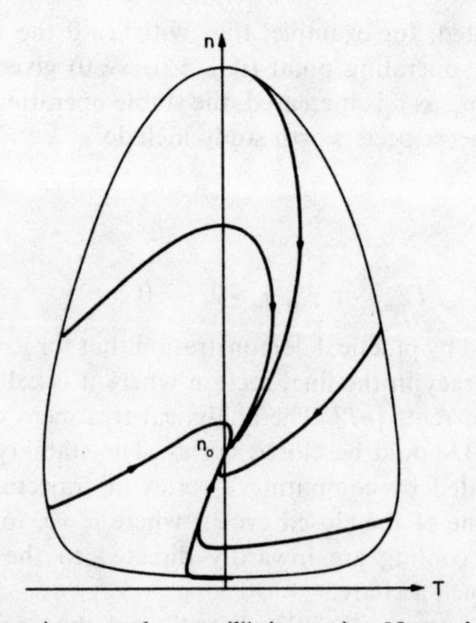

FIG. 9.23. Trajectories around an equilibrium point: Newton's law of cooling.

Conclusion

The introduction to the practice of analogue computing and the presentation of examples in enough detail for reproduction by the novice should enable him to develop some confidence in his ability to use this versatile tool in studies of reactor behaviour, particularly operational behaviour, optimisation and safety transients. The series of problems that follow are designed to exercise that skill further in the straightforward use of analogue computers. More advanced elements, using logic and digital/analogue hybrids, are discussed in the references.

Problems for Chapter 9

9.1. *Mass–spring oscillator.* Add viscous damping to the mass–spring system so that the state equation is

$$\ddot{x} + 2\zeta\omega\dot{x} + \omega^2 x = 0$$

where ζ is the viscous damping factor $0<\zeta$.

Investigate the behaviour for $\zeta=0.4$, 1 and 1.4, making any suitable adjustments to the time scaling. Include consideration of phase plots, i.e. $\dot{x}(t)$ versus $x(t)$.

9.2. Extend the consideration of problem 9.1 to include a *forcing* function $f(t)$ in the damped system so that

$$\ddot{x} + 2\zeta\omega\dot{x} + \omega^2 x = f(t)$$

Choose $f\ t)=a \sin \omega_0 t$. Study the response of the system to different driving frequencies ω_0. Since the response for ω_0 close to the natural frequency ω will be large (resonance) it will be necessary to reduce the constant a in this range. Draw the transfer function for the second order damped system, i.e. the relative amplitude of response to input and the relative phase, for a range of values of ω/ω_0 and ζ.

9.3. Consider the circuit diagram below and derive its equation. What use can be made of this in low power reactor kinetics studies?

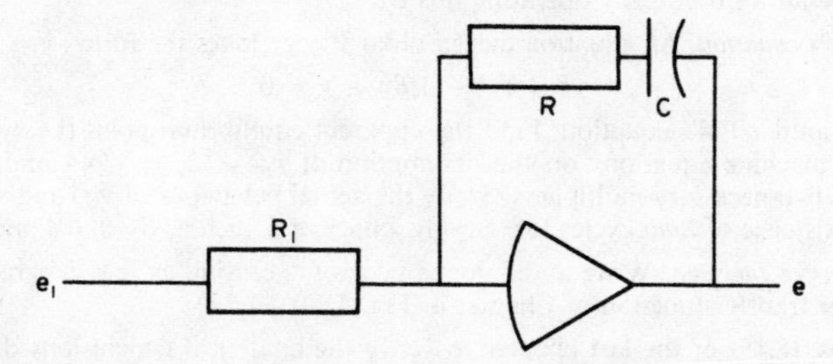

9.4. Consider the various circuits in the diagram below containing a multiplier and derive the relation of e to e_x and e_y.

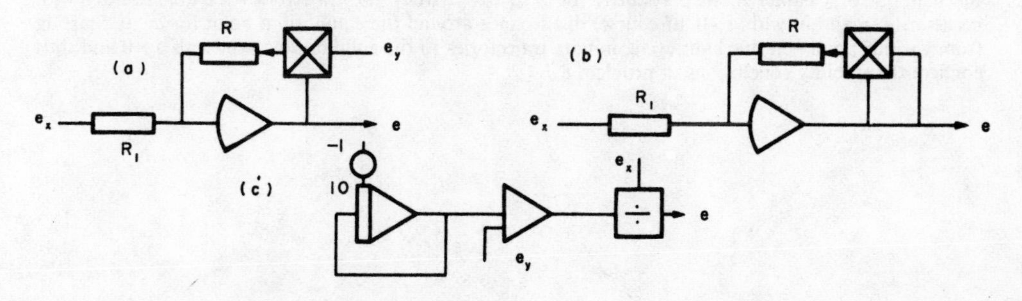

9.5. Modify the one group precursor model program to follow the prompt jump approximation, i.e. where the integration of the rate of change of density is replaced by an algebraic solution for $n(t)$ in terms of $c(t)$. Run similar cases and observe the difference (if any) for various values of step and ramp reactivity functions.

9.6. Modify the xenon poisoning program to accommodate (a) the direct yield of xenon in fission, and (b) the case where the flux is adjusted so as to maintain the xenon concentration constant with time (as required in the discussion of Chapter 8) at a given level, X_0 say.

9.7. Samarium builds up in a reactor from the decay of the fission product promethium without subsequent decay, leading to a permanent poisoning increase or reactivity loss. System equations and values suggested by the *Reactor Handbook* for both samarium explicitly and general fission product poisoning are as follows:
Samarium: fission→^{149}Pm (54 hr $T_{\frac{1}{2}}$)→^{149}Sm (stable)
$\sigma = 5\times10^{-24}$ m^2; γ (^{235}U) $= 0.0113$; $\lambda = 3.56\times10^{-6}$ s^{-1} $= 0.0128$ hr^{-1}

$$\frac{dP}{dT} = \gamma\Sigma^f\phi - \lambda P; \quad \frac{dS}{dT} = \lambda P - \phi\sigma S$$

General fission products:

$$\frac{dZ_i}{dT} = \gamma_i\Sigma^f\phi - Z_i\sigma_i\phi$$

where for ^{235}U

i	γ_i	σ_i (10^{-24} m^2)
1	0.809	0.852
2	0.122	0.1137
3	−0.054	−0.0129

NB: negative values arise from an artificial fitting of more complex data to three terms only.

Write an analogue program to simulate the growth of poisoning and consequent loss of reactivity.

9.8. Justify eqn. (9.28). Calculate the ratio of maximum reactivity rates of increase in startup and shutdown as a function of the steady operating flux ϕ_0.

9.9. *Van der Pol's equation.* An equation met in radar theory takes the form

$$\ddot{y} + (y^2 - 1)\beta\dot{y} + y = 0$$

and is known as van der Pol's equation. Find the apparent equilibrium point (i.e. where y is in steady state). Derive the machine equations on the assumption of $y_{\max}=2$, $\dot{y}_{\max}=4$ and give the patching program using the two necessary multipliers. Study the actual behaviour of $y(t)$ and $\dot{y}$ versus y and thus demonstrate the existence of *limit* cycles for varying values of β such as $\beta=0$, 0.4 and 2.0.

9.10. *Boiler transfer function.* Write an analogue programme using as few integrators as possible to represent the boiler transfer function of Chapter 4 (Fig. 4.20).

9.11. Study eqns. (8.18) of the last chapter to verify the analytical conclusions drawn in Chapter 8 for the precursor-prompt jump approximation.

9.12. Carry out the studies indicated in the text for Newton's law of cooling and thus supplement the material of Chapter 8. In particular, for zero $\rho_0-\beta$ study the trajectories for an arbitrary (not necessarily small but with $n>0$, of course) disturbance around the equilibrium point for $\mu=0$. Starting from various points on this limit cycle, initiate trajectories to the equilibrium point with $\mu>0$ and thus confirm the stability conclusions of problem 8.7.

References for Chapter 9

1. D. I. Rummer, *Introduction to Analog Computer Programming*, Holt, Rinehart & Winston, New York, 1969.
2. A. S. Jackson, *Analog Computation*, McGraw-Hill, New York, 1960.
3. C. A. Steward and R. Atkinson, *Basic Analogue Computer Techniques*, McGraw-Hill, New York, 1967.
4. F. H. Raven, *Automatic Control Engineering* (2nd edn.), McGraw-Hill, New York, 1968. A general text with a good section on analogue computing.
5. R. Tomovic and W. J. Karplus, *High Speed Analogue Computers*, Wiley, 1962 and Dover, 1970. A specialised monograph.
6. *EAI*-180 *Operators Reference and Maintenance Manual*, EAI-Electronics Associates, Australia, 1972.
7. *Basics of Parallel Hybrid Computers*, EAI, Australia, 1968.
8. M. A. Schultz, *Control of Nuclear Reactors and Power Plants* (2nd edn.), McGraw-Hill, New York, 1955. Special applications to nuclear engineering.
9. *Reactor Physics Constants*, *ANL*-5800 (2nd edn.), USAEC, Washington, 1963.
10. M. Benedict and T. H. Pigford, *Nuclear Chemical Engineering*.
11. R. P. Silvernnoinen, *Reactor Core Fuel Management*, Pergamon, Oxford, 1976.
12. A. Henry, *Nuclear-Reactor Analysis*, MIT, Boston, 1975.

Index